GLOBALIZATION, BIOSECURITY, AND THE FUTURE OF THE LIFE SCIENCES

Committee on Advances in Technology and the Prevention of Their Application to Next Generation Biowarfare Threats

Development, Security, and Cooperation
Policy and Global Affairs Division

Board on Global Health
Institute of Medicine

INSTITUTE OF MEDICINE ᴬᴺᴰ
NATIONAL RESEARCH COUNCIL
OF THE NATIC...

THE NATIONAL ACADEMIES PRESS
Washington, D.C.
www.nap.edu

THE NATIONAL ACADEMIES PRESS 500 Fifth Street, N.W. Washington, DC 20001

NOTICE: The project that is the subject of this report was approved by the Governing Board of the National Research Council, whose members are drawn from the councils of the National Academy of Sciences, the National Academy of Engineering, and the Institute of Medicine. The members of the committee responsible for the report were chosen for their special competences and with regard for appropriate balance.

This study was supported by contracts between the National Academies and the Department of Homeland Security, the Centers for Disease Control and Prevention, the Food and Drug Administration, the National Institute of Allergy and Infectious Diseases, the National Science Foundation, and the Intelligence Technology Innovation Center. The views presented in this report are those of the National Research Council and Institute of Medicine Committee on Advances in Technology and the Prevention of Their Application to Next Generation Biowarfare Threats and are not necessarily those of the funding agencies.

International Standard Book Number 0-309-10032-1 (BOOK)
International Standard Book Number 0-309-65418-1 (PDF)
Library of Congress Control Number 2006925454

Additional copies of this report are available from the National Academies Press, 500 Fifth Street, N.W., Lockbox 285, Washington, DC 20055; (800) 624-6242 or (202) 334-3313 (in the Washington metropolitan area); Internet, http://www.nap.edu.

THE NATIONAL ACADEMIES
Advisers to the Nation on Science, Engineering, and Medicine

The **National Academy of Sciences** is a private, nonprofit, self-perpetuating society of distinguished scholars engaged in scientific and engineering research, dedicated to the furtherance of science and technology and to their use for the general welfare. Upon the authority of the charter granted to it by the Congress in 1863, the Academy has a mandate that requires it to advise the federal government on scientific and technical matters. Dr. Ralph J. Cicerone is president of the National Academy of Sciences.

The **National Academy of Engineering** was established in 1964, under the charter of the National Academy of Sciences, as a parallel organization of outstanding engineers. It is autonomous in its administration and in the selection of its members, sharing with the National Academy of Sciences the responsibility for advising the federal government. The National Academy of Engineering also sponsors engineering programs aimed at meeting national needs, encourages education and research, and recognizes the superior achievements of engineers. Dr. Wm. A. Wulf is president of the National Academy of Engineering.

The **Institute of Medicine** was established in 1970 by the National Academy of Sciences to secure the services of eminent members of appropriate professions in the examination of policy matters pertaining to the health of the public. The Institute acts under the responsibility given to the National Academy of Sciences by its congressional charter to be an adviser to the federal government and, upon its own initiative, to identify issues of medical care, research, and education. Dr. Harvey V. Fineberg is president of the Institute of Medicine.

The **National Research Council** was organized by the National Academy of Sciences in 1916 to associate the broad community of science and technology with the Academy's purposes of furthering knowledge and advising the federal government. Functioning in accordance with general policies determined by the Academy, the Council has become the principal operating agency of both the National Academy of Sciences and the National Academy of Engineering in providing services to the government, the public, and the scientific and engineering communities. The Council is administered jointly by both Academies and the Institute of Medicine. Dr. Ralph J. Cicerone and Dr. Wm. A. Wulf are chair and vice chair, respectively, of the National Research Council.

www.national-academies.org

Preface

The most significant recognized biological attack launched in the United States in recent times came shortly after the devastating events of 9/11 in late 2001. Combined, these events propelled the nation into what many perceived to be a new era, fraught with the real hazards of global terrorism abetted by the exploitation of common everyday technologies—in the one case, an efficient and highly automated national postal service, and in the other, the enormous latent energy carried by a commercial airliner set for a transcontinental flight. That the biological agent used in the postal attacks was anthrax, a "classic" choice of those intent on waging biological warfare, undoubtedly contributed to the nature of the government's response and the biodefense research priorities that evolved subsequently with a nearly exclusive focus on well-recognized, "traditional" biowarfare agents. Such a focus is dangerously narrow, although past successes in weaponizing anthrax and the potentially devastating consequences of a smallpox release within an immunologically naïve population cannot be ignored. Conventional threats must be addressed by any successful biodefense plan. However, it would be dangerous to ignore the ingenuity displayed in the past by those who are intent on disrupting and perhaps destroying our society. Smart and well-informed terrorists who seek to use against us the technologies we have developed and upon which we have come to rely so heavily at the beginning of the 21st century have unparalleled opportunities to do harm. We also need to be concerned about the unintended consequences from those who use technologies in an irresponsible or ill-informed manner. These dangers are nowhere more evident than in the life sciences, where biotechnology and our understanding of the biological processes that define our very being are advancing at extraordinary rates.

Concerns about how new developments in the life sciences, including their convergence with other rapidly advancing fields such as nanotechnology and materials science, may enable the creation and production of wholly new threats of biological origin led to the formation of the Committee on Advances in Technology and the Prevention of Their Application to Next Generation Biowarfare Threats, an ad hoc committee of the National Research Council and the Institute of Medicine. The committee's charge has been to examine current trends and future objectives of research in the life sciences that may enable the development of a new generation of future biological threats. In taking on the charge to define a horizon of 5 to 10 years, the committee has sought to identify ways to anticipate, identify, and mitigate these dangers to society. While this has been far from easy, and while the committee may worry about how successful it has been in fulfilling its charge as this report goes to press, several conclusions stand out with startling clarity.

First, the future is now. Even in the short time since the creation of the committee, we have seen the phenomenon of RNA interference capture the collective consciousness of the life sciences community, providing entirely new insights into how human genes are normally regulated and how this regulation might be disrupted for malevolent purposes by those intent on doing harm. Similarly, "synthetic biology," an approach embraced and discussed by few at the time the committee was formed, has now been redefined and promoted on the cover of one of the most widely read scientific journals. Neither of these developments could have been foretold even a few years back, pointing to the futility of trying to predict with accuracy what will come in the next few years. This leads to the second conclusion, that the task of surveying current technology trends in order to anticipate what new threats may face us down the road will be never ending. This report, published in early 2006, will in some respects be out of date by 2007.

These considerations led to two of the major recommendations adopted by the committee: the need to survey the threat horizon continually for what we may face in the future and, in order to do this effectively, the need to enhance in a significant manner the scientific expertise of those charged with this task. Thus, rather than laying forth a list of threats as perceived at the end of 2005, the committee has endeavored to describe a process and set of organizing principles, a method by which technological advances might be assessed and future risks for their malevolent use considered. Such a contribution is likely to be more lasting than any specific list, although the process itself must be continuously reassessed in light of advancing knowledge.

In addressing our charge, committee members have been blessed by a large committee, well endowed with expertise in a number of diverse scientific fields and with several international members consistent with the imperative that these issues be addressed in a global context. Our discussions highlighted many different perspectives held by members of the committee—differences that stem from past experiences in the very different fields represented by the members, which include biological discovery, global emerging infections, nuclear physics, bioethics, law enforcement, and international arms control, to list a few. However, every member was challenged by the committee's charge.

Paradigms for threat reduction that may have worked reasonably well for controlling nuclear arms proliferation, information control, materials inventory, and so forth, may have limited relevance to the control of biological weapons proliferation. This is especially true given the wide dispersion of biological information and the mechanisms in place that support this globally, the capacity of the relevant materials to replicate, and the lack of any readily apparent "global bargain" resembling the "Atoms for Peace" initiative of the past. Yet it is increasingly important that life scientists, and the funding agencies and editors who support their activities, take every possible step to ensure that the fruits of their work are not exploited in a malevolent fashion, to the detriment of society. This will require that those working in the life sciences achieve a much greater appreciation of the dangers than that now held by most and a greater willingness to shoulder this responsibility. A new ethos is required, and it must be achieved on a global scale. This was apparent to all committee members, although a clear path toward achieving this goal was apparent to none of us—our recommendations are but first steps in this direction.

In many ways, this committee has worked in the shadows of the groundbreaking National Research Council report *Biotechnology Research in an Age of Terrorism* (2004), commonly known as the Fink report. However, there is a clear difference between that report and the present one. Our focus has been on advances in the life sciences and related and convergent technologies that are likely to alter the biological threat spectrum over the next 5 to 10 years. To a greater extent than was the case with the previous effort, we have attempted to take a global perspective and consider how future threats might be anticipated. In contrast, the Fink report focused primarily on the regulatory oversight of research employing biotechnology and the flow of scientific knowledge derived from the use of biotechnology, mostly within the United States.

It is an unfortunate reality that almost all advances in life sciences technology pose potential "dual-use" risks. But better science is the best protection against potential threats. This is not to advocate the creation of a biological arms race, but to recognize the simple fact that better vaccines, better drugs, and better countermeasures in general, not to mention anticipation of potential threats, will stem from such a flow of information.

The committee has broadly considered ways to prevent or mitigate the consequences of malevolent exploitation (either by state actors, nonstate groups, or individuals) or naïve misuse of these technologies. To the extent that we do cover some ground trod earlier by the Fink report, for example, in considering how regulations and policies may have utility in addressing these risks—the conclusions here are much the same: The committee recognizes and emphasizes the counterproductive nature of efforts to control the flow of biological information. Given the widening threat spectrum, the best means of future protection comes from the exploitation of science, paradoxically the very advances in technology about which there is so much concern. It is imperative that we (defined broadly as free societies) keep ahead scientifically and remain technologically advantaged over potential opponents. Such protection can only come from a robust scientific enterprise, which in turn depends on the free exchange of biological data among scientists.

Committee members have been struck by the often "self-centered" and limited perspective taken by some in the United States charged with addressing these critically important issues. Although we cannot pretend to understand completely the forces that govern advances in science and technology and the need for regulating these activities in diverse regions of the globe, it is clear that different societies may have vastly different perspectives on these issues and may adopt divergent paths while aiming to achieve similar goals. To succeed in reducing the threats posed by these advancing technologies will require an appreciation of these differences and an understanding that science does not stop at our borders.

The futile nature of attempting to predict the future accurately and which of the myriad scenarios a set of terrorists or malevolent state actors might choose leads to the committee's final conclusion—which is perhaps the most obvious, the most important, but the least novel and therefore, unfortunately, the least likely to be heeded. The best anticipatory practices, thoughtful predictions, and preventive actions are unlikely to be completely successful in preventing a future significant biological attack, whether with a conventional "classic" biological weapons agent or a newly engineered weapon of biological origin. Thus, we must be prepared; the best preparation will be to strengthen the nation's fractured public health infrastructure and the lack of coordination that exists among the myriad federal and state agencies that will be called on in such an event. Unfortunately, many of the same considerations likely apply to all nations of the world. Ironically, substantial returns from these investments are guaranteed. Even in the absence of a deliberate attack, a robust and agile public health system and a biodefense strategy informed by advancing science and technology greatly enhances our ability to address the ever-present and constantly evolving threats to health from nature, as should be clear from the avian flu threat and other emerging infectious diseases. Although short on sizzle, such efforts are imperative. The costs will be high if we fail to make such investments.

The committee is indebted to the many individuals who provided their unique perspectives on the issues it faced and who through formal presentations and discussions ensured that the committee possessed the information it needed to inform its findings and recommendations. These individuals include Robert Carlson, James B. Petro, Pim Stemmer, Charles Rice, Drew Endy, Herb Lin, Sonia Miller, John Steinbruner, Barry Kellman, Michael Moodie, Terence Taylor, David Lipman, Charles Jennings, Phillip Campbell, Jonathan Tucker, Gerald Epstein, Jerrold Post, David Banta, Decio Ripandelli, Charles Arntzen, Miguel Gomez Lin, Luis Herrera-Estrella, Rosiceli Barreto Goncalves Baetas, Jacques Ravel, Patrick Tan Boon Ooi, Abdallah Daar, Gerardo Jimenez-Sanchez, Tibor Toth, Amy Sands, Robert Mathews, Jerome Amir Singh, Peter Herby, Nadrian Seeman, Michael Morgan, Kathryn Nixdorff, and Elliott Kagan. The committee also greatly appreciates the role played by the *Instituto Nacional de Salud Pública* (National Institute of Public Health) in Cuernavaca, Mexico, in hosting a workshop convened by the committee in September 2004 and recognizes in particular the head of the Institute, Mauricio Hernandez, and his staff.

This report has been reviewed in draft form by individuals chosen for their diverse perspectives and technical expertise, in accordance with procedures approved by the National Academies' Report Review Committee. The purpose of this independent review is to provide candid and critical comments that will assist the institution in making its published report as sound as possible and to ensure that the report meets institutional standards for objectivity, evidence, and responsiveness to the study charge. The review comments and draft manuscript remain confidential to protect the integrity of the process.

We wish to thank the following individuals for their review of this report: Ronald Atlas, University of Louisville; Edouard Brezin, French Academy of Science; Robert Carlson, University of Washington; Malcolm Dando, Bradford University; Drew Endy, Massachusetts Institute of Technology; Gerald Epstein, Center for Strategic and International Studies; David Franz, Midwest Research Institute; Gerald Fink, Whitehead Institute for Biomedical Research; Alistair Hay, Leeds University; James Hughes, Emory University; Stephen Johnston, University of Texas; Gigi Kwik Grönvall, Johns Hopkins University; Frederick Murphy, University of California, Davis; and Mark Wheelis, University of California, Davis.

Although the reviewers listed above provided many constructive comments and suggestions, they were not asked to endorse the conclusions or recommendations, nor did they see the final draft of the report before its release. The review of this report was overseen by Gilbert Omenn, of the University of Michigan. Appointed by the National Academies, he was responsible for making certain that an independent examination of this report was carried out in accordance with institutional procedures and that all review comments were carefully considered. Responsibility for the final content of this report rests entirely with the authoring committee and the institution.

Most important, this study and this report benefited greatly from the efforts of Eileen Choffnes, study director, who played a critical role in the committee's work and deliberations. Among her many contributions, she provided invaluable insight into previous efforts by the National Academies that related to many of the issues the committee was charged to address, helpful feedback on the committee's progress toward its goals, an effective organizational structure, constant encouragement (and occasional prodding), and great dedication to the project. For all of this the committee is most grateful. Other members of the National Research Council/Institute of Medicine staff also contributed substantially to the committee's work, including Stacey Knobler, who made important and thoughtful contributions during the early phases of the study, and Kate Skoczdopole and Katherine McClure, who provided extensive assistance in orchestrating committee meetings, organizing the international workshop, and keeping information flowing between committee members. Additional assistance with the writing of this report was provided by Leslie A. Pray. The committee expresses its profound gratitude to all of these highly talented individuals, with whom it has been a pleasure and a privilege to work during the past two years.

Stanley M. Lemon,
Co-chair

David A. Relman,
Co-chair

Contents

Executive Summary

Knowledge, materials, and technologies with applications to the life sciences enterprise are advancing with tremendous speed, making it possible to identify and manipulate features of living systems in ways never before possible. On a daily basis and in laboratories around the world, biomedical researchers are using sophisticated technologies to manipulate microorganisms in an effort to understand how microbes cause disease and to develop better preventative and therapeutic measures against these diseases. Plant biologists are applying similar tools in their studies of crops and other plants in an effort to improve agricultural yield and explore the potential for the use of plants as inexpensive manufacturing platforms for vaccine, antibody, and other products. Similar efforts are underway with animal husbandry. Scientists and engineers in many fields are relying on continuing advances in the life sciences to identify pharmaceuticals for the treatment of cancer and other chronic diseases, develop environmental remediation technologies, improve biodefense capabilities, and create new materials and even energy sources.

Moreover, other fields not traditionally viewed as biotechnologies—such as materials science, information technology, and nanotechnology—are becoming integrated and synergistic with traditional biotechnologies in extraordinary ways enabling the development of previously unimaginable technological applications. It is undeniable that this new knowledge and these advancing technologies hold enormous potential to improve public health and agriculture, strengthen national economies, and close

1

the development gap between resource-rich and resource-poor countries. However, as with all scientific revolutions, there is a potential dark side to the advancing power and global spread of these and other technologies. For millennia, every major new technology has been used for hostile purposes, and most experts believe it naive to think that the extraordinary growth in the life sciences and its associated technologies might not similarly be exploited for destructive purposes.

This is true despite formal prohibitions against the use of biological weapons and even though, since antiquity, humans have reviled the use of disease-causing agents for hostile purposes. In its most recent unclassified report on the future global landscape, the National Intelligence Council predicted that a major terrorist attack employing biological agents will likely occur by 2020, although it suggested that most future (i.e., over the course of the next 15 years) terrorist attacks are expected to involve conventional weapons. Official U.S. statements continue to cite around a dozen countries that are believed to have or to be pursuing a biological weapons capability. In addition to the efforts by terrorists or states with malevolent intent, we must be concerned about the grave harm that may result from misuse of the life sciences and related technologies by individuals or groups that are simply careless or irresponsible.

The continuing threat of bioterrorism, coupled with the global spread of expertise and information in biotechnology and biological manufacturing processes, has raised concerns about how advancing technological prowess could enable the creation and production of new threats of biological origin possessing unique and dangerous but largely unpredictable characteristics. The Committee on Advances in Technology and the Prevention of Their Application to Next Generation Biowarfare Threats, an ad hoc committee of the National Research Council and the Institute of Medicine, was constituted to examine current trends and future objectives of research in the life sciences, as well as technologies convergent with the life sciences enterprise from other disciplines, such as materials science and nanotechnology, that may enable the development of a new generation of biological threats over the next five to ten years, with the aim of identifying ways to anticipate, identify, and mitigate these dangers.

Specifically, the charge to the committee was to:

1.   Examine current scientific trends and the likely trajectory of future research activities in public health, life sciences, and biomedical and materials science that contain applications relevant to the development of "next generation" agents of biological origin five to ten years into the future.

2.   Evaluate the potential for hostile uses of research advances in ge-

netic engineering and biotechnology that will make biological agents more potent or damaging. Included in this evaluation will be the degree to which the integration of multiple advancing technologies over the next five to ten years could result in a synergistic effect.

3. Identify the current and potential future capabilities that could enable the ability of individuals, organizations, or countries to identify, acquire, master, and independently advance these technologies for both beneficial and hostile purposes.

4. Identify and recommend the knowledge and tools that will be needed by the national security, biomedical science, and public health communities to anticipate, prevent, recognize, mitigate, and respond to the destructive potential associated with advancing technologies.

This report is part of a larger body of work that the National Academies has undertaken in recent years on science and security and the contributions that science and technology could make to countering terrorism, beginning with *Scientific Communication and National Security* in 1982 and continuing with *Chemical and Biological Terrorism: Research and Development to Improve Civilian Medical Responses* (1999), *Firepower in the Lab: Automation in the Fight Against Infectious Diseases and Bioterrorism* (2001), *Making the Nation Safer: The Role of Science and Technology in Countering Terrorism* (2002), *Biological Threats and Terrorism: Assessing the Science and Response Capabilities* (2002), and *Countering Agricultural Terrorism* (2002). Most recently and of particular relevance to this report is the National Research Council report *Biotechnology Research in an Age of Terrorism* (2004). The principal difference between that report and the present report is that the former revolves around issues pertaining to the regulatory oversight of research employing biotechnology and the flow of scientific knowledge derived from the use of biotechnology, with a focus on the United States. In contrast, this report adopts a more global perspective, addressing the increasing pace of advances in the life sciences and related convergent technologies likely to alter the biological threat spectrum over the next five to ten years and broadly considering ways to prevent or mitigate the consequences of malevolent exploitation or naïve misapplication of these technologies.

While many readers might hope to find a well-defined, prioritized list or set of lists of future threats, the pace of research discovery in the life sciences is such that the useful lifespan of any such list would likely be measured in months, not years. Instead, the committee sought to define more broadly how continuing advances in life sciences technologies could contribute to the development of novel biological weapons and to develop a logical framework for analysts to consider as they evaluate the evolving technology threat spectrum. The committee concluded that there are classes or categories of advances that share important features relevant to their potential to contribute to the future development of new biological weapons. These shared characteristics are based on common purposes, common conceptual underpinnings, and common technical enabling platforms. Thinking of technologies within this framework should help in evaluating the potential they present for beneficial and destructive applications or technological surprise(s).

The committee classified new technologies according to a scheme organized around four groupings: (1) technologies that seek to acquire novel biological or molecular diversity; (2) technologies that seek to generate novel but pre-determined and specific biological or molecular entities through directed design; (3) technologies that seek to understand and manipulate biological systems in a more comprehensive and effective manner; and (4) technologies that seek to enhance production, delivery, and "packaging" of biologically active materials. This classification scheme highlights commonalities among technologies and, by so doing, draws attention to critical enabling features; provides insight into some of the drivers behind life sciences-related technologies; facilitates predictions about future emerging technologies; and lends insight into the basis for complementarities or synergies among technologies and, as such, facilitates the analysis of interactions that lead to either beneficial or potentially malevolent ends.

To a considerable extent, new advances in the life sciences and related technologies are being generated not just domestically but also internationally. The preeminent position that the United States has enjoyed in the life sciences has been dependent upon the flow of foreign scientific talent to its shores and is now threatened by the increasing globalization of science and the international dispersion of a wide variety of related technologies. The increasing pace of scientific discovery abroad and the fact that the United States may no longer hold a monopoly on these leading technologies means that this country is, as never before, dependent on international collaboration, a theme that is explored in depth in Chapter 2.

Foreign scientific exchange is an integral and essential component of the culture of science. The training of scientists from other countries in the United States has played an important role in fostering these interactions and has contributed substantially to the productivity of the American scientific enterprise. It has, however, been threatened recently by increased scrutiny of visa applications as well as the growing attractiveness of science and technology training opportunities outside of the United States. As technological growth becomes increasingly dependent on the global commons, international scientific exchanges and collaborations become an ever more vital component of U.S. technological capacity, including biodefense technological capacity. Weakening this link by prohibiting or discouraging bi-directional foreign scientific exchange—including the engagement of foreign students and scientists in U.S. laboratories, meetings, and business enterprises—could impede scientific and technological growth and have counterproductive, unintended consequences for the biodefense research and development enterprise.

Although this Report is concerned with the evolution of scientific and technological capabilities over the next five to ten years with implications for next-generation threats, it is clear that today's capabilities in the life sciences and related technologies have already changed the nature of the biothreat "space." The accelerating pace of discovery in the life sciences has fundamentally altered the threat spectrum. The immune, neurological, and endocrine systems are particularly vulnerable to disruption by manipulation of bioregulators. Some experts contend that bioregulators, which are small, biologically active compounds, pose an increasingly apparent dual-use risk. This risk is magnified by improvements in targeted delivery technologies that have made the potential dissemination of these compounds much more feasible than in the past.

The viruses, microbes, and toxins listed as "select agents" or "category A/B/C agents" and on which U.S. biodefense research and development activities are so strongly focused today are just one aspect of the changing landscape of threats. Although some of them may be the most accessible or apparent threat agents to a potential attacker, particularly one lacking a high degree of technical expertise, this situation is likely to change as a result of the increasing globalization and international dispersion of the most cutting-edge aspects of life sciences research.

The committee concluded that a broad array of mutually reinforcing actions are required to successfully manage the threats that face society. These must be implemented in a manner that engages a wide variety of communities that share stakes in the outcome. As in fire prevention, where the best protection against the occurrence of and damage from catastrophic fires comprises a multitude of interacting preventive and mitigating actions (e.g., fire codes, smoke detectors, sprinkler systems, fire trucks, fire hydrants, and fire insurance) rather than any single "best" but impractical or improbable measure (e.g., stationing a fire truck on every block), the same is true here. The committee, therefore, envisions a broad-based, intertwined network of steps—a *web of protection*—for reducing the likelihood that the technologies discussed in this report will be used successfully for malevolent purposes. It believes that the actions suggested in its recommendations (Box ES-1), taken in aggregate, will likely decrease the risk of inappropriate application or unintended misuse of these increasingly widely available technologies.

BOX ES-1 Recommendations

1. The committee endorses and affirms policies and practices that, to the maximum extent possible, promote the free and open exchange of information in the life sciences.

1a. Ensure that, to the maximum extent possible, the results of fundamental research remain unrestricted except in cases where national security requires classification, as stated in National Security Decision Directive 189 (NSDD-189) and endorsed more recently by a number of groups and organizations.

1b. Ensure that any biosecurity policies or regulations implemented are scientifically sound and are likely to reduce risks without unduly hindering progress in the biological sciences and associated technologies.

1c. Promote international scientific exchange(s) and the training of foreign scientists in the United States.

2. The committee recommends adopting a broader perspective on the "threat spectrum."

2a. Recognize the limitations inherent in any agent-specific threat list and consider instead the intrinsic properties of pathogens and toxins that render them a threat and how such properties have been or could be manipulated by evolving technologies.

2b. Adopt a broadened awareness of threats beyond the classical "select agents" and other pathogenic organisms and toxins, so as to include, for example, approaches for disrupting host homeostatic and defense systems and for creating synthetic organisms.

3. The committee recommends strengthening and enhancing the scientific and technical expertise within and across the security communities.

3a. Create by statute an independent science and technology advisory group for the intelligence community.

3b. The best available scientific expertise and knowledge should inform the concepts, plans, activities, and decisions of the intelligence, law enforcement, homeland security, and public policy communities and the national political leadership about advancing technologies and their potential impact on the development and use of future biological weapons.

3c. Build and support a robust and sustained cutting-edge analytical capability for the life sciences and related technologies within the national security community.

3d. Encourage the sharing and coordination, to the maximum extent possible, of future biological threat analysis between the domestic national security community and its international counterparts.

4. **The committee recommends the adoption and promotion of a common culture of awareness and a shared sense of responsibility within the global community of life scientists.**

4a. Recognize the value of formal international treaties and conventions, including the 1972 Biological and Toxin Weapons Convention (BWC) and the 1993 Chemical Weapons Convention (CWC).

4b. Develop explicit national and international codes of ethics and conduct for life scientists.

4c. Support programs promoting beneficial uses of technology in developing countries.

4d. Establish globally distributed, decentralized, and adaptive mechanisms with the capacity for surveillance and intervention in the event of malevolent applications of tools and technologies derived from the life sciences.

5. **The committee recommends strengthening the public health infrastructure and existing response and recovery capabilities.**

5a. Strengthen response capabilities and achieve greater coordination of local, state, and federal public health agencies.

5b. Strengthen efforts related to the early detection of biological agents in the environment and early population-based recognition of disease outbreaks, but deploy sensors and other technologies for environmental detection only when solid scientific evidence suggests they are effective.

5c. Improve the capabilities for early detection of host exposure to biological agents, and early diagnosis of the diseases they cause.

5d. Provide suitable incentives for the development and production of novel classes of preventative and therapeutic agents with activity against a broad range of biological threats, as well as flexible, agile, and generic technology platforms for the rapid generation of vaccines and therapeutics against unanticipated threats.

Recommendation 1

The committee endorses and affirms policies and practices that, to the maximum extent possible, promote the free and open exchange of information in the life sciences.

Overall, society has gained from advances in the life sciences because of the open exchange of data and concepts. The many ways that biological knowledge and its associated technologies have improved and can continue to improve biosecurity, health, agriculture, and other life sciences industries are highlighted in Chapter 2. Conversely, restrictive regulations and the imposition of constraints on the flow of information are not likely to reduce the risks that advances in the life sciences will be utilized with malevolent intent in the future. In fact, they will make it more difficult for civil society to protect itself against such threats and ultimately are likely to weaken national *and* human security. Such regulations and constraints would also limit the tremendous potential for continuing advances in the life sciences and its related technologies to improve health, provide secure sources of food and energy, contribute to economic development in both resource-rich and resource-poor parts of the world, and enhance the overall quality of human life.

The potential to develop effective countermeasures against biological threats is strongly enhanced by the nation's leadership position in the life sciences. However, implementation of the regulatory regime imposed by the PATRIOT and Bioterrorism Response acts on the life sciences community has raised concerns that qualified individuals may be discouraged from conducting biomedical and agricultural research of value to the United States for a variety of reasons. Moreover, many features of these statutes are considered unlikely to be effective in accomplishing their desired effect—limiting access to select agents by would-be terrorists—and may, in fact, lead to unintended consequences.

Recommendation 2

The committee recommends adopting a broader perspective on the "threat spectrum."

U.S. national biodefense programs currently focus on a relatively small number of specific agents or toxins, chosen as priorities in part because of their history of development as candidate biological weapons agents by some countries during the 20th century. The committee believes that a much broader perspective on the "threat spectrum" is needed. Recent advances in understanding the mechanisms of action of bioregulatory compounds, signaling processes, and the regulation of human gene expression—combined with advances in chemistry, synthetic biology, nanotechnology, and other technologies—have opened up new and exceedingly challenging frontiers of concern.

The limitations of the current select agent lists, and indeed any list, point to the need for a broadened awareness of the threat spectrum. Mechanisms must be put in place to ensure regular and deliberate reassessments of advances in science and technology and identification of those advances with the greatest potential for changing the nature of the threat spectrum. The process of identifying potential threats needs to be improved. This process needs to incorporate newer scientific methodologies that permit more rigorous assessment of net overall risks. Rather than adopting a static perspective, it will be important to identify and continually reassess the degree to which scientific advances or current or future biological "platforms" hold the potential for being put to use by potential adversaries. This will require the engagement of the scientific community in new ways and an expansion of the science and technology expertise available to the intelligence community.

Recommendation 3

The committee recommends strengthening and enhancing the scientific and technical expertise within and across the security communities.

A sound defense against misuse of the life sciences and related technologies is one that anticipates future threats that result from misuse, one that seeks to understand the origins of these threats, and one that strives to preempt the misuse of science and technology. It would be tragic if society failed to consider, on a continuing basis, the nature of future biological threats, using the best available scientific expertise, and did not make a serious effort to identify possible methods for averting such threats. Interdiction and prevention of malevolent acts are far more appealing than treatment and remediation. The committee, therefore, urges a proactive, anticipatory perspective and action plan for the national and international security communities.

There are several existing problems within the national security community and national political leadership related to the task of anticipating future biological threats. First, these groups have not developed the kinds of working relationships with the "outside" (non-governmental) science and technology communities that are needed (and are feasible). Second, "inside" groups (national security community and national political leadership) have been unable to establish and maintain the breadth, depth, and currency of knowledge and subject matter expertise in the life sciences and related technologies that are needed. The number of analysts in the national security community that have professional training in the life sciences and related technologies is small and insufficient; these analysts lose touch with the cutting edge of science and technology over time and tend to be moved from position to position, preventing them from developing any particular depth of expertise and experience. To the degree that the right kinds of expertise do exist in the analysis sectors, they do not adequately penetrate the intelligence collection process, and the expertise is distributed unevenly across these inside communities without sufficient coordination and integration. Moreover, intelligence assessments are not always shared among the different member agencies of the national security community. Finally, historical, political, and cultural barriers have prevented the national security community from working closely with counterparts from other nations and regions of the world. Yet the life sciences and related technologies are globally distributed in a seamless fashion, and future threats that arise from this science and technology will be globally distributed as well.

The committee, therefore, recommends the creation of an independent advisory group that would work closely with the national security community for the purpose of anticipating future biological threats based on an analysis of the current and future science and technology landscape, and current intelligence. In proposing the creation of this group, the committee supports Recommendation 13.1 of The Commission on the Intelligence Capabilities of the United States Regarding Weapons of Mass Destruction (March 31, 2005) that suggests the creation of a similar group, which they named the Biological Sciences Advisory Group. While the committee is mindful of the recent creation of the National Science Advisory Board for Biosecurity (NSABB) by the secretary of the U.S. Department of Health and Human Services, the current charter of the NSABB does not provide for the critical anticipatory and analytical functions that the committee envisions this new advisory group should provide to the intelligence community.

While the exact structure and specific charge of the entity that might fill this role are beyond the purview of this committee, the committee believes that the features of the advisory group, described in more detail in Chapter 4, will address critical unmet needs.

Recommendation 4

The committee recommends the adoption and promotion of a common culture of awareness and a shared sense of responsibility within the global community of life scientists.

The 1972 Biological and Toxin Weapons Convention (BWC) and the 1993 Chemical Weapons Convention (CWC) serve as cornerstones of the global biological-chemical regime, which has expanded to include rules and procedures rooted in measures ancillary to the two treaties. The biological-chemical regime as it currently exists—including the BWC, CWC, Australia Group, Security Council Resolution (SCR) 1540, and other measures—must be recognized for its positive contributions and placed within the overall array of measures taken to prevent biological warfare. Such international conventions should not be considered the solution to the issues society confronts today with respect to potential harmful use of advances in the life sciences, nor should they be cast aside and ignored. Despite their limitations, the committee appreciates their value in articulating international norms of behavior and conduct and suggests that these conventions serve as a basis for future international discussions and collaborative efforts to address and respond to the proliferation of biological threats.

The committee also appreciates the potential for codes of conduct or codes of ethics to mitigate the risk that advances in the life sciences might be applied to the development or dissemination of biological weapons. The committee concluded that the primary effect of such codes would be to create an enabling environment that would facilitate the recognition of potentially malevolent behavior (i.e., experiments aimed at purposefully developing potential weapons of biological origin) or potentially inappropriate experiments that might unwittingly promote the creation of a more dangerous infectious agent. The committee also recognized that such codes could generally be expected to achieve their desired effect only when reinforced by a substantial educational effort and appropriate role modeling on the part of scientific leaders. The "informal curriculum" probably drives what students learn and emulate more powerfully than the formal curriculum. Identifying, celebrating, and rewarding senior scientists who through word and deed serve as role models in preventing the malicious application of advances in biotechnology is perhaps the most important element in creating an environment that enables ethical and appropriate behavior.

The committee also envisions the establishment of a decentralized, globally distributed, network of informed and concerned scientists who have the capacity to recognize when knowledge or technology is being used inappropriately or with the intent to cause harm. This network of scientists and the tools they use would be adaptive in the sense that the capacity for surveillance and intervention would evolve along with advances in technology. Such intervention could take the form of informal counseling of an offending scientist when the use of these tools appears unwittingly inappropriate or reporting such activity to national authorities when it appears potentially malevolent in intent. While decentralized and adaptive solutions are potentially limited in effectiveness, they are nonetheless of substantial interest. Their usefulness may be limited to their ability to engender public opprobrium, but active steps to promote the development of distributed, decentralized networks of scientists will at the least heighten awareness while potentially enhancing surveillance. A good example of such a network is the Program for Monitoring Emerging Diseases, which hosts the ProMED-mail Web site. A similar instrument could be useful in establishing a shared culture of awareness and responsibility among life scientists. Such a distributed reporting and response network would be directed primarily at the community of legitimate scientists, its aggregate aim being to stimulate both creativity in anticipating activity that could be malicious, and vigilance in detecting and reporting such activity.

Recommendation 5

The committee recommends strengthening the public health infrastructure and existing response and recovery capabilities.

The committee recognizes that all of its recommended measures, taken together, provide no guarantee that continuing advances in the life sciences—and the new technologies they spawn—will not be used with the intent to cause harm. No simple or fully effective solutions exist where there is malevolent intent, even in cases where only minimal resources are available to individuals, groups, or states. Thus, its recommendations recognize a critical need to strengthen the public health infrastructure and the nation's existing response and recovery capabilities. In keeping with the focus of this report, the committee urges that the insights and potential benefits gained through advances in the life sciences and related technologies be fully utilized in the development of new public health defenses. Although many of the concepts and suggestions embodied in these recommendations were articulated in the 2002 National Research Council report, *Making the Nation Safer: The Role of Science and Technology in Countering Terrorism* ("Intelligence, Detection, Surveillance, and Diagnosis," Chapter 3, pp. 69-79), they remain as relevant and needed today as they were then.

An effective civil defense program will require a well-coordinated public health response, and this can only occur if there is strong integration of well-funded, well-staffed, and well-educated local, state, and federal public health authorities. Despite substantial efforts since September 11, 2001, few if any experts believe that the United States has achieved even a minimal level of success in accomplishing this goal, which is as important for responses to naturally-emerging threats, such as pandemic influenza, as for a deliberate biological attack. Current efforts to accomplish these aims have been woefully ineffective and have not provided the nation with the infrastructure it needs to deal rapidly, effectively, and in a clearly coordinated manner when faced with a catastrophic event such as an overwhelming tropical cyclone, a rapidly spreading pandemic, or a large-scale bioterrorism attack. These efforts need to be enhanced and expanded.

Early and specific diagnosis, even prior to the onset of typical signs and symptoms, should be the goal of research and development efforts. While it is reasonable to hope that improved diagnostic tests will be developed as a result of current federal biodefense research efforts, it is not clear that adequate attention, prioritization, or investment have been devoted to this important area or that all of the potentially useful approaches (e.g., comprehensive monitoring of host-associated molecular biological markers) have been adequately explored. There is a similar need for early recognition and diagnosis of animal and plant diseases. Equally important is the development of broadly active vaccines or biological response modifiers capable of providing protection against large classes of agents. To date, well-established companies in the pharmaceutical and vaccine industries have had little financial incentive to develop new vaccines or therapeutics for biological threat agents for which the market is extremely uncertain and dependent ultimately on government procurement decisions. Continued efforts must be taken to address this failure of the market to produce the countermeasures needed.

CONCLUSION

Because its members believe that continuing advances in the life sciences and related technologies are essential to countering the future threat of bioterrorism, the committee's recommendations affirm policies and practices that promote the free and open exchange of information in the life sciences. The committee also affirms the need to adopt a broader perspective on the nature of the threat spectrum and to strengthen the scientific and technical expertise available to the security communities so that they are better equipped to anticipate and manage a diverse array of novel threats. Given the global dispersion of life sciences knowledge and technological expertise, the committee recognizes the international dimensions of these issues and makes recommendations that call for the global community of life scientists to adopt a common culture of awareness and a shared sense of responsibility, including specific actions that would promote such a culture.

It remains unclear how the country's response to a future biological attack will be managed. How will the responses of many different federal departments (e.g., Departments of Homeland Security, Health and Human Services, Justice, and Defense and the myriad agencies within them) be effectively integrated, and who will control operations and ensure they are adequately interfaced with local and state governments and public health agencies? Although well beyond the scope of the committee's charge, the development of an effective means of integrating the responses by multiple government agencies would provide the nation with perhaps the most necessary of "tools" with which to meet any future challenge.

1

Framing the Issue

In these early years of the 21st century, scientific discovery and understanding are playing an important and growing role in meeting the challenges—environmental, human health, economic—facing societies everywhere. At the forefront are advances in biology. Indeed, it is reasonable to say we are entering the Age of Biology, paralleling in many ways the Age of Physics in the first half of the 20th century.[1]

For many thousands of years, humans have been manipulating plant and animal stocks—first by accident and later selectively—to meet changing societal and environmental needs. But the discovery of the structure of DNA in 1953, followed by the invention of DNA recombinant technology two decades later, paved the way for the powerful potential to manipulate genes directly and in such a way that the "nature" of an organism can be altered with precision in a single generation. In 2001, scientists finished the initial draft of the human genome sequence, representing a shift in the way biology is studied and opening a portal to vast post-genomic possibilities—from RNA interference (RNAi) therapeutics to DNA nanotechnology. This rapid pace of technological growth in the life sciences research enterprise reflects a revolutionary change in the way people interact with biological systems and a growing capacity to manipulate such systems. Such advancing technologies offer great promise for improving the quality of human life: promoting health, preventing disease, and ensuring adequate food and even the possibility of new energy sources. However, as with all technological advances, there is a potential dark side, the ability for these technologies to be used, either purposefully or negligently, in ways that cause harm to humans. Devising optimal approaches for preventing this has been the overarching aim of this committee.

This chapter provides an overview of recent growth in the life sciences and its associated technologies—with an emphasis on the rapid and shifting nature of this growth. It defines key terms that are used throughout this report and explores the broad-based nature of the threat posed by the rapid, unpredictable growth, and widespread dissemination of life sciences knowledge and associated technologies. This overview takes into account contemporary understanding of how naturally emerging pathogens cause disease and recently developed technologies that have opened up novel approaches to engineer potentially more harmful agents from both pathogenic and nonpathogenic microbes or viruses. In reviewing this material, the committee developed a heightened awareness of the tremendous potential of the benefits to be derived from the advancement of knowledge and technological growth in the life sciences. At the same time, committee members came to appreciate the magnitude of what hangs in the balance should society fail to address the potential for these technologies to be exploited to cause harm or, by overreacting and imposing unduly restrictive measures on activities in the life sciences, unwittingly muzzle the ability of the life sciences to contribute to future human good.

COMMITTEE CHARGE AND PROCESS

As discussed above and in more detail throughout the report, life sciences knowledge, materials, and technologies are advancing with tremendous speed, making it possible to identify and manipulate features of living systems in ways never before possible. On a daily basis and in laboratories around the world, biomedical researchers are using sophisticated technologies to manipulate microorganisms in an effort to understand how microbes cause disease and to develop better preventative and therapeutic measures against infectious disease. Plant biologists are applying similar tools in their studies of crops and other plants in an effort to improve agricultural yield and explore the potential for the use of plants as inexpensive platforms for vaccine, antibody, and other product manufacturing. Similar efforts are underway with animal husbandry. Scientists and engineers in many fields are relying on continuing advances in the life sciences to identify pharmaceuticals for the treatment of cancer and other chronic diseases, develop environmental remediation technologies, improve biodefense capabilities, and create new materials.

Moreover, other fields not traditionally viewed as biotechnologies—such as materials science, information technology, and nanotechnology—are converging with biotechnology in unforeseen ways and thereby enabling the development of previously unimaginable technological applications. It is undeniable that this new knowledge and these advancing technologies hold enormous potential to improve public health and agriculture, strengthen national economies, and close the development gap between resource-rich and resource-poor countries. However, as with all scientific revolutions, there is a potential dark side to the advancing power and global spread of these and other technologies. Every major new technology has been used for hostile purposes, and many experts believe it is naive to think that the extraordinary growth in the life sciences and its associated technologies might not be similarly exploited for malevolent purposes.[2]

This is true despite formal prohibitions against the use of biological weapons and even though, since antiquity, humans have reviled the use of disease for hostile purposes. In its most recent unclassified report on the future global landscape, the National Intelligence Council argued that, although most future (i.e., over the course of the next 15 years) terrorist attacks are expected to involve conventional weapons, a bioterrorist attack will likely occur by 2020.[3] Official U.S. statements continue to cite around a dozen countries that are believed to have or to be pursuing biological weapons capabilities.[4]

The threat of bioterrorism, coupled with the global spread of expertise in biotechnology and biological manufacturing processes, raises concerns about how this advancing technological prowess could enable the creation and production of new biological weapons and agents of biological terrorism possessing unique and dangerous but largely unpredictable characteristics. The Committee on Advances in Technology and the Prevention of Their Application to Next Generation Biowarfare Threats, an ad hoc committee of the National Research Council and the Institute of Medicine, was constituted to examine current trends and future objectives of research in the life sciences, as well as technologies convergent with the life sciences enterprise from other disciplines, such as materials science and nanotechnology, that may enable the development of a new generation of biological threats over the next five to ten years, with the aim of identifying ways to anticipate, identify, and mitigate these dangers.

As part of its study, the committee convened a workshop in September 2004 at the *Instituto Nacional de Salud Pública* (National Institute of Public Health) in Cuernavaca, Mexico. The purpose of this information gathering workshop was to sample global perspectives on the current advancing technology landscape. Experts from different fields and from around the world presented their diverse outlooks on advancing technologies and forces that drive technological progress; local and regional capacities for life sciences research, development, and application (both beneficial and nefarious); national perceptions and awareness of the risks associated with advancing technologies; and strategic measures that have been taken or could or should be taken to address and manage the potential misapplication of technology(ies) for malevolent purposes. The results of this workshop helped inform the committee as it developed this report.

The committee was charged to:

1.   Examine current scientific trends and the likely trajectory of future research activities in public health, life sciences, and biomedical and materials science that contain applications relevant to the development of "next generation" agents of biological origin 5 to 10 years into the future.
2.   Evaluate the potential for hostile uses of research advances in genetic engineering and biotechnology that will make biological agents more potent or damaging. Included in this evaluation will be the degree to which the integration of multiple advancing technologies over the next 5 to 10 years could result in a synergistic effect.
3.   Identify the current and potential future capabilities that could enable the ability of individuals, organizations, or countries to identify, acquire, master, and independently advance these technologies for both beneficial and hostile purposes.
4.   Identify and recommend the knowledge and tools that will be needed by the national security, biomedical science, and public health communities to anticipate, prevent, recognize, mitigate, and respond to the destructive potential associated with advancing technologies.

In interpreting its charge the committee sought to examine current trends and future objectives of research in public health and the life and biomedical sciences that contain applications relevant to the development of new types of biological weapons or agents of bioterrorism, with a focus on five to ten years into the future. It is recognized that the global technology landscape is shifting so dramatically and rapidly that any attempt by the committee to devise a formal risk assessment of the future threat horizon exploiting dual-use technologies by state actors, non-state actors, or individuals could be an exercise in futility. Given that within just the past few years the global scientific community has already witnessed the unexpected emergence of some remarkable new technologies, such as RNA interference and nanobiotechnology, biological threats in the next five to ten years could extend well beyond those that can be predicted today.

Rather than a formal risk assessment, the committee has proposed a conceptual framework for how to think about the nature of the future threat landscape. Indeed, as the world becomes more competent and sophisticated in the biological sciences, it is vitally important that the national security, public health, and biomedical science communities have the necessary knowledge and tools to address the present and future applications of advances in the life sciences.

This report is part of a larger body of work that the National Academies has undertaken in recent years on science and security and the contributions that science and technology could make to countering terrorism, beginning with *Scientific Communication and National Security* in 1982 and continuing with *Chemical and Biological Terrorism: Research and Development to Improve Civilian Medical Responses* (1999), *Firepower in the Lab: Automation in the Fight Against Infectious Diseases and Bioterrorism* (2001), *Making the Nation Safer: The Role of Science and Technology in Countering Terrorism* (2002), *Biological Threats and Terrorism: Assessing the Science and Response Capabilities* (2002), and *Countering Agricultural Terrorism* (2002). Most recently, and of particular relevance to this report, is the National Research Council report *Biotechnology Research in an Age of Terrorism* (2004). The principal difference between that report and the present report is that the former revolves around issues pertaining to research oversight and the flow of scientific knowledge, with a focus on the United States, whereas this report adopts a more global perspective and broadly considers the use and applications of such knowledge.

EMERGING TECHNOLOGIES IN THE LIFE SCIENCES

Heralded by *Science* magazine as the 2002 "Breakthrough of the Year,"[5] RNA interference (RNAi) has emerged as a promising therapeutic approach for the treatment of a wide range of diseases, including cancer.[6] Yet just a year before it earned its breakthrough title, RNAi was met with doubt and criticism.[7] RNAi therapy involves using small interfering RNA molecules (siRNAs) to cleave and destroy sequence-specific RNA and, in so doing, silence endogenous genes that participate in the pathway of human disease. The technology is expected to prove particularly valuable in cases where the targeted RNA encodes genes and protein products with activities that cannot be modulated today by conventional drugs. Several recent experiments indicate that investigators are well on their way to overcoming the clinical challenges of delivering effective RNAi therapy.[8] In October 2004, Acuity Pharmaceuticals (Philadelphia, PA) announced that it was beginning a Phase I clinical trial of an investigational drug known as Cand5, making Cand5 the first RNAi therapeutic to enter clinical trial. Cand5 is an siRNA that turns off the expression of proteins contributing to vision loss in patients with age-related macular degeneration.

In addition to its therapeutic applications, RNAi has emerged as a key basic research tool for use in functional genomics; by blocking the expression of a particular gene, one can create a phenotype that yields clues about the function of that gene. RNAi technology is forecast to grow at an annual average rate of just over 30 percent between 2003 and 2010.[9] Although European and U.S.-based companies currently dominate the market (i.e., there are about 50 U.S. and European companies active in the RNAi market, most of their revenues coming from RNAi reagents and research tools),[10] this may change over the course of the next several years as Asian companies begin specializing in RNAi applications.

Touted alongside RNAi in Massachusetts Institute of Technology's (MIT) 2004 *Technology Review* as one of the top 10 emerging technologies that "will change your world," synthetic biology is the assemblage of gene networks—or circuits (i.e., analogous to silicon circuits)—that can guide the construction of novel, synthetic proteins and direct cells to perform assigned tasks.[11] By assembling genes into circuits that direct cells to perform assigned tasks, synthetic biologists have taken genetic engineering to a level so profoundly different from recombinant technology that, in an October 2004 *Nature* news article, the latter was referred to as "old hat."[12] DNA synthesis applications are now largely limited to places like the MIT's Independent Activities Period (IAP) course, where students design DNA circuitry, send their designs via the Internet to Blue Heron Biotechnology, Inc. (Bothell, WA), and then introduce the resulting synthetic DNA molecules into *E. coli* strains.[13] Because it is in its early growth phase, the future industrial potential of synthetic biology is unclear.[14] Meanwhile, research scientists are using the technology to design unique genomes and test novel hypotheses and models.

In just five years, nanotechnology has catapulted from being a specialty of a handful of physicists and chemists to a worldwide scientific and industrial enterprise.[15] The U.S. government estimates that the nanotech economy will be worth $1 trillion by 2012, and the White House recently requested $1 billion for fiscal 2006 to develop nanotechnology (up from $442 million in 2001). In April 2005, the National Academies Keck Futures Initiative announced that it had awarded a total of $1 million to 14 interdisciplinary research projects in nanoscience and nanotechnology. The awards, which are considered seed money to allow recipients to develop research approaches and position themselves competitively for other project funding, will be used for a variety of projects ranging from an examination of the interactions of nanoparticles with biosystems to the development of a new approach for capturing solar energy.

Nanoparticles are already being used in a variety of commercial products, like sunscreen, paint, inkjet paper, stain-resistant trousers, and highly durable engine parts.[16] Some industry analysts predict that by lowering drug toxicity and the cost of treatment (among other benefits), nanotechnology-enabled drug delivery systems will probably be among the first biomedical markets to evolve and to provide significant business revenue opportunities.[17] For example, Elan Corporation (Dublin, Ireland) has developed a proprietary technology known as NanoCrystal, which transforms poorly water-soluble drugs into nanometer-sized particles that can be used to create any of a variety of more soluble common dosage forms for both parenteral and oral administration. There are several NanoCrystal-based therapeutics already on the market or in development.[18]

Nanobiotechnology—also known as DNA nanotechnology—refers to the convergence of nanotechnology with molecular biology.[19] In fact, most of the examples in the preceding paragraph fall within its domain. Nanobiotechnology and nanobiotech start-up companies constitute nearly 50 percent of the venture capital invested in nanotechnology.[20] Scientists are increasingly reporting discoveries with implications for potential applications of nanobiotechnology. For example, in January 2005, in a paper published in *Physical Review Letters*, researchers from the University of California, Los Angeles, described a nanoscale mechanism for externally controlling protein function, a technological advance that could ultimately lead to a generation of targeted "smart" drugs that are active only when certain DNA is present or a certain gene is expressed.[21] In February 2005, in a paper published in the *Proceedings of the National Academy of Sciences*, Northwestern University researchers described a nanoparticle-based assay for detecting the onset of Alzheimer's disease.[22] Also in February 2005, an Illinois-based company, Nanosphere, Inc., announced plans to expand and market the application of the same assay to a variety of other diseases, including cancer.[23]

While new tools, like RNAi therapeutics and nano-based drug delivery are emerging, already proven tools such as the polymerase chain reaction (PCR) and DNA sequencing, are becoming more versatile, more affordable, and faster. For example, real-time, or quantitative PCR (qPCR), which is arguably one of the fastest growing PCR technologies, allows users to quantitatively monitor the amplification process as copies of DNA accumulate (unlike "traditional" PCR, which provides only an end product, a "yes/no" answer, and a qualitative measure of the abundance of the target material).[24] In 2004, the least expensive qPCR thermocycler on the market was listed in the mid-$20,000 range. In spring 2005, Bio-Rad Laboratories (Hercules, CA) launched a "personal" qPCR machine that sells for about $16,500 and is one of the smallest machines on the market (i.e., in terms of size and the number of samples it can accommodate).

Moreover, it should not be forgotten that PCR itself was not widely anticipated before its arrival on the scene.[25] And it is instructive to remember how it developed, first as a relatively straightforward concept in which DNA synthesis was recycled through a series of cyclic thermal manipulations.[26] This resulted in a doubling of the product each thermal cycle with an exponential amplification of the product over many thermal cycles of annealing, extension, and denaturation, with the DNA polymerase enzyme being destroyed during the denaturation step. However, it was not until a thermally-resistant DNA polymerase was isolated from nature that the process became widely available and widely utilized.

Now PCR is as indispensable a "tool" for many 21st-century biologists as a microscope was to a 19th-century microbe hunter. Its impact on accelerating the velocity of life sciences research is readily appreciated by anyone in the field, as most biotechnologists today would have difficulty accomplishing their aims without this technique. Its importance overall to the life sciences is reflected in the relatively unusual actions of the Norwegian Nobel Committee, conferring its award on the inventor of PCR, Kary Mullis, only a few years after the technique was first reported. Parallels to the thinking that went into PCR are seen today in an unrelated field—the investigation of spongiform encephalopathies, like "mad cow disease," where an analogous cycling technique has been reported recently for in vitro amplification of prions, putative infectious agents that lack genes (i.e., DNA or RNA) and that consist of a protein with "infectious" capacity to initiate misfolding of similar proteins.[27] This series of events in the development of PCR recapitulates a theme in the life sciences: the sudden arrival of a new technique, followed by its technological exploitation, further refinement, and subsequent extension to other related fields. Similar scenarios have accompanied the discovery of restriction endonucleases and the development of recombinant DNA, and are unfolding now with RNAi technology or recently described multiplex DNA synthesis capabilities.

The speed of DNA sequencing, DNA synthesis, and protein structural analysis—each a different measure of biotechnological power—has increased practically exponentially over the past 15 years.[28] Indeed, progress in the life sciences, rather than being "linear," is often marked by periodic and unpredictable major breakthroughs in our understanding of the living world that consequently radically transforms the growth and development of advances in disparate disciplines.[29] At present, the 10 plant and animal genomes and the approximately 100 microbial genomes that are sequenced every year are done so, largely, at a small number of factory-like DNA sequencing centers. It has been estimated that if technological developments continue to improve the efficiency of DNA sequencing as they have up to this point, by 2010 a single lab worker will be able to sequence (or synthesize) about 10^{10} bases in one day (there are 3×10^9 bases in the human genome).[30]

The future of DNA synthesis is likely to follow a similarly rapid trajectory, with scientists being able to synthesize complete microbial genomes by 2010 if not sooner.[31] In December 2004 Harvard University's George Church and colleagues published an article in *Nature* describing a new microchip-based technology for the multiplex synthesis of long oligonucleotides.[32] The researchers used the new technology to synthesize all 21 genes that encode proteins of the *E. coli* 30S ribosomal subunit. This technological advance is coupled with falling prices. In 2000, sequence

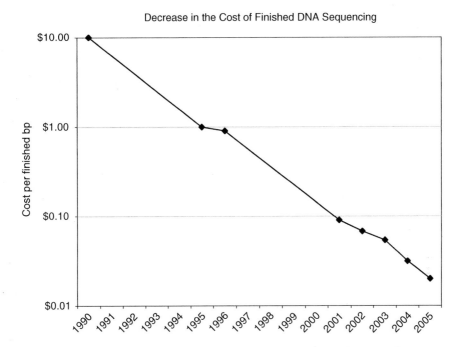

FIGURE 1-1 The plunging cost of DNA sequencing has opened new applications in science and medicine.
SOURCE: Reprinted with permission. Service, RF. 2006. Gene sequencing: The race for the $1,000 genome. *Science* 311(5767):1544-1546. Available online at www.sciencemag.org/cgi/content/full/311/5767/1544.

assembly cost about $10 to $12 per base pair. By the beginning of 2005, the cost had dropped to about $2 per base pair (e.g., Blue Heron offers a special price of $1.60 for new customers[33]), and it is expected to fall to 1 cent per base pair within the next couple of years[34] (see Figure 1-1).

This has had real and practical consequences. For example, when the first successful autonomously replicating RNA replicons for hepatitis C virus were described by the Bartenschlager laboratory in 1999,[35] several other groups immediately synthesized the entire ~7,000 nucleotide-long complementary DNA sequence of this RNA so as to be able to access this technology. De novo chemical synthesis was judged to be a more rapid, or less expensive means to acquire the technology than working through Materials Transfer Agreements, etc., with those who first described the replicons. The DNA synthetic "muscle" for this was readily available on a contract basis, even five years ago. Such an exercise would be trivial to-day, however, given recent advances in DNA synthetic capacities.[36]

Similar predictions about feasibility, rapidity, and affordability can be made for the structural analysis of proteins and other biologically important molecules. It is not unreasonable to expect that, before long, scientists will develop and have access to computer programs that simulate in detail the molecular processes in cells, so that the interaction of cells with pathogenic microbes and molecules can be fully anticipated and understood.

Notable Features of Technological Growth in the Life Sciences

Technological growth in the life sciences is characterized by several notable features. These are critically important to recognize if a reasonable estimate is to be made of what is or is not possible in predicting its future.

First, as described above, progress in biology has been marked repeatedly by successive serendipitous discoveries and applications that over time have lead to the widespread adoption of new technologies with independent scientific and economic impacts. Indeed, the rapid growth of bio- and other relevant technologies over the past 30 years has been driven by two processes working together: a quantitative increase in performance coupled with a decrease in the cost of existing technologies (such as template independent DNA synthesis) and instruments, as explained in the previous section, and sudden and occasionally dramatic qualitative changes (paradigm shifts) resulting from unanticipated new inventions, unexpected discoveries, and insights, all of which may be significantly enhanced by the occurrence of unforeseen, historically significant events that impact significantly on human society and its everyday concerns. In addition to recombinant DNA technology (which sparked the biotech revolution back in the 1970s), prominent new inventions and discoveries in recent history include PCR (i.e., which originated in the mid-1980s as described above), the transfer of nuclei from cell to cell (i.e., cloning, also known as somatic cell nuclear transfer, or SCNT), the advent of RNAi technology (as described above), and the introduction of new techniques for parallel DNA synthesis capable of greatly accelerating the rate at which genes can be created de novo. New inventions and discoveries like these are a precondition for the rapid growth of technology. They result in the capacity to reduce the development costs associated with new and potentially very useful products, such as the recombinant hepatitis B vaccine, one of the early "fruits" of the recombinant DNA era, or to genetically engineer crops with intrinsic resistance to pests.

Equally important, however, are both public and political support for these efforts. Such support can, in turn, drive the availability of government or venture capital funding required to fuel the advancement of research and development activities in the life sciences. Current levels of government support in the life sciences can be attributed in part to unforeseen historical events, such as the political decision to declare a "war on cancer" in the 1970s, the occurrence of the HIV/AIDS pandemic in the 1980s, and the 2001 anthrax mailings, which, in part, contributed to the current "war on terror." On the other hand, the public perception of risks can readily derail the expansion of this technology, as evidenced by the impact of the "green" movement in Europe on the acceptance of genetically engineered crops by the public.

This constantly changing and rapidly growing global technological landscape, marked as it is by the seemingly stochastic arrival of new paradigm-shifting concepts, makes it extremely difficult, if not impossible, to predict specific future trends. Just a year before it earned its "Breakthrough of the Year" title by *Science* magazine,[37] RNAi was met with doubt and criticism. Self-assembling nano-devices, such as the DNAzyme (a device that can bind and cleave RNA molecules one by one) developed in 2004 by Purdue University researcher Chengde Mao, were unimaginable just a couple of years ago.[38] About the only thing one can predict is that the life sciences will continue to advance quickly, in a variety of directions, and that new and previously unanticipated paradigm shifts are very likely to occur in the future.

Second, as difficult as it is to predict what kind of technological or scientific breakthroughs might occur next, it is practically impossible to know where in the world these breakthroughs might happen. As discussed in greater detail in Chapter 2 of this report and in an earlier workshop summary report from this committee, a number of countries around the world are investing heavily in life sciences technologies.[39] Indeed, several countries that are not commonly viewed as being technologically sophisticated, or that have not been considered technologically savvy in the past, are making remarkable progress in biotechnology and are well-positioned to become regional or global leaders in the near future. Importantly, the rapid global dispersion of life sciences materials, knowledge, and technologies is not limited to technologies with proven therapeutic and market value. While India is currently strong in generic and bulk biopharmaceutical manufacturing, several factors, including its growing technological expertise and its 2005 accession to the World Trade Organization, are contributing to its greater capacity for innovation and research and development of novel products. South Korea is rapidly gaining global prominence for its breakthrough contributions to stem cell research, although some of these "breakthroughs" are now in dispute.[40] Meanwhile, Singapore has identified biotechnology as a central pillar of its future economy. Biotechnology is no longer the restricted playing field of a few privileged nations, but is truly a global enterprise.[41]

Third, the number of known biologically active molecules, and potential genetically engineered organisms, that could cause harm to humans through inadvertent use, inappropriate use, or as a result of purely malevolent intent such as in the development of a weaponizable biological or chemical agent, is increasing rapidly. This stands in sharp contrast to the still relatively small number of nuclear materials that could potentially be used for malign intent. This is evident in the increasing pace of research activity in the life sciences, as reflected in the number of biotech drug approvals (i.e., as opposed to large pharma drug approvals), which grew from fewer than 5 in 1982 (and none in 1983) to more than 30 in 2000.[42] According to the Biotechnology Industry Organization (BIO), there are approximately 370 biotech drug products and vaccines currently in clinical trials targeting more than 200 diseases.[43] This growing number of potential and approved drugs is due in part to a fundamental shift in the drug discovery process.

New technologies—genomics, microarrays, proteomics, structural biology, combinatorial chemistry, toxicogenomics, and database mining—allow drug developers to identify likely molecular targets early in the discovery process and then screen large numbers of compounds that bind to and affect the targets. Moreover, purely "*in silico*" screening approaches are becoming more common. There has thus been a shift in drug discovery methodology from pure empiricism to more rationally based drug design. In addition, new methods for synthesizing chemical libraries have led to the aggregate generation of several hundred million new potential ligands, while the same discovery process has identified thousands of potentially toxic compounds each year.[44] Until now, most of the databases produced by these efforts have been proprietary and jealously guarded by the companies that generated them.[45] However, the Chemical Genomics Center network, recently established by the National Institutes of Health (NIH), will make this type of information much more accessible (see Box 1-1). Without any knowledge of the underlying biological mechanism, it will be possible to mine this vast chemical database to unearth structural relationships between desirable targets and the chemical compounds known to interact with them. While originally conceived as leading ultimately to a "roadmap" of the functions of the myriad proteins expressed by the human genome, it is also possible that this novel program could become a "roadmap" to new generations of very efficient poisons. Thus, as will become evident throughout this report, almost any effort to advance knowledge in the life sciences, such as the NIH Roadmap (Box 1-1), brings with it the potential for malevolent use as well as beneficial impact.

Finally, one can imagine a future where, as biotechnology continues to change radically, rather than becoming big and centralized, the life sciences and related applications may become increasingly domesticated and accessible. An example of this is the recent pet store appearance of genetically modified tropical fish with new and brilliant colors (see Box 1-2). The fish were developed as a commercial product by the Taikong Corporation of Taiwan; they first appeared on the commercial market in Taiwan, Japan, Hong Kong, and Malaysia in 2003. More importantly, just as computer technology was transformed over the course of a few decades to a point where computers were small enough and cheap enough to be used in homes (e.g., to prepare income tax returns or homework) and then to a point where computer games and toys became a dominant feature of children's lives, biotechnology may similarly be transformed.[46]

Definitions

The life sciences are defined broadly in this report to include any field of science that is leading to or has the potential to lead to an enhanced understanding of living organisms, especially human life. These sciences include, for example, branches of mathematics and computational science, as these are now being applied in efforts to effectively model a wide variety of biological systems, or materials science, as it is applied to the manipulation of biological systems. Here "associated technology" refers to the development and application of tools, machines, materials, and processes based on knowledge derived within or applied to the life sciences: genetic engineering, synthetic biology, aerosol technology, combinatorial chemistry, and nanotechnology are just a few of these technologies. The future problem in some respects resembles the current threat; for every biological threat agent there must be a means for delivery to the target (aerosols are usually thought to be the most efficient mode for agents that are not contagious). The efficient delivery of medicines by aerosol (insulin) or nanotechnology is intensely relevant "enablig technologies" to future problems, even if they are not inherently "biology." In addition, the future holds threats that are so different from the classical biothreats that they fall into entirely new categories and will require highly innovative ways to detect and counter them. The aim of this report, therefore, is to be as inclusive as possible in looking at technologies—including those not traditionally viewed as biotechnologies—but that may lead to the creation and production of new biological weapons and biological warfare threats.

The terms "weapon" and "bioweapon" are also used broadly, and include any biological agent or biologically active molecule or other entity[47] that is used or developed and/or stockpiled for use in an effort to cause harm to humans, plants, or animals. In some cases a bioweapon

---

BOX 1-1
The NIH Roadmap: Where Will It Lead Us?

Even seemingly "benign" or solely beneficial activities such as those envisioned by the achievements of the National Institutes of Health (NIH) Roadmap could serve as a new source of potential dual-use information— that is, information that can be used inappropriately or for malicious purposes as well as for the beneficial intent for which it is designed. The NIH Roadmap is designed to identify major opportunities and gaps in biomedical research that no single NIH institute could tackle alone but that, by relying on an interdisciplinary research approach, would make the biggest impact on medical research progress in the coming century. It is highlighted here simply to point out that the potential dual-use nature of the information to be generated may not be fully appreciated. The challenge is to devise a strategy for allowing this necessary and beneficial research to move forward while preventing inappropriate or malicious use.

There are three NIH Roadmap "themes." The first, "New Pathways to Discovery," among other goals, addresses the need to know more precisely the combination of molecular events that lead to disease and involves establishing a library of chemical molecules for use in identifying potential targets for new therapies and other purposes. The second, "Research Teams of the Future," involves stimulating new ways of combining skills and disciplines in the physical and biological sciences, for example, by encouraging public-private partnerships and encouraging investigators to conduct research that is high-risk but also high-payoff. The third, "Re-engineering the Clinical Research Enterprise," addresses the need for new partnerships and networks between and among the scientific and clinical communities in order to better translate research discoveries into drugs, treatments, and preventative methods.

Each of these themes comprises several or more groups and initiatives, the details of which can be viewed on the NIH Roadmap Web site:

---

may be purposefully modified to enhance its ability to be delivered or to otherwise cause harm. However, it is not necessary for a biological agent to be specifically "weaponized" for it to be used as a weapon, as, for example, a routine culture of a bacterial pathogen might simply be added to food or drinking water.

The term "dual-use" refers to the capacity or potential for biological agents, information, materials and supplies, or technologies to be used for either harmful or peaceful purposes. This definition needs to be distinguished from a common use of the term within defense circles, wherein "dual-use" implies a potential military application for a civilian technology.

nihroadmap.nih.gov/overview.asp. One example of an initiative that may eventually lead to potent new dual-use information is the Chemical Genomics Center, established in June 2004 by the Molecular Libraries and Imaging Implementation Group, as part of the New Pathways to Discovery theme. This center will be part of a consortium of chemical genomics screening centers to be located across the country whose purpose will be to identify small molecule inhibitors of every important human cellular protein or signaling pathway. Part of the rationale for the chemical genomics initiative(s) is that, in contrast to researchers in the pharmaceutical industry, many academic and government scientists do not have easy access to large libraries of small molecules (i.e., organic chemical compounds that are smaller than proteins and that can be used as tools to modulate gene function). The database will give academic and government researchers an opportunity to identify useful biological targets and thereby contribute more vigorously to the early stages of drug development. With plans to screen more than 100,000 small-molecule compounds within its first year of operation, one of the goals of the Chemical Genomics Center network is to explore the areas of the human genome for which small molecule chemical probes have yet to be identified. Data generated by the network will be deposited in a comprehensive database of chemical structures (and their biological activities). The database, known as PubChem, will be freely available to the entire scientific community. In addition to screening and probe data, it will list compound information from the scientific literature. Should this come to pass, it will offer enormous opportunities for industry and academic scientists alike to pursue novel "drugable" targets in a search for small molecule inhibitors of certain pathways that could offer substantial clinical benefit. However, the availability of information and reagents that enable one to disrupt critical human physiological systems has profound implications for the nature of the future biological and chemical threat spectrum.

As used in this report, the terms "inappropriate use" and "malicious use" or "malevolent use" capture a continuum of potentially dangerous activities that are largely distinguished on the basis of *intent.* Thus, (1) the use of technology without the intent to cause harm but with unanticipated dual-use consequences, including experiments or other activities conducted with inadequate oversight or without an awareness of the consequences of certain outcomes would be considered *inappropriate* use, while, (2) the deliberate use of technology for the creation, development, production, or deployment of biological weapons is considered *malevolent* or *malicious* use, with malevolent indicating the intent to cause death or

BOX 1-2
One Fish, Two Fish, Three Fish—GloFish™

The GloFish™, a fluorescent red zebrafish sold as a novel pet, has become the first transgenic animal sold to U.S. consumers. Its sale has produced regulatory controversies, a lawsuit, and profits for its proponent, Yorktown Technologies (Austin, TX). With the market plan calling for sales in a widening number of countries, continuing controversy seems likely.

What Is a GloFish™?

The GloFish™ is a trademarked transgenic zebrafish (*Danio rerio*) expressing a red fluorescent protein from a sea anemone under the transcriptional control of the promoter from the myosin light peptide 2 gene of zebrafish.[a] Produced and patented by a group at the National University of Singapore,[b] exclusive rights for international marketing were purchased by Yorktown Technologies in 2002. Yorktown produces GloFish™ through contracts with 5-D Tropical (Plant City, FL) and Segrest Farms (Gibsonton, FL) and began marketing the fish in the United States in December 2003.

Issues Posed

The prospect of commercial sales of GloFish™ has raised a number of issues. Among them was the issue of whether GloFish™ pose an environmental hazard. Zebrafish, a tropical species native to south Asia, are sensitive to low temperatures. Despite decades of production and use in the United States, zebrafish have not established self-sustaining populations in this country. Laboratory tests showed that viability, reproductive success, and temperature tolerance of transgenics were equal to or somewhat less than those of the wild type.[c] While preliminary, results supported the expectation that the modification would not increase invasiveness and that environmental risk was small.

Commercialization of the GloFish™ in the United States poses regulatory uncertainty because existing biotechnology policy bases oversight on use of the product. Sales of ornamental fishes are not federally regulated. The Food and Drug Administration asserts jurisdiction over genetically modified animals using the New Animal Drug Application process.[d] After a brief internal review and interagency consultation, the FDA's Center for Veterinary Medicine determined that "because tropical aquarium fish are not used for food purposes, they pose no threat to the food supply. There is no evidence that these genetically engineered zebra fish pose any more threat to the environment than their unmodified counterparts which have long been widely sold in the United States. In the absence of any clear risk to the public health, the FDA finds no reason to regulate these particular fish."[e] Alan Blake, CEO of Yorktown Technologies, also made contact with the U.S. Department of Agriculture, the U.S. Fish and Wildlife Service, and the Environmental Protection Agency, which expressed no regulatory concerns regarding GloFish.™

Future Prospects
Future prospects for the GloFish™ include marketing additional color lines in a wider range of markets. Not only red, but also green and yellow fluorescent proteins have been introduced into stable transgenic lines, yielding green, yellow, and orange fish.

Commercialization of fluorescent zebrafish has gone forward in several countries and is stymied in others. Fluorescent green zebrafish developed in Taiwan have been sold in Taiwan, Malaysia, and Hong Kong. Singapore confiscated attempted imports of the fish. Despite this, Yorktown Technologies is considering other markets, including parts of Asia and Latin America. Extensive information requirements suggest that GloFish™ will not be marketed in Canada or the European Union in the near future. Despite these regulatory challenges, according to Blake, "The GloFish™ venture is a profitable one, and the company looks forward to continuing to provide a safe and enjoyable product for many years to come."[f]

---

[a]Gong, W., et al. 2003. Development of transgenic fish for ornamental and bioreactor by strong expression of fluorescent proteins in the skeletal muscle. *Biochemical and Biophysical Research Communications* 308:58-63.

[b]National University of Singapore. 2004. Zebrafish as pollution indicators. Available online at www.nus.edu.sg/corporate/research/gallery/research12.htm [accessed January 4, 2006].

[c]U.S. Food and Drug Administration. 2003. FDA statement regarding Glofish. Available online at www.fda.gov/bbs/topics/NEWS/2003/NEW00994.html [accessed January 4, 2006].

[d]FDA asserts its authority to regulate transgenic animals under the "new animal drug application" authorities it has under the Federal Food, Drug, and Cosmetics Act (Title 21 of the Code of Federal Regulations.)

[e]Schuchat, S. 2003. Why GloFish won't glow in California. *San Francisco Chronicle* (December 17):A29. Available online at sfgate.com/cgi-bin/article.cgi?f=/c/a/2003/12/17/EDGQV3KOLB1.DTL [accessed January 4, 2006].

[f]Adapted from Hallerman, EM. 2004. Glofish, The First GM Animal Commercialized: Profits Amid Controversy. (June). Available online at www.isb.vt.edu/articles/jun0405.htm [accessed January 4, 2006].

serious injury, and malicious a lesser degree of intended damage. These latter terms do not include the deliberate use of technology to create potentially harmful materials or other disease-causing agents for defensive research purposes in the absence of any intent to cause harm (e.g., the equivalent of computer hacking[48]). The committee recognizes, however, the controversial and troubling aspects of such approaches, and their potential to add to, as well as potentially help mitigate, the threat of bioterrorism or biological warfare.

The term "bioterrorist" refers to individuals or groups, usually non-state actors, that develop and/or use biological agents with the intent to cause harm. On the other hand, "biological warfare" refers to the intentional use of such weapons by state actors, regardless of whether they are deployed against civilian or military targets, or on either a large or small scale.

Finally, the term "biosecurity" is used to refer to security against the inadvertent, inappropriate, or intentional malicious or malevolent use of potentially dangerous biological agents or biotechnology, including the development, production, stockpiling, or use of biological weapons as well as natural outbreaks of newly emergent and epidemic diseases. Although it is not used as often as it is in other settings, to refer to a situation where adequate food and basic health are assured,[49] there may be significant overlap in measures that guarantee "biosecurity" in either sense.

20TH CENTURY GERM-BASED BIOWARFARE

History has demonstrated that research in biology, even when conducted without any military application in mind, may still contribute to the production of biological weapons. Indeed, people figured out how to intentionally spread illnesses long before naturalists discovered that germs cause disease.[50] But it was only after the discovery of the germ theory of disease in the late 19th century that infectious diseases were seriously considered, on a continual basis, as tools of war. Biologists were able for the first time to identify, isolate, and culture disease-causing microbes under controlled conditions and use them to intentionally induce disease in a "naïve" host.

For example, one of the first attempts to use anthrax during warfare was in World War I, when the Germans reportedly attempted to ship horses and cattle inoculated with *B. anthracis* (as well as *Burkholderia mallei*, the bacterium that causes glanders in livestock) from U.S. ports to Allies.[51] In 1917, German spies were caught allegedly trying to spread *B. anthracis* among reindeer herds in northern Norway, near the Russian border.[52] These charges were confirmed when anthrax-laced sugar cubes, obtained from a Swedish-German-Finnish aristocrat arrested as a German agent in 1917, were recently found to still be viable after being stored in the archives of a Norwegian museum for the past 80 years.[53]

During World War II, every major combatant had a biological weapons program in place (including the United Kingdom, the United States, Canada, France, the Soviet Union, Germany, and Japan).[54] The U.S. offensive biological weapons program originated in 1942, at Camp Detrick, in Frederick, Maryland. Its focus during WWII was on biological warfare research on the causative agents of anthrax, botulism, and many other human, animal, and plant pathogens.[55] The Special Projects Division of the Army Chemical Warfare Service, which was primarily responsible for carrying out the program, had at its peak approximately 3,900 personnel, including about 2,800 Army and 1,000 Navy personnel and 100 civilians. Although Camp Detrick remained the parent research and pilot plant center, field-testing facilities were established in 1943 and 1944 in Mississippi and Utah, respectively, and production plants were constructed in Indiana and Pine Bluff, Arkansas in 1944. After WWII, until the program was dismantled in 1969, it developed and perfected offensive weapons capabilities for the Department of the Army and certain weapons capabilities for the Air Force, Navy, and the Central Intelligence Agency, using a variety of human, animal, and plant pathogens. All work was conducted under the strictest secrecy. [56]

Japan's secret biological warfare program, Imperial Unit 731 (hereinafter Unit 731), which was officially known as the Army Anti-Epidemic Prevention and Water Supply Unit, studied, cultured, and developed a large number of biological agents, including B. anthracis and C. perfringens, which reportedly were used on prisoners of war.[57] There were at least four operational units of Japan's secret biological warfare complex: Unit 731, located in Ping Fan; Unit 100 in Changchun; Unit 9420 in Singapore; and Unit Ei 1644 in Nanking. There is also some evidence that the Japanese had an "epidemic prevention center"—a euphemism for biological weapons research on tropical diseases—in Rangoon, Burma. Each unit had 10 to 15 individual facilities located within and outside mainland China.[58] During the Sino-Japanese War (1937-1945), Japan repeatedly attacked China with the plague-causing bacterium Yersinia pestis, reportedly targeting over eleven cities. At least 700 Chinese reportedly died from plague alone, although the actual morbidity and mortality associated with Unit 731's germ warfare "experiments" against Chinese nationals and others is likely to be several orders of magnitude higher.[59]

Even after the Biological and Toxin Weapons Convention (BWC) was opened for signature in 1972, the Soviet Union retained and expanded an extensive secret biological weapons program that involved tens of thousands of workers—the largest biological weapons complex ever created. On April 3, 1979, an accidental release of anthrax was believed to have occurred at the Soviet Institute of Microbiology and Virology in Sverdlovsk. A reported 67 people died from inhalation anthrax, and another 33, perhaps more, were reported to have been infected with *B. anthracis*, the causative agent of the disease.[60] For years the Soviet government officially maintained the cover story that the outbreak was gastrointestinal anthrax and was due to ingestion of contaminated beef, denying that the incident had anything to do with an accidental release of anthrax from an upwind military research facility. In 1992, the Russian press reported that President Boris Yeltsin officially acknowledged that the 1979 incident had in fact been an accidental airborne release of anthrax spores from a military research facility, although many Russian scientists continued to steadfastly deny the occurrence of such an accident. The incident reinforced U.S. suspicions that the Soviets had a biological weapons program, despite having signed the BWC in 1972 and upon ratification, making the statement: "The Soviet Union does not possess any bacteriological agents and toxins, weapons, equipment, or means of delivery." The quantity of spores released at Sverdlovsk was recently estimated at less than one gram, but the basis for this estimate is speculative.[61]

The nature and extent of the former Soviet Union's biological weapons program became known to western governments after Vlademir Pasechnik and then Ken Alibek, chief and deputy chief (respectively) of Biopreparat, defected to the United Kingdom[62] and the United States in the late 1980s and early 1990s.[63] Biopreparat, an ostensibly state-owned pharmaceutical organization was, in reality, carrying out a secret offensive and defensive biological weapons program that operated from 1972 until at least 1992.[64] It was the most sophisticated biological weapons program in the world, and its size and scope were enormous. By the early 1990s, more than 60,000 people were involved in the research, development, and production of biological agents for use in weapons, and the complex had the capability to stockpile hundreds of tons of material containing anthrax spores and dozens of tons of material containing other pathogens, including smallpox and plague agents.[65] Many state programs were involved in various aspects of this effort. The Ministry of Defense and its research facilities, of course, played a central role in setting requirements for the program and in program implementation. Components of the Ministry of Health and the Ministry of Agriculture and selected institutes of the Soviet Academy of Sciences were also involved. The KGB developed the capability to deliver biological weapons through clandestine systems.[66] These activities were carried out despite Soviet assurance set forth in international agreements not to develop a biological weapons capability and open declarations to the United Nations that it was not developing such capacity.

South Africa's clandestine program, Project Coast, a chemical and biological weapons program that existed from the 1980s until 1993-1994, when the government announced that it was dismantling all weapons of mass destruction programs, serves as another example of a recent state-level, clandestine bioweapons program. The extent of Project Coast was not publicly known until 1998-1999, when the Truth and Reconciliation Commission offered immunity to many scientists in exchange for disclosure of their involvement with the project.[67] The now-transparent history serves as a dramatic example of how science can be subverted to undermine entire communities and how scientists can be persuaded to participate in a clandestine state-level biological weapons program. At the time of the project, research conducted in the national interest was considered the most important research in the country.[68] As recommended by the international community, the South African government has attempted to keep many experts in this area employed under its watch rather than have them take their expertise elsewhere.[69]

Beating Nature: Is It Possible to Engineer a "Better" Pathogen?

The rapid, unpredictable, and widespread growth of the life sciences and biotechnology has raised concerns that, while such growth benefits national development and enriches the quality of life for millions of people worldwide, it also creates new opportunities for inappropriate or malicious use. The question then becomes, what type of biological agent, or bioweapon, poses the greater threat, and do human-engineered bioweapons pose a greater or lesser threat than naturally emerging infectious disease agents?

Natural Threats

It has been argued by some that nature serves as the most potent reservoir of biological threats to humans, animals, and plants and as a source of biothreat diversity. This issue deserves further discussion here, given the relevance of the counter-arguments (synthetic or engineered agents may be as potent or more potent—at least in the short-term—than "natural pathogens") that are based on the potential impact of advancing technologies. This argument posits that deliberate efforts to create novel biological threat agents will not succeed in constructing agents more (or even, as) potentially harmful than those that have or can arise through natural means, because of the broad spectrum of natural mechanisms that give rise to biological diversity and the competitive and selective pressures brought to bear on these natural agents. In considering this argument, however, it is important to consider the principles underlying "pathogenicity" and to recognize that the capacity to injure humans does not, by

TABLE 1-1 Cases and Deaths of Emerging Infectious Diseases in the Past and Present

Historic Pandemics[a]	Causative Agent	Cases	Deaths
Justinian Plague, 6th Century (First recorded outbreak of bubonic plague)	bacterium *Yersinia pestis*	142 million (based on an estimated 70% mortality rate)	~100 million
The "Black Death"	bacterium *Yersinia pestis*		25 million
China Plague (or "Third Pandemic"), 1896-1930	bacterium *Yersinia pestis*	30 million	12 million
Spanish Flu 1918-1919	influenza A virus	200 million	50 million - 100 million

Current Pandemics[b]	Causative Agent	Cases	Deaths
Malaria	*Plasmodium* parasites	300 million to 500 million per year	1.5 million to 2 million per year
Tuberculosis	*Mycobacterium tuberculosis*	8 million to 10 million per year	2 million per year
Hepatitis C	Hepatitis C virus (HCV)	~170 million (cumulative)	10,000 per year (U.S. only)
HIV/AIDS	Human Immunodeficiency Virus-1 (HIV-1)	more than 60 million (cumulative)	more than 20 million (cumulative)

itself, provide any virus, bacterium, or other infectious agent with a selective survival advantage. Rather, injury or "disease" occurs as an incidental effect of mechanisms evolved by the infectious agent to promote its multiplication and long-term survival. To illustrate the devastation that natural biological agents can cause, Table 1-1 provides a snapshot of cases and deaths of emerging infectious diseases in the past and present. In addition to those listed, many other infectious diseases have emerged, reemerged, or developed drug resistance over the past couple of decades and across the globe; every hour an estimated 1,500 people die from an infectious disease.[70]

Influenza virus is considered by many to be the greatest natural infec-

TABLE 1-1 Continued

Recent and Current Outbreaks[c]	Causative Agent	Cases	Deaths
Marburg hemorrhagic fever (in Angola; as of May 17, 2005)	Marburg virus	337	311
Avian influenza (in Asia and Eurasia) beginning January 2003; as of April 18, 2006)	H5N1 Influenza A virus	194	109
Meningococcal disease (in Burkina Faso, from January 1 to April 20, 2003)	N. meningitidis	7,146	1,058
Severe Acute Respiratory Syndrome (Worldwide, from November 1, 2002, to July 31, 2003)	SARS-associated coronavirus (SARS-CoV)	8,096	774

[a]Information on historic pandemics is adapted from "Killer Diseases through Time," *The Scientist* 17(11), 2003:16; updated information from Institute of Medicine. 2005. *The Threat of Pandemic Influenza: Are We Ready?* Washington, DC: The National Academies Press; Osterholm, M. 2005. Preparing for the Next Pandemic. *Foreign Affairs*, 84, (4): 24-37; McNeill, W. 1998. Plagues and Peoples. Anchor Book published by Doubleday Press: New York.

[b]Institute of Medicine. 2003. *Microbial Threats to Health: Emergence, Detection, and Response.* Washington, DC: The National Academies Press.

[c]Information on recent outbreaks is from the World Health Organization.

tious disease threat faced by the world today. However, it is but one example of a potentially devastating natural threat. The magnitude of the threat posed by influenza reflects several different features of the virus: its ability to be readily transmitted among humans, to cause significant tissue injury, and to circumvent preexisting immunity within a population by the rapid acquisition of novel surface antigens that are not recognized by antibodies elicited by prior influenza infections. This latter feature of influenza illustrates one of several natural mechanisms by which viruses and microbes create genetic diversity in their populations. However, influenza virus does this in an ongoing fashion and at a dizzying pace, at times making fantastic genetic leaps. Many scientists consider an influenza pandemic—one that could conceivably kill tens or even hundreds of millions of people worldwide—to be imminent.[71] With the spread of the highly pathogenic avian influenza (H5N1) in Asia and Eurasia, politicians and the general public have recently begun to realize the danger.[72]

Sometimes referred to as a "continually emerging" infectious disease agent (as opposed to an emerging one), influenza viruses cause epidemics annually, in part aided by a phenomenon known as "antigenic drift." Antigenic drift reflects the fact that the virus constantly accumulates genetic mutations (errors in its genome) and that over time, this eventually results in significant antigenic changes in its surface proteins that lessen their ability to be recognized by virus-neutralizing antibodies prevailing in the host population. Less often, a more dramatic change in the antigenic structure of the virus takes place through a process of reassortment of its segmented genome. This occurs through mixing of gene segments from different influenza viruses co-infecting the same host, producing a new influenza strain with a different complement of gene segments. Reassortment between avian and human influenza viruses is thought to occur in intermediate hosts, such as swine, and can lead to the appearance of a novel human virus with potentially heightened virulence as well as complete resistance to preexisting immunity ("antigenic shift").[73] Such an event is thought to have led to the emergence of a pandemic strain of influenza virus three times in the past century: in 1918 ("Spanish" influenza, H1N1); in 1957 ("Asian" influenza, H2N2); and in 1968 ("Hong Kong" influenza, H3N2).[74] The 1918 influenza A pandemic (H1N1), which may have claimed as many as 50 million to 100 million lives worldwide in less than a year, ranks as one of the worst disasters in human history.[75] Isolated cases and small outbreaks of disease due to highly pathogenic avian influenza have become more frequent over the past decade. The current epizootic of H5N1 avian influenza in Asia and Eurasia is unprecedented in its scale, geographic distribution, and economic loss. Tens of millions of birds have died of influenza, and hundreds of millions more have been culled to protect humans.[76] According to information provided by the World Health Organization, between January 2003 and April 21, 2006, there were 204 confirmed human cases and 113 deaths of avian influenza A (H5N1), spread across 9 countries: Viet Nam (93 cases, 42 deaths), Thailand (22 cases, 14 deaths), Cambodia (6 cases, 6 deaths), China (17 cases, 12 deaths), Indonesia (32 cases, 24 deaths), Turkey (12 cases, 4 deaths), and Iraq (2 cases, 2 deaths).[77] Evidence suggests that the currently circulating H5N1 virus has accumulated mutations that have made it increasingly infectious and deadly in multiple bird species, as well as in mammals.[78] Thus far there has been little evidence of human-to-human transmission of the H5N1 virus, but many experts are concerned that the virus may need to accumulate only a limited number of mutations in order for it to acquire the ability to be efficiently spread between people.

The influenza virus genome is composed of RNA, placing it among a group of human pathogens that evolve relatively quickly, even within a single host. Other RNA viruses, such as HIV and hepatitis C, are also particularly prone to this behavior. These agents generate significant degrees of genetic variability and appear to have "sampled" (or to be sampling) a large proportion of, if not all possible, gene sequence possibilities ("sequence space") as they replicate and spread in host populations. This is due to the lack of proofreading capabilities in the polymerases responsible for amplification or copying their RNA genomes, a feature that distinguishes these RNA-dependent RNA polymerases and reverse transcriptases from the DNA polymerases that copy bacterial or protozoan genomes. This lack of proofreading allows for rapid generation of sequence diversity and, coupled with very efficient replication schema (a typical, chronically infected human, produces about 10^{12} new hepatitis C virus particles per day and a typical HIV-infected person about 10-fold less), promotes a process of accelerated "natural selection" that optimizes the ability of the virus to sustain a successful interaction with the host, which generally means to be able to multiply and spread to the next host.

Despite the fact that their genomes are copied with much greater fidelity, bacteria are also capable of generating genetic diversity by sharing mobile genetic elements, such as plasmids, or by receiving exogenous genes via bacterial virus infection. In fact, genes that confer virulence on a bacterium tend to be carried on mobile genetic elements.[79] These "accessory" genes and functional potential allow an organism to compete more successfully in their interactions with a host and are dispensable to those microbes that either choose a different (exogenous) habitat or temporarily adopt a nonpathogenic lifestyle in the host. Complex regulatory systems recognize cues indicative of the host environment and modulate expression of virulence-associated genes accordingly. Some DNA viruses, such as the herpesviruses, appear to have "picked up" genes from their hosts, probably through a process of DNA recombination, and have modified these for their own purposes, thus increasing their genetic diversity and potential for survival.

Given the clear capability of at least some microbes and viruses to evolve quickly, acquire new genes, and alter their behavior, it might seem reasonable that over hundreds of thousands of years all conceivable biological agents have been "built" and "tested" and that the agents seen today are the most "successful" of these. Thus, is there any reason to think that it might be possible to artifically create a more successful biological agent? Possibly not, but it is important to understand that "successful" in this context means the most able to survive within, on, or near human populations over time. "Success" does not necessarily equate with virulence, or pathogenicity, the ability to cause disease or injury.

The Evolution of Pathogenicity: What Does It Take to Cause Disease?

Early views of pathogenicity and virulence were based on the assumption that these characteristics were intrinsic properties of microorganisms, although it was recognized that pathogenicity was neither invariant nor absolute.[80] Over the course of the last century, as increasing numbers of viral and microbial pathogens were identified and the pathogenesis of multiple infectious diseases was characterized, the complexity and individuality of host-pathogen relationships became evident, while the general definitions of pathogenicity and virulence became increasingly qualified and cumbersome. Viral pathogenicity reflects two fundamental features of a viral infection: the ability of the virus to cause direct injury to tissue (i.e., its cytopathogenicity) and the amount of injury associated with either an effective or ineffective immune response to the presence of the virus. More subtle forms of pathogenicity also arise, such as when a virus such as a papillomavirus causes malignant transformation of a cell or when a differentiated cellular function (i.e., insulin secretion or T-helper cell function) is lost along with the targeted destruction of a special, differentiated cell type. Bacterial, fungal, and multicellular parasites are somewhat different in that they can survive on host mucosal surfaces or skin, and do not necessarily require invasion of a cell to multiply and survive. Pathogenicity may be equated in many, but not all, bacterial pathogens with the inherent ability to cross host cell barriers—a property conferred by the expression of virulence factors, many of which are encoded by discrete DNA segments known as pathogenicity islands. On the other hand, from the host's perspective, disease occurs only when the presence of a microbe (whether protozoan, bacterial, or viral) results in damage—whether that damage is actually mediated by the pathogen itself or by the host's immune response to it. [81]

Virulence—defined broadly as the ability of an infectious agent to cause disease in a host—is a relatively rare trait even among those microbes capable of survival within a host, such as humans. All but a tiny fraction of the microbes that have thus far been found on the planet are incapable of replication under human physiologic conditions. Of those that spend a significant fraction of their existence in a human or other mammalian host, most are on or near a mucosal surface, or on the skin, in competition with a wide variety of other microbial strains and species. The human body has been estimated to contain approximately 10^{14} cells, 90 percent of which are microbial![82] Some human endogenous sites, such as the skin and mucosal surfaces, have evolved with a robust microbial community in attendance. Successful colonization of these sites by so-called commensals does not precipitate a strong enough response to result in damage. In fact, most microbes that reside in the gut—such as *Lactobacillus spp.* and *Bacteroides spp.*—may actually serve a protective, not pathogenic, role.[83]

Rather than producing overt illness in their hosts, the vast majority of microbes establish themselves as persistent colonists: either low-impact parasites (organisms that cause asymptomatic infections), commensals (organisms that "eat from the same table," deriving benefit without harming their hosts), or symbionts (microbes that benefit their hosts).[84] These states, while separate, represent a section of a continuum—one that extends to pathogenicity and disease—which is occupied by various microbial species at various times depending on environmental, genetic, and host factors.[85] Persistent colonization of a host by a microbe is rarely a random event; such coexistence depends on a relationship between host and microbe that can be characterized as a stable equilibrium.[86] Pathogenic microbes acquire genes that enable them to exploit their hosts, but they generally have evolved to do so in ways that allow both the hosts and pathogens to persist. In certain cases (e.g., when microbes cause persistent, asymptomatic infection), this equilibrium can be disrupted by physiological or genetic changes in either the host or microbe, shifting the relationship toward pathogenesis and resulting in illness and possibly death for the host.[87] The selective forces controlling evolution of the microbe are determined by its survival on the planet, not necessarily what it does to its host. However, there are often no direct positive benefits derived by the microbe per se in causing disease or killing its host.[88]

Of the several thousand species estimated to inhabit the body, only a handful are capable of causing disease on a routine basis, while only a modest additional number are capable of causing disease when host defenses become impaired. Those that regularly cause disease in unimpaired hosts employ a strategy for replication and survival that involves colonization of a highly protected anatomic site that is usually off-limits to microbes; the strategy includes mechanisms for resisting or subverting host defenses. The net result of this strategy and the ensuing host response is damage to the host and disease. Among the different viral and bacterial species that routinely cause human disease, there are multiple and diverse strategies for gaining access to the appropriate habitat, adhering to the relevant receptors, overcoming host defenses, replicating and/or persisting.[89] Sometimes the strategy involves a long-term association with the host and more subtle disruptions of host physiology. Hepatitis C virus is a prime example, causing a persistent infection of the liver that typically remains asymptomatic for decades and causes significant disease (cirrhosis or liver cancer) only in a minority of infected persons. Another is *Helicobacter pylori*, which infects half the world's population but causes gastric disease in only one out of five carriers. *H. pylori* is an example of a potentially pathogenic (parasitic) microbe that more often assumes the role of commensal or symbiont.[90]

Recent research indicates that viral as well as bacterial pathogens that infect or colonize animals share broadly common strategies with those that infect plants.[91] Both can express proteins that mimic, suppress, or modulate host cell-signaling pathways and enhance pathogen fitness, and both are recognized by similarly sophisticated host surveillance systems. Striking architectural similarities between surface appendages of plant and animal pathogenic bacteria suggest common mechanisms of infection, while structural differences—most notably the presence versus absence of a cell wall—reflect the profound differences between plant and animal cells. Studies of "interkingdom" pathogens, such as *Pseudomonas aeruginosa*, which can infect both humans and plants, such as *Arabidopsis thaliana*, reveal common features that permit a wide host range.[92]

Pathogenic bacteria are relatively restricted in their phylogenetic distribution across the bacterial domain. To date, only 7 of the more than 80 divisions of bacteria contain well-recognized pathogens, and within these 7 divisions the distribution of pathogens is focal. Clearly, some microbes are inherently capable of adapting to life within or on humans, while others are not. Notably, there is not a single known organism within the domain *Archaea* that is capable on its own of causing disease in humans (see Figure 1-2).[93]

A few methanogens are common inhabitants of the human intestinal tract, and they have recently become strongly implicated in the common gum disease chronic periodontitis, where their role is believed to be indirect, as partners in syntrophic relationships with other bacteria.[94] However, the true spectrum of archaeal associations with disease and archaeal virulence mechanisms has barely been explored, in part because of the difficulty in detecting and characterizing these organisms. Similarly, in recent years there has been a growing awareness that viral agents also inhabit and replicate robustly in humans in the absence of disease expression, such as the DNA TT viruses[95] that persistently infect the majority of some well-studied human populations, or GB virus C, a distantly-related, non-hepatotropic cousin of the RNA hepatitis C virus that thus far is not recognized to cause any specific disease.

In discussing the known diversity of pathogens and the considerable microbial community diversity that has yet to be characterized, it is important to recognize that the pressures that have guided the evolution of these specialized microbes over long periods of time impose a number of critical constraints. All organisms, including pathogens, have been se-

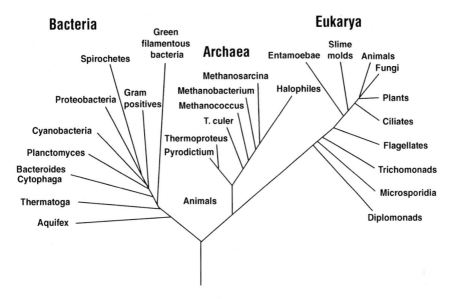

FIGURE 1-2 *Archaea* represents one of the three domains of life, the other two being *Bacteria* and *Eukarya*.
SOURCE: Todar, K. 2004. Major Groups of Prokaryotes. University of Wisconsin-Madison Department of Bacteriology. Available online at www.bact.wisc.edu/Bact303/MajorGroupsOfProkaryotes [accessed January 4, 2006].

lected for their ability to persist and survive on the planet. There is, after all, a fitness cost associated with being too virulent—not just for the host but for the parasite as well. If a pathogen kills or inflicts too much damage on its host, it may impact its own survivability and transmissibility, and could end up effectively committing suicide. Most successful pathogens maintain an evolutionary middle ground with respect to the amount of damage they exact on their host; to survive in the privileged anatomic niche they have chosen and to be transmitted to a new susceptible host, they may need to inflict some degree of injury but not so much that they hinder the fitness of their host as an optimal partner in attaining these goals: pathogen survival, persistence, and transmission.

An interesting and potentially serious anomaly is provided by those infectious agents, such as some arthropod-borne viruses like West Nile virus, that infect humans "accidentally." For such pathogens, human infection is not a necessary part of its essential life cycle, as for example West Nile virus usually cycles between avian and mosquito species with only occasional forays into mammals such as humans. When this happens, disease (tissue injury) can be catastrophic, as the host-pathogen interaction has not been tempered by evolution. It is interesting that most pathogens on the category A select agent list fall into this category, among them anthrax, plague, tularemia, and the viral hemorrhagic fevers.

Recent studies have suggested that some natural human pathogens are not nearly as virulent as they could be. Genes associated with "hyper-virulence" have been identified: Some encode products that down-modulate virulence (mutations that lead to depressed virulence); other genes that are found to be inactive or missing in some pathogens encode factors that might enhance virulence in some hosts.[96] Hyper-virulence is thought to reflect an exaggerated form of behavior that might be deleterious to the microbe over long periods of time, and a behavior against which natural selective forces tend to act; however, over shorter periods of time, or in more restricted portions of the natural environment, these behaviors or capabilities might not be so detrimental or even relevant. For example, in extant strains of disease-causing *Salmonella enterica*, the product of the *pcgL* gene, a D-Ala-D-Ala dipeptidase, facilitates growth in nutrient-poor conditions, presumably found outside the human host.[97] Naturally occurring disruptions of this gene in some clinical strains cause the organism to be hyper-virulent, but at the cost of impaired survival in the external environment. Some pathogens occasionally increase their virulence to overcome a different type of disadvantage they face in the environment, such as poor vector competence.[98]

Thus, the setting and conditions under which nature judges the "success" of a pathogen may limit our appreciation for the kinds of virulence properties that might be possible in a biological agent and cause us to arrive at false conclusions concerning our ability to create new pathogenic agents. The overall survival strategy of a pathogen may involve adaptation to an external environment and transmission among hosts over thousands of years; the result is often attenuation of virulence for the human host.[99] Nature rewards long-term survival of an organism whereas, long-term survival is not an important requirement for an organism to be capable of causing disease and disrupting human populations in the short-term (e.g., months, years). This is particularly so if the infectious agent is aided and abetted in its production and distribution by a motivated and knowledgeable actor. Such reflections suggest that the spectrum of virulence, both in terms of severity and phylogenetic diversity, observed among natural pathogens may not reflect the spectrum that is possible and that might be achieved, albeit for only short periods of time, if one were to explore additional realms of genetic diversity and genomic arrangement (see Chapter 3, "sampling biological diversity"). Thus, Mother Nature may not be "the mother of all terrorists" after all, and it is reasonable to anticipate that humans are capable of engineering infectious agents with virulence equal to or perhaps far worse than any observed naturally.

The Importance of the Host Response

Our growing understanding of host-microbial interactions has led to an increased awareness of the host as a key component and determinant of the host-microbe outcome. While this greater awareness and understanding of host defense may lead to new strategies for the recognition, prevention, treatment, and prediction of outcome of microbial disease, it also broadens the knowledge base from which bioterrorists could design new forms of biological weapons that disrupt host homeostatic systems,[100] with diverse and potentially devastating consequences.

Irrespective of whether caused by a virus or bacterium, infectious diseases typically result either from direct tissue injury caused by an infectious agent or from the host's response to it. How, then, does the host differentiate between pathogenic and benign microbes? Evidence has accumulated, over the past decade, to suggest that the host's "innate" immune system recognizes the presence of molecular "danger signals" in the form of molecular patterns that are primarily associated with most microorganisms and viruses, regardless of whether it is a pathogen or commensal.[101] For example, in mammals, considerable research on the epithelial cells lining the gut and other mucosal surfaces indicates that this cell layer serves as a key interchange in a signaling network that transmits signals between microbes, and adjacent and underlying immune and inflammatory cells.[102] Commensal microorganisms constantly stimulate pattern recognition receptors (e.g., Toll-like receptors (TLR)) at a low level and produce stereotyped responses that are protective for the gut mucosa.[103] Thus, in addition to providing a mechanical barrier between the host and its environment, mucosal epithelial cells sense the makeup of the intestinal microflora and provide signals to the host that affect the growth, development, and function of nearby cells, including the activation of inflammatory and immune responses. Macrophages and other cells of the immune system continuously patrol the inner tissues of the body, expressing similar molecular pattern receptors and constantly on the search for similar stimuli where they should not be. Parenchymal cells also express Toll-like receptors and, while much less well studied, are likely to contribute to this constant watch. The distinction between host responses to pathogens and commensals is thus likely to relate to the location, as well as the timing, duration, and intensity of the stimuli that initiate these stereotyped responses. The complexion of the responses signaled by the activation of these molecular pattern receptors can be modified through the action of secreted virulence factors (e.g., toxins) or in the case of many viruses by direct interference with the intracellular signaling pathways. Either an inappropriate dampening or induction of signaling by the host's pattern recognition receptors, or the provocation of signaling responses in cells that do not typically encounter microbes, can produce damage and disease.

The ability to measure host responses in terms of genome-wide fluctuations in gene transcript or protein abundance, using DNA microarrays or mass spectroscopy, respectively, has raised the possibility of new approaches for early diagnosis and outcome prediction in infectious diseases. Patterns of host transcript or protein abundance may reveal the nature of the causative factor in microbial disease and help classify infected hosts based on future clinical course, as well as help elucidate disease mechanisms.

Comprehensive analyses of host responses will also help define signal transduction pathways and the regulatory mechanisms elicited by biological agents and may lead to new therapeutic approaches for controlling inflammatory and immune responses, as well as for regulating the growth and development of epithelial and other cell types. For example, recent findings indicate that Crohn's disease—a chronic inflammatory bowel disease—results from mutations in a recently recognized mammalian microbe-associated molecular pattern (MAMP) receptor known as Nod2, which recognizes as ligand a bacterial-derived muramyl peptide.[104] These mutations appear to result in defective regulation of the MAMP receptor's response to either commensal or pathogenic bacteria, with resultant aberrant downstream signaling.

Thus elucidation of the wiring diagram of the host and its programs for responding to pathogens, as well as detailed descriptions of the varied strategies used by pathogens for disruption of these programs, will be critically important for the design of improved diagnostics, therapeutics, and preventives. However, it will also provide new and potent opportunities for the would-be malefactor. Deliberate efforts to disrupt host response systems could target critical nodes in the host cell wiring diagram directly. Importantly, the effectors (i.e., weapons) might not necessarily be traditional "threat agents."[105]

Advancing Technologies Will Alter the Future Threat Spectrum

Although this report is concerned with the evolution of science and technology capabilities over the next 5 to 10 years with implications for next-generation threats, it is clear that today's capabilities in the life sciences and related technologies may have already changed the nature of the biothreat "space." In a 1996 Department of Defense (DOD) report, it was argued that advances in biotechnology and genetic engineering had provided the means to modify agents in very specific ways and had consequently facilitated the development of a new generation of biological warfare agents that could potentially be more dangerous than classical agents. The report identified five "potential types of novel biological agents"—all microorganisms: benign microorganisms that have been genetically altered to produce a toxin, venom, or bioregulator; engineered microorganisms that are resistant to antibiotics, standard vaccines, and therapeutics; microorganisms with enhanced aerosol and environmental stability; immunologically-altered microorganisms able to defeat standard identification, detection, and diagnostic methods; and combinations of any of the above four types coupled with improved delivery systems.[106] In a 1997 study, six imagined future bioweapon constructs were described, conveying an even stronger sense of the broad range of possibilities that could be created by advances in life science technologies: designer genes, viruses, and other life forms (e.g., organisms or life forms designed to be drug-resistant); designer diseases; binary biological weapons (i.e., one involving infection with one agent that has little initial pathogenic consequence until a subsequent co-infection from a second organism, or an environmental trigger, activates the pathogenic aspect of the original infection); gene therapy as a weapon; stealth viruses (i.e., cryptic viral infections); and host-swapping (zoonotic) diseases.[107] This dynamic biological "threat space"—past, present, and future—is illustrated in Figure 1-3. This timeline depicts the relative threat level presented by traditional

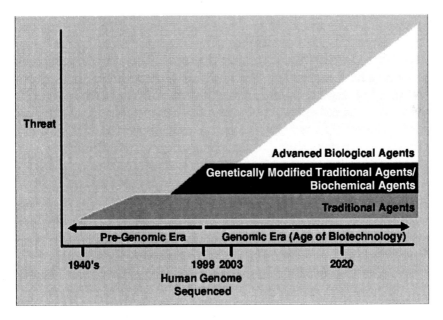

FIGURE 1-3 Timeline describing impact of biotechnology on biological warfare threat. Reprinted with permission from *BioSecurity and Bioterrorism: Biodefense Strategy, Practice, and Science* 1(3), 2003.

agents (e.g., naturally occurring bacterial or viral agents), genetically modified traditional agents (e.g., antibiotic-resistant bacteria), and advanced biological agents (novel biological weapons agents created by using biotechnological applications).

Returning to influenza as an illustrative example, advances in technology have led to the possibility that, even if a new lethal influenza A virus does not emerge in nature in the near future, one could be artificially generated through reverse genetic engineering (see Chapter 3 for a description of the technology).[108] Although not possible until recently with negative-strand RNA viruses, in October 2004 researchers at the University of Wisconsin used reverse genetic engineering techniques to partially reconstruct the highly virulent strain of influenza responsible for the 1918-1919 pandemic;[109] and the following year the complete sequence and characterization of the 1918-1919 influenza A virus was reconstructed.[110] Although the knowledge, facilities, and ingenuity to carry out this sort of experiment are beyond the abilities of most non-experts at this time, this situation is likely to change over the next 5 to 10 years.[111]

Some experts argue that bioregulators, which are small, biologically active organic compounds, may pose a more serious dual-use risk than had been previously perceived, particularly as improved targeted delivery technologies have made the potential dissemination of these compounds much more feasible than in the past. This shift in the perceived magnitude of the risk posed by bioregulators highlights the fact that the materials, equipment, and technology necessary for disseminating and delivering biological agents to their intended recipient(s) may be equally, if not more, important than the agents themselves in determining the risk they pose. However, growing understanding of how these macromolecules function and interact in mammalian systems has also increased their dual-use potential.[112]

The immune, neurological, and neuroendocrine systems are particularly vulnerable to bioregulator modification. In fact, the capacity to develop bioweapons that can be aimed at the interaction of the immune and neuroendocrine systems again points to a shift in focus from the agents to, in this case, how a range of agents can be exploited (or created) to affect the human body in targeted, insidious ways. In addition to bioregulators, large numbers of potentially toxic compounds are being generated, and identified as such, as a result of high-throughput drug discovery efforts, reflecting important changes in the drug discovery process and the expanded application of new combinatorial chemistry and other high-throughput technologies. Small interfering RNAs, described in detail earlier in this chapter, also fall into this category of biologically active small molecules.

The growing concern regarding bioregulators and other agents of biological origin does not diminish the importance of naturally-occurring or engineered pathogenic organisms. But it does mandate the need to adopt a broader perspective in assessing the threat, focusing not on a narrow list of pathogens, but on a much wider spectrum that includes biologically active chemical agents. Bioregulators are not currently included in the select agents list, which comprises those agents currently considered to be the greatest biosecurity threats.[113] The threat spectrum is broad and evolving—in some ways predictably, in other ways unexpectedly. The viruses, microbes, and toxins listed as select agents are just one aspect of the continually changing, complex threat landscape.[114, 115] In the future, genetic engineering and other technologies may lead to the development of pathogenic organisms with unique and unpredictable characteristics, and biochemical compounds that target homeostatic and physiological processes will likely pose a greater threat than they do now.

The Development and Use of Biological Weapons

Human history seems to suggest that, as technology advances, malevolent use is the rule, not the exception. As Matthew Meselson observed, "Every major technology—metallurgy, explosives, internal combustion, aviation, electronics, nuclear energy—has been intensively exploited, not only for peaceful purposes, but also hostile ones."[116] The rapidly growing and constantly shifting global knowledge base and technology landscape may lessen delivery and other challenges to developing bioweapons capabilities—at both the state and individual actor levels. With respect to the former, based on unclassified U.S. government information, the media, foreign governments, and academic reports, there are about a dozen countries with suspected, likely, or probable biological weapons capabilities.[117] Although biological weapons are outlawed by the 1972 BWC,[118] the prohibitions of the BWC do not extend to basic and applied research nor to the actions of individuals—only state parties. In the 30 years since the BWC came into effect, the acquisition and engineering of biological agents has become easier and cheaper for both state and nonstate actors, and for individuals. State actors that have not been involved with biological weapons in the past (e.g., because such weapons have not been considered accessible, or particularly useful) may start developing new bioweapons programs, either secretly or under the cover of biodefense research programs. Likewise, the technologies and tools required to develop bioweapons capabilities are becoming increasingly accessible and affordable to individuals.

National security experts and even biologists themselves are concerned that rogue scientists could create new biological weapons—such as deadly viruses that lack natural foes. They also worry about innocent mistakes: organisms that could potentially create havoc if allowed to reproduce outside the lab. In July 2000, the topic of *Scientific American* magazine's "The Amateur Scientist" was "PCR at Home." Gazing into the not too distant future, it seems likely that not only will the fruits of biotechnology—whether recombinant vaccines or GloFish™—become increasingly common features of everyday life, but the technology itself, like computer technology, will become increasingly accessible to interested amateurs. If and when biotechnology becomes as user-friendly as computer technology is currently, plant and animal breeding hobbyists and professionals, as well as gardeners, pet owners, and "bio-Unabombers"[119] will likely access and use these technologies for good or evil. Sooner or later, it is reasonable to expect the appearance of "bio-hackers,"[120, 121] mirroring the computer hackers that repeatedly cause mischief today through the creation of a succession of more and more sophisticated computer "viruses."

Since the mid-1990s, when the specter of bioterrorism emerged, only a small number of credible incidents of bioterrorism have been experienced—such as the mailing of anthrax spores through the U.S. Postal Service in the fall of 2001. Additionally, a few criminal events have occurred that have involved the use of disease-causing microorganisms or biologically-derived toxins, as can be seen in Table 1-2. The data presented below exclude accidental releases, such as happened in Sverdlovsk, USSR, in April 1979, resulting in the documented deaths of 67 people from inhalational anthrax as well as purposeful, experimental releases, such as the

TABLE 1-2 Authenticated Acts of Biological Warfare or Terrorism Directed Against People, 1940 to 2004

Date	Location	Details of Episode	Key Information Sources
1940-1941	China: Hangzhou and Nanjing	Japanese aircraft drop packages containing fleas infected with *Yersinia pestis*. There are reports of several other such episodes later.	Recent testimony in a Tokyo court by one of the aircraft pilots.
1957-1963	Brazil: Mato Grosso	Introduction of smallpox, influenza, tuberculosis, and measles into Indian tribal populations via contaminated gifts and *mestizos* in furtherance of large-scale land takeovers.	The Figueiredo report (1968), which led to the indictment of 134 employees of the Government Service for the Protection of Indians

TABLE 1-2 Continued

Date	Location	Details of Episode	Key Information Sources
1981	United Kingdom: Chemical Defence Establishment (CDE) Porton Down, Wiltshire	"Dark Harvest Commandos" deposit outside a defense research facility a parcel of soil containing anthrax bacteria taken from a former bioweapons proving ground.	The perpetrators' own account plus CDE's soil analysis confirming presence of *B. anthracis* at less than 10 org/gram.
1984	United States: The Dalles, Oregon	Rajneeshee cultists seek to influence local elections by infecting voters with salmonellosis by contaminating neighborhood restaurants.	Subsequent medical investigation of 751 sick persons.
1989	Namibia: near Windhoek	Covert operation by a South African government agency, the Civil Cooperation Bureau (CCB), to contaminate the water supply of a refugee camp with cholera bacteria.	Perpetrator's testimony during recent trial of Brig. Dr. Wouter Basson.
1990-1993	Japan: Tokyo	Aum Shinrikyo cultists, prior to their 1994-1995 sarin attacks, had sprayed biological agents, including anthrax bacteria, against several U.S. and other facilities in and around the city, but with no discernible effect.	Confessions and other information contained in leaked police reports.
2001	United States	U.S. Postal Service used by still-unknown perpetrator(s) to spread anthrax spores contained in letters addressed to individuals in the media and the U.S. Senate: 22 people catch anthrax in its cutaneous or inhalational forms: 5 die; 10,286 receive postexposure prophylaxis; and many more have their daily lives disrupted.	Medical investigations coordinated through the Centers for Disease Control and Prevention.

SOURCE: Compiled from published and documentary sources held in the Sussex Harvard Information Bank, University of Sussex, United Kingdom.

"hot agent" field trials conducted by the United Kingdom, the United States, and others. Episodes involving non-harmful simulants, such as the biological hoaxes whose proliferation accelerated in 1997, are excluded from the table, as are the myriad recent instances of false alarms, in which substances initially suspected of being harmful were subsequently found not to be so.

However, with the increasing availability of and accessibility to pertinent knowledge and technology, it is not at all unreasonable to anticipate that biological threats will be increasingly sought, threatened, and used for warfare, terrorism, and criminal purposes and by increasingly less sophisticated and resourced individuals, groups, or nations. One can envision that the ratio of credible threats and actual attacks relative to hoaxes could actually increase substantially in the years to come.

Biological Weapons Are Fundamentally Different from Other "Weapons of Mass Destruction"

Clear thinking about threats posed by and appropriate responses to biological, chemical, and nuclear weapons is often confused by the vocabulary commonly used in discussing these threats. In particular, analysts, policymakers, journalists, and even scientists often refer casually to weapons of mass destruction, or WMD, as if biological, chemical, and nuclear weapons were each merely variants of some common type of weapon. But, in fact, these weapons differ greatly in their proliferation potential, the challenges they pose for deterrence, and the effectiveness of defensive measures (each of these areas of difference are discussed in detail below).[122] Moreover, the WMD label fails to capture the disparate future trajectories of the technologies underlying biological, chemical, and nuclear weapons, so it will likely become ever more misleading over time. Box 1-3 provides the United Nations' definition of weapons of mass destruction. Some communities, such as the DOD, have tried to reconcile the "terminology problem" by introducing yet another term "weapons of mass effect."[123]

This does not mean there is no common ground or any lessons to be applied from experience dealing with one class of weapons to the others. Moreover, there is no question that from a political and strategic point of view these weapons are often strongly coupled, for example, in the connection between biological or chemical weapons and state doctrines regarding "no first use" of nuclear weapons. Of note, one striking similarity between nuclear technology and biotechnology is their dual-use nature. Nuclear technology was seen from the beginning as offering peaceful benefits, which today include research reactors, power reactors, and radioisotope production. The range of benefits is not nearly as great as the media

---

BOX 1-3
What Is a Weapon of Mass Destruction?

In 1948, the United Nations defined "weapons of mass destruction" as "atomic explosive weapons, radio-active material weapons, lethal chemical and biological weapons, and any weapons developed in the future which have characteristics comparable in destructive effect to those of the atomic bomb."

SOURCE: United Nations Security Council, Commission for Conventional Armaments. 1948. Resolution Adopted by the Commisssion at Its Thirteenth Meeting, 12 August 1948, and a Second Progress Report of the Commisssion. S/C.3/32/Rev.1; August 12:2.

---

hype predicted back in the mid-1950s when President Eisenhower presented his Atoms for Peace program to the General Assembly of Nations (e.g., automobiles that would run for a year on vitamin-sized nuclear pellets).[124] Yet despite this similarity, there is a demarcation between weapons-related and nonweapons-related nuclear science and technology. From the beginning, the civilian uses of nuclear energy were cordoned off from weapons developments through a large investment in security classification, international diplomacy, and a discourse that insists that nuclear weapons are special. Quite the opposite is true of the life sciences.

Clear thinking on the issue must proceed from an understanding of the significant differences among these weapons. Although there are lessons to be learned from the history of and our experience with nuclear weapons technology, many of the differences between the nuclear and biological realms are too great to adopt a similar mix of nonproliferation, deterrence, and defense. Effective strategies for anticipating, identifying, and mitigating the dangers associated with advancing and emerging life sciences technologies demand a clear understanding of the varied and unique nature of the biological threat spectrum.

Proliferation Potential

One of the most pronounced differences between biological weapons and other weapons of mass destruction is the proliferation potential of the former. Imagine a line that begins with nuclear weapons at the left, continues through chemical, radiological and biological weapons, and terminates with cyber "weapons" on the right. Moving from left to right along

TABLE 1-3 Characteristcis of Fissile Materials and Pathogens

Fissile Materials	Biological Pathogens
Do not exist in nature	Generally found in nature
Nonliving, synthetic	Living, replicative
Difficult and costly to produce	Easy and cheap to produce
Not diverse: plutonium and highly enriched uranium are the only fissile materials used in nuclear weapons	Highly diverse: more than 20 pathogens are suitable for biological warfare
Can be inventoried and tracked in a quantitative manner	Because pathogens reproduce, inventory control is unreliable
Can be detected at a distance from the emission of ionizing radiation	Cannot be detected at a distance with available technologies
Weapons-grade fissile materials are stored at a limited number of military sites	Pathogens are present in many types of facilities and at multiple locations within a facility
Few nonmilitary applications (such as research reactors, thermoelectric generators, and production of radioisotopes)	Many legitimate applications in biomedical research and the pharmaceutical/ biotechnology industry

SOURCE: Tucker, J.B. 2003. Preventing the misuse of pathogens: The need for global biosecurity standards. *Arms Control Today*, June.

this continuum, an effective nonproliferation regime becomes increasingly difficult. To make these differences clear, it is useful to begin with a comparison of the characteristics of fissile materials and pathogens, as depicted in Table 1-3.

Nuclear nonproliferation policy seeks to limit the number of states with nuclear weapons and to keep such weapons out of the hands of nonstate groups. The Nuclear Nonproliferation Treaty (NPT) established a near-global verification regime, carried out by the International Atomic Energy Agency (IAEA), intended to prevent the diversion of fissile material from civilian use to weapons programs. The regime employs inspections, audits, and surveillance cameras and instrumentation, monitoring some 1,100 facilities and installations worldwide. Despite important shortcomings, the NPT inspection regime has been largely successful, in part because the facilities needed to produce weapons-usable uranium or plutonium are, for the most part, big and hard to hide.[125] As the NPT verification regime has faced new challenges, the inspection regime has been modified in response so as to continue playing an important and credible nonproliferation role. The regime has lagged behind technical developments, but the pace of technical evolution has been slow enough that the regime has, with delay, still been useful.

The roughly forty nations of the Nuclear Suppliers Group (NSG) further pursue nonproliferation objectives by adhering to consensus guidelines for nuclear and nuclear-related dual-use exports. These guidelines are intended to supplement the NPT by controlling the transfer of listed items without hindering legitimate international nuclear cooperation. United Nations Security Council Resolution 1540 aims to extend many of these nonproliferation measures to all states, not just members of the NSG. Through Cooperative Threat Reduction, the United States and other members of the Group of Eight industrialized nations (G-8) act to impede the theft of nuclear material or the transfer of personnel with nuclear weapons-relevant knowledge from the former Soviet Union and other states. Diplomatic pressure and security guarantees have also played crucial roles, and intelligence has been vital throughout.

Despite its challenges, including difficulties created by uncooperative nations that may possess capable delivery systems, the nuclear nonproliferation regime has been reasonably successful in part because the production of nuclear weapons-usable plutonium or uranium has substantial technical requirements (reactors or enrichment plants, respectively) that create conspicuous bottlenecks for any would-be weapons program. Declared facilities that need to be monitored are few, and undeclared illegal facilities are at risk of discovery by intelligence or other means. Moreover, the special nuclear materials used in nuclear weapons do not occur naturally. Any detection of such materials outside of legitimate, declared facilities would automatically be a matter of grave concern. The theft of nuclear weapons or weapons-grade nuclear material may be the only way to avoid these bottlenecks, making the prevention of nuclear theft an extremely high priority.

At the far right end on the continuum of unconventional weapons are cyber weapons. At this extreme, a traditional nonproliferation regime seems nearly insurmountable, if not impossible. Exports of certain high-end computers, components, and codes are certainly controlled, but these neither prevent cyber attacks nor reduce the use of these technologies for illicit or malicious purposes. Such attacks may in principle be launched from any of over 100 million computers all over the world that already have access to the Internet—and that number is growing rapidly. Implementing a nonproliferation on-site verification regime analogous to that used to monitor nuclear facilities would therefore seemingly require unannounced inspections or the monitoring of hundreds of millions of residences and businesses. Cybersecurity poses a *reductio ad absurdum* for a traditional inspection regime.

Chemical, biological, and radiological weapons fall between the nuclear and cyber extremes of the continuum of unconventional weapons described above in the context of nonproliferation control regimes. The Organization for the Prohibition of Chemical Weapons (OPCW), established under the Chemical Weapons Convention (CWC) oversees an international verification regime for chemical weapons that is more challenging than that for nuclear weapons because of the larger number of relevant facilities and dual-use materials. An entire industrial sector and well over 5,000 inspectable facilities must be monitored by the OPCW. Nonetheless, under the CWC, chemical weapons stocks have been declared and verified, the destruction of these stocks has begun, even if fitfully, and hundreds of inspections of dual-use chemical plants have taken place. For example, Albania, India, the Republic of Korea, Libya, Russia, and the United States have declared over 70,000 metric tons of chemical agents. Except for Albania, these six countries, along with Bosnia and Herzegovina, China, France, Iran, Japan, Serbia and Montenegro, and the United Kingdom have declared post-1945 chemical weapons production facilities (Japan's declaration referred to facilities that had belonged to the Aum Shinrikyo). Completion of destruction of the declared stocks falls due in 2012. Once the disarmament phase is complete, the dual-use monitoring component will dominate verification, permitting increased attention to the challenges of new science and technology. The CWC regime is further supplemented by the Australia Group of nations that, analogously to the NSG, establishes consensus guidelines restricting the export of weapons-relevant materials.

Biological weapons pose greater challenges to a nonproliferation regime than do chemical weapons, so they lie closer to the cyber extreme along this continuum. There are some formal analogies to the nuclear and chemical cases, but the analogies only go so far. For example, while the maintenance of inventories of critical materials is of considerable use in monitoring the proliferation of nuclear weapons and, while less so, still of use in monitoring the proliferation of chemical weapons, there is little basis for such efforts in monitoring or preventing the proliferation of biological weapons. Due to the inherent replicative properties of bacteria and viruses, large stocks of biological agents can readily be produced from very small quantities of stolen agent, so small that their absence would not be missed from even the most carefully inventoried collections. Moreover, DNA synthesis technology is the equivalent of a matter compiler for genetic material—the transition from "matter" to "information" pushes the control of biology closer to the cyber extreme.

The 1925 Geneva Protocol forbade the use of biological and chemical weapons,[126] and the BWC, which entered into force in 1975, forbids the development, production, and stockpiling of biological weapons. Article I of the treaty reads:

Each State Party to this Convention undertakes never in any circumstances to develop, produce, stockpile or otherwise acquire or retain: (1) Microbial or other biological agents, or toxins whatever their origin or method of production, of types and in quantities that have no justification for prophylactic, protective or other peaceful purposes; (2) Weapons, equipment or means of delivery designed to use such agents or toxins for hostile purposes or in armed conflict.

Like the CWC, but unlike the NPT, the BWC prohibits an entire class of weapons for every country in the world, and in that sense is perhaps better characterized as an arms control treaty than a "nonproliferation" treaty. But unlike the NPT and CWC, the BWC established no inspection and monitoring regime, and there is no agency analogous to the IAEA or OPCW, respectively, that is responsible for technical monitoring of compliance with the treaty. Monitoring compliance with this treaty is especially problematic since, under the BWC, ambiguities may arise over what level of effort is consistent with "prophylactic, protective or other peaceful purposes" and, to a considerable extent, intent becomes an extremely important issue. It may prove very difficult to determine what levels of defensive efforts are legitimate and which serve as camouflage for an illicit offensive program.

After the discovery of the Iraqi and Soviet offensive biological weapons programs in the early 1990s, an international effort began in 1994 to negotiate a compliance protocol to the BWC. The protocol would have required all countries to declare certain facilities, and it would have established on-site visits for facilities meeting certain criteria and the possibility of targeted investigations in the event of suspicions regarding either particular facilities or disease outbreaks. An Organization for the Prohibition of Biological Weapons, analogous to the OPCW, was to have been created. In July 2001, however, the United States rejected the draft protocol, stating that the BWC was unverifiable; that the proposed protocol would threaten both private-sector proprietary information and U.S. government biodefense secrets; that it would threaten existing export controls; and that it risked giving a false stamp of legitimacy to violators of the BWC by failing to detect their violations but nevertheless declaring them to be in compliance with the BWC.[127]

There are other nonproliferation measures in place that are reminiscent of those established for the BWC, NPT, and CWC. For example, the Australia Group adopts consensus national export controls to impede the transfer of biological agents and technology where possible. Nevertheless, the biological nonproliferation regime faces intrinsically greater challenges than does its nuclear or even chemical counterpart, because many of the relevant materials, technologies, and knowledge are far more widespread in the biological case. As this report documents, these technologies are rapidly becoming "democratized" in that they are far more available and accessible to individuals as well as states and will become ever more so in the years to come.

Deterrence and Dissuasion

Deterrence through the threat of retaliation was a central nuclear weapons strategy during the Cold War. Deterring any form of terrorism may prove difficult, since some terrorist groups may be unconcerned about retaliation or may hope to remain unidentified. This conclusion may hold regardless of the type of weapon being used. However, biological weapons pose a unique challenge, as some infections may incubate without symptoms for days or even weeks, making it especially difficult to trace an attack back to its perpetrators.

Biological attacks may also be dissuaded through means other than the threat of retaliation (dissuasion by deterrence), such as the force of law. In the initiatives now being pressed forward nationally and internationally to criminalize the activities of individuals (i.e., activities that the BWC and the CWC prohibit at the state level), dissuasion may grow from the possibility of responsible individuals being held accountable in courts of law, provided intelligence and nonproliferation create a significant probability of detection and assuming, too, the existence of determined and empowered prosecutors.[128] Dissuasion of a different kind may also reside in the obloquy of detection that is made public, and the political or other consequences that could follow from it.

Defense

So far, biological weapons have been based on agents that also cause naturally occurring disease outbreaks. This aspect of these weapons has no good analog in the realm of nuclear weapons and only partial ones in the realm of chemical weapons. Many of the same tools that address natural disease threats will be needed to respond to an attack using biological weapons, or to prevent such an attack from succeeding, and this is likely to remain true even in a future case involving a genetically engineered pathogen. In the biological case, therefore, there is the opportunity to ensure that many of the steps required to improve biodefense will benefit public health even if major acts of bioterrorism never occur. "Defense" in the case of biological security means, above all, improvements in domestic and international disease surveillance and response and strengthened public health systems. The recent evolution of biomedical defense programs has largely focused on detection and vaccine development. This "dual-use" aspect of much of biodefense is very different from the nuclear case, though it bears some similarity to the chemical case where preparation for accidental chemical spills or explosions is relevant to consequence management for terrorism as well.

The "Arms Race" Metaphor and the Difficult Issue of Secrecy

The "arms race" metaphor must be used with caution; it too is in danger of calling up misleading analogies to the Cold War nuclear arms race that are not relevant to the biological case. We are transitioning from the classical biowarfare/bioterrorism mode in which naturally occurring agents are the threat weapons. They may have minor modifications such as induction of antibiotic resistance, but generally most countermeasures will be expected to be active, and the list of candidate agents is only debated in a small group of organisms. In this situation, production of vaccines and countermeasures is not inherently destabilizing provided transparency and interchange are maintained. (It is worth noting that we are far from solving defenses against these agents at this time.) However, in tomorrow's world of proliferating threats that may be quite different from the 20th century agents and indeed may not even exist at this time, there will be very different considerations. The primary driver of offensive capabilities in that biological arms race is not mainly a particular adversary but rather the ongoing global advance of biotechnology and microbiological and biomedical research. That is, protective measures are in a race with the malevolent application of potentially beneficial basic research, rather than primarily against technologies being developed in weapons programs of other countries.[129]

However, there is a legitimate concern that defensive research undertaken in one country's program could be misperceived as offensive, or potentially offensive, in character and drive other nations to pursue offensive research as well. For example, would it be legal and wise to have classified biodefense research activities produce modified pathogens that no adversaries are believed to have yet created in order to ostensibly understand and more robustly defend against them? On the one hand, defense against such potential threats that we can already anticipate would seem to prompt an answer of "yes." On the other hand, such classified research risks making the responsible government a driver of the biological arms race, as other nations may misperceive such research as offensive.

Yet "open" biodefense research risks providing information to individuals or groups with malicious or malevolent intent. Strategic decisions must be made about what, if any, biological weapons research will be conducted in the name of biodefense, how much of this research will be classified, and how that program will be publicly described or even performed under a regime of transparency. This too is very different from the nuclear weapons case and the NPT, under which it is legally permissible, at least for the time being, for the United States and certain other NPT states to stockpile and conduct research in nuclear weapons. Under the BWC, no nation may develop or stockpile biological weapons. However, the situation is complicated by the reality that a vast reservoir of technical information that could be used to develop bioweapons capabilities is freely available in the public domain, unlike the smaller number of closely held secrets associated with nuclear weapons technology. Moreover, much of biotechnology, unlike nuclear technology, is increasingly accessible even to minimally trained individuals.

The risks associated with open versus closed biodefense research activities highlight the difficult questions surrounding the issue of secrecy and whether there is certain biological information that should be kept secret and, if so, where the boundary that defines this information lies and what circumstances dictate a need for secrecy. Other questions about secrecy pertain to the role of "sensitive" information and whether information should only be categorized as either classified or not, and the challenges associated with imposing a global secrecy structure.[130]

The Need to Strike a Balance: Benefits of Technological Growth

Much of this report focuses on the potential misuse of technology. Indeed, the rationale for conducting this study was based on the growing threat of misuse posed by dual-use advances in science and technology. Importantly, there are also tremendous benefits to be gained from the very same scientific and technological advances. The purpose of this section is to enumerate some of these benefits. A more detailed discussion of these benefits, particularly for the developing world, may be found in an earlier workshop summary report from this committee, *An International Perspective on Advancing Technologies and Strategies for Managing Dual-Use Risks: A Workshop Report.*[131]

There is no question that many populations have benefited greatly from advances in biotechnology and applications of related technologies (e.g., nanotechnology and informatics) to biomedicine and agriculture. Health biotechnology holds out promises for improved nutrition, a cleaner environment, a longer and healthier lifespan, and cures for many once-formidable diseases. Even older technologies, such as classic vaccine technology, have enabled the eradication or reduction of many once-dreaded diseases, such as smallpox, polio, diphtheria, tetanus, and whooping cough. Newer reverse genetic technologies for RNA viruses may make possible rapid, rational development of vaccines against newly recognized pathogens, such as SARS, avian influenza, Nipah, and many others for which current technologies are too slow, hit-or-miss, or too resource intensive for timely vaccine development.

In the developing world, broader application of biotechnology may make it economically feasible for resource-limited countries to produce inexpensive vaccines to protect their own populations against emerging infections that most afflict them. However, the potential applications of life sciences technologies extend far beyond more affordable vaccines. A technology foresight study conducted by the University of Toronto Joint Centre for Bioethics (JCB) identified the ten biotechnology-related developments that are likely to improve human health in developing countries within the next five to ten years:[132] molecular diagnostics, recombinant vaccines, drug and vaccine delivery systems, bioremediation, sequencing pathogen genomes, female-controlled STI protection, bioinformatics, enriched genetically modified crops, recombinant drugs, and combinatorial chemistry.

In addition to improved health, world agriculture stands to benefit greatly from new discoveries in the life sciences and growing technological capabilities. Many staple crop plants represent virtual monocultures—that is, specific strains that were selected and propagated to give high yield under certain specific conditions.[133] Biotechnology could enable agriculturalists to develop plants and livestock that are more resistant to disease, pests, or harsh environmental conditions. Such modified agricultural stock could be both an assurance for the future—against the possibility of a disease wiping out a major part of the food supply, as happened in the Irish potato famine of the mid-19th century—and a boon to resource-poor farmers in developing countries.

Importantly, while posing risks, U.S. biodefense efforts stand to benefit as well from global cooperation in the spread of biotechnology. Efforts to expand disease surveillance, improve detection and diagnostic capabilities, and develop new vaccines and therapeutics—all of which are crucial for a rapid, effective response in the event of either a deliberately introduced or a naturally occurring biological attack—will continue to rely on excellent scientific research and technological growth. Much of the latter in particular relies, in turn, on international collaboration, a theme explored in depth in Chapter 2. It has been argued that domestic regulations designed to strengthen biodefense but that either indirectly and/or directly restrict international collaboration may, ironically, "have an unintended negative influence on the current U.S. effort to develop enhanced biodefense capabilities, because it creates incentives to shift away from international collaboration in biodefense research and production."[134] This is particularly alarming with respect to vaccine research, in which international biotechnology companies play a vital role. Of the six category A biological threats on the U.S. Department of Health and Human Services' (DHHS) select agents list (i.e., anthrax, botulinum toxin, plague, smallpox, tularemia, and viral hemorraghic fevers), vaccines for all but botulinum toxin are currently being developed in cooperation with international biotechnology companies located in Canada, Japan, Russia, and throughout Europe. Global cooperation, including the training of foreign nationals in the United States, also provides opportunities to introduce foreign scientists to the concept of dual-use risk and the need to pursue a culture of awareness and responsibility in the global scientific community.

The Dual-Use Dilemma

It can be expected that tension between the potential beneficial and malevolent uses of technology will increase in the future, as science and technology increasingly empower users to manipulate the materials and processes of life itself. The same reverse genetic technologies that can be used to develop new vaccines against RNA viruses could also be used to construct modified viruses, including possibly viruses that express heterologous virulence factors that result in more lethal disease. The Fink report summarized the details and implications of three recent examples of "contentious research" in the life sciences—experiments that resulted in the creation of new infectious agents or knowledge with dual-use potential: the 2001 ectromelia virus (mousepox) experiment, in which Australian researchers engineered a recombinant virus that expressed the mouse interleukin-4 (IL-4) gene and, in so doing, inadvertently created a lethal virus that kills mice genetically resistant or recently vaccinated against mousepox;[135] the 2002 announcement that researchers from the State University of New York, Stony Brook, had artificially synthesized a virulent poliovirus from scratch, using mail-order segments of DNA and a viral genome map freely available on the Internet;[136] and a 2002 study on the difference in virulence between the *Variola major* virus, which causes smallpox and has a 30 to 40 percent mortality rate, and the *vaccinia* virus, which is used to vaccinate humans against smallpox and does not cause disease in immunocompetent persons.[137]

The following serve as additional illustrative examples of scientific publications that pose dual-use dilemmas (see also Figure 1-4):

• Publication of the complete DNA sequences of human pathogens. This information, which is widely available on public databases, could potentially facilitate the development and production of novel biological weapons agents. According to one assessment: "The ever-expanding microbial genome databases now provide a parts list of all potential genes involved in pathogenicity and virulence, adhesion and colonization of

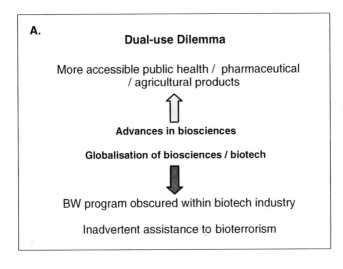

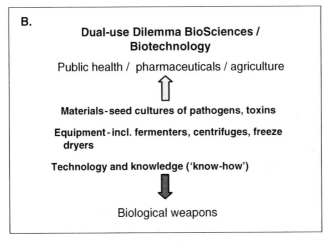

FIGURE 1-4 A and B The Dual-Use Dilemma.
SOURCE: B. Mathews, presentation at committee workshop, Cuernavaca, Mexico.

host cells, immune-response evasion and antibiotic resistance, from which to pick and choose the most lethal combinations."[138]
   • A method for the construction of "fusion toxins," derived from two distinct toxins, for the purpose of killing cancer cells.[139] This technique might be redirected to develop novel toxins that could target the normal cells of almost any tissue when introduced into a human host.

- Development of a genetically engineered strain of *Bacillus anthracis* containing an inserted gene for a foreign toxin, potentially rendering the agent resistant to the existing anthrax vaccine.[140]
- Development of "stealth" viruses that can evade the human immune system and serve as molecular vectors to introduce curative genes into patients with inherited diseases, or alternatively express unwanted proteins such as toxins.[141]
- Publication of molecular details of two highly virulent strains of influenza, the 1997 Hong Kong flu[142] and the 1918 Spanish flu.[143] Other labs are trying to sequence and publish all of the genes in the 1918 influenza strain,[144] which killed 20 million to 40 million people, so as to be prepared should it reemerge.[145] Such work may help control influenza, but it also makes it easier to recreate the highly pathogenic 1918 virus for malevolent purposes.
- Genetic engineering of the tobacco plant to produce subunits of cholera toxin, making it theoretically possible to produce large quantities of this toxin cheaply and relatively easily, for good or bad.[146]
- Efforts to do away with insulin injections for diabetes by developing new technologies for delivering drugs by aerosol spray.[147]
- The deliberate release of botulinum toxin into the U.S. milk supply could poison several hundred thousand individuals, but early detection measures, such as in-process testing, could eliminate this threat. Researchers have explored the effects of a deliberate release of botulinum bacteria at various points in the milk supply chain, including raw milk silos and tanker delivery trucks, with varied assumptions for key variables.[148] This work helps biodefense and public health workers pinpoint vulnerabilities so they can be appropriately addressed, but it also could help direct a terrorist to the most vulnerable points in the milk supply.

In the past, dual-use concerns have focused on pathogens and on the challenges associated with controlling dangerous pathogens. As already emphasized, this committee's deliberations have indicated that the problem will be far broader and more profound in the future. For example, advances in neurobiology may make it possible to manipulate behavior and thought processes, while gene therapy and gene expression technologies just now coming to fruition will make it possible to activate endogenous molecules in the body—with possibly wide-ranging and devasting effects. Advances in synthetic biology and nanotechnology will offer similar rich opportunities for dual use. Nanodevices that may be used to unplug blocked arteries could instead be employed to interfere with circulatory function. Advanced drug delivery technologies and pharmacogenomics knowledge could be used to develop and deliver

with greater efficiency new bioweapons, perhaps even selectively targeting certain racial or ethnic groups.

COMMITTEE PROCESS

In considering the rapid and unpredictable advance of life sciences
technologies, it became apparent to the committee that the possibility of
inappropriate or malevolent use could never be completely eliminated
without fundamentally undermining the vitality of the scientific enterprise and endangering the tremendous benefits this enterprise brings.
However, we can also learn from past experiences. For example, recombinant DNA was a new enabling technology some 30 years ago, and the
possibilities of this powerful new technology led to widespread concern
in both the scientific and the political communities. The issue was discussed at the now-famous Asilomar Conference of 1975, when scientists
gathered to discuss the safety of manipulating DNA from different species and when many of the safeguards now in place were originally developed.[149] As scientists developed more experience with the technology,
many of the original stringent safeguards could gradually be softened.
Similarly, this committee has considered how possibilities for inappropriate and malicious use could be greatly reduced or mitigated, as discussed
in greater detail in Chapter 4.

In creating the ad hoc Committee on Advances in Technology and the
Prevention of Their Application to Next Generation Biowarfare Threats,
the National Research Council (the operating arm of The National Academies) and the Institute of Medicine selected committee members representing a broad spectrum of backgrounds, expertise, and interests. Areas
of expertise included public health, molecular and cellular biology, biophysics, clinical medicine, drug and vaccine discovery, national security
and law enforcement, bioethics, and sociology (see Appendix C for bio-
graphical information of committee members). In addition, the committee
relied on the expertise and advice of representatives of the executive
branch of the U.S. government, international governmental and nongovernmental organizations, national governmental and nongovernmental
technical and policy experts, and educators and private consultants. Information available from the open literature and materials submitted by
experts were reviewed and considered during the committee's deliberations (see Appendix B).

REPORT ROADMAP

Chapter 2 of this report reviews the current global dispersion of tools
and technologies used in the life sciences enterprise both domestically

and globally. This global dispersion is being driven by a multitude of economic, social, and political forces. Chapter 3 provides an overview and perspective on the breadth and types of technologies that will—directly or indirectly—have an impact on how the life sciences enterprise will evolve in the near-term future. Finally, Chapter 4 presents the committee's conclusions and recommendations about the ways in which the adoption of a "web of prevention" approach might enhance our collective abilities to mitigate or minimize the negative consequences of inadvertent, inappropriate or purposeful malevolent applications of any of these technologies in the decades to come.

ENDNOTES

[1] Cracraft, J. 2004. Editorial: A new AIBS for the age of biology. *BioScience* (November). Available online at www.aibs.org/bioscience-editorials/editorial_2004_11.html [accessed January 4, 2006].

[2] Meselson, M. 1999. The problem of biological weapons. Presentation given to the 1818th Slated Meeting of the American Academy of Arts and Sciences, Cambridge, MA, January 13.

[3] National Intelligence Council. 2004. Mapping the Global Future, Report of the National Intelligence Council's 2020 Project. Available online at www.cia.gov/nic/NIC_globaltrend2020.html#contents [accessed May 3, 2005].

[4] National Research Council. 2004. *Biotechnology Research in an Age of Terrorism.* Washington, DC: The National Academies Press.

[5] Couzin, J. 2002. Breakthrough of the year: small RNAs make big splash. *Science* 298(5602):2296-2297. Available online at http://www.sciencemag. org/cgi/reprint/298/5602/2296.pdf [accessed January 4, 2006].

[6] Cancer Bulletin, National Cancer Institute. 2005. "Cancer biologists are using RNAi to do everything from investigating individual genes to running high-throughput screens for new drugs to developing therapeutics," says Dr. Natasha Caplen, head of the Gene Silencing Section in NCI's Center for Cancer Research. Available online at www.nci.nih.gov/ncicancerbulletin/NCI_Cancer_Bulletin_032905/page4 [accessed January 4, 2006].

[7] 10 emerging technologies that will change your world. *Technology Review* (February 2004). Available online at www.lib.demokritos.gr/InTheNews/emerging0204.htm [accessed January 4, 2006].

[8] Scherr, M. et al. 2003. Inhibition of GM-CSF receptor function by stable RNA interference in a NOD/SCID mouse hematopoietic stem cell transplantation model. *Oligonucleotides* 13:353-363; Song, E. et al. 2003. RNA Interference targeting Fas protects mice from fulminant hepatitis. *Nature Medicine* 9(3):347-351; Soutschek, J. et al. 2004. Therapeutic silencing of an endogenous gene by systemic administration of modified siRNAs. *Nature* 432(7014):173-178.

[9] Frost & Sullivan. 2004:B349.

[10] Ibid.

[11] 10 emerging technologies that will change your world *Technology Review*

(February 2004). Available online at www.lib.demokritos.gr/InTheNews/emerging0204.htm [accessed January 4, 2006].

[12] Ball, P. 2004. Synthetic biology: Starting from scratch. *Nature* 431 (7009): 624-626. Available online at www.nature.com/nature/journal/v431/n7009/pdf/431624a.pdf [accessed January 4, 2006].

[13] Morton, O. 2005. Life, reinvented. *Wired* 13.01 (January). Available online at www.wired.com/wired/archive/13.01/mit.html [accessed January 4, 2006].

[14] June 29 /PRNewswire/—Synthetic Genomics, Inc., a new company that will develop and commercialize synthetic biology, was launched today. J. Craig Venter, Ph.D., is the founder, chairman, and chief executive officer of the company. Synthetic Genomics, Inc. was founded in 2005 to develop and commercialize synthetic biology. The company is applying recent scientific advances, including newly discovered genetic sequences of novel photosynthetic and metabolic pathways, to execute various life functions within a synthetically devised organism. These breakthroughs present unprecedented opportunities that could restructure and revolutionize industries including energy, industrial organic compounds, pharmaceuticals, CO_2 sequestration, fine chemicals, and environmental remediation. Synthetic Genomics, Inc.'s initial focus will be on ethanol and hydrogen production. The company will engineer modular "cassette" based systems to execute specific functions using reprogrammed cells as bio-factories. After leveraging enormous archives of genomic sequence data, the company will integrate novel processes to design, build, and test desired outputs from synthetic organisms. openwetware.mit.edu/index.php?title=Craig_Venter's_Synthetic_Genomics_plans_to_%22program_cells%22_to_produce_hydrogen [accessed January 4, 2006].

[15] Service, R.F. 2004. Nanotechnology grows up. *Science* 304 (5678): 732-1734. Available online at www.sciencemag.org/cgi/reprint/304/5678/1732.pdf [accessed January 4, 2006].

[16] Chang, K. 2005. Tiny is beautiful: translating 'Nano' into practical. *New York Times* (February 22); N. Munro. 2005. How fast can nanotechnology go? *National Journal* 37(8).

[17] Moradi, M. 2005. Six opportunities in nano-enabled drug delivery systems. NanoMarkets, LC (February 23).

[18] The first was Wyeth's first solid-dose formulation of the immunosuppressant Rapamune® (sirolimus), which received marketing approval from the U.S. Food and Drug Adminstration (FDA) in August, 2000. Rapamune, which was developed to reduce organ rejection in patients who receive kidney transplants, had previously been available only as an oral solution which required refrigeration and mixing with water or orange juice prior to administration. The NanoCrystal-based tablet allows for more convenient storage and administration. As a more recent example of the utilization of this technology, Johnson & Johnson Pharmaceutical Research & Development, L.L.C., is currently conducting a Phase III clinical study of a NanoCrystal-based formulation of the drug paliperidone palmitate for use in patients with schizophrenia.

[19] Fortina, P. et al. 2005. Nanobiotechnology: The promise and reality of new approaches to molecular recognition. *Trends in Biotechnology* 23(4):168-173; C.M. Niemeyer and C.A. Mirkin, eds. 2004. *Nanobiotechnology: Concepts, Applications and Perspectives*, (Wiley).

[20] Paull, R. et al. 2003. Investing in nanotechnology. *Nature Biotechnology* 21(10):1144-1147.

[21] Choi, B. et al. 2005. Artificial allosteric control of maltose binding protein. *Physical Review Letters* 94(3):038103.

[22] Georganopoulou, D.G. et al. 2005. Nanoparticle-based detection in cerebral spinal fluid of a soluble pathogenic biomarker for Alzheimer's disease. *Proceedings of the National Academy of Sciences* 102 (7):2273-2276.

[23] www.nanosphere-inc.com/3_media/1_pr/020105.html [accessed February 23, 2005].

[24] First described in the mid-1980s, PCR has become the workhorse of biological laboratories worldwide. Researchers and clinicians use the technology to multiply, or copy, specific regions of genomes for use in various types of downstream analyses (e.g., to detect the presence of a specific DNA sequence).

[25] Mullis, K. 1990. The unusual origin of the polymerase chain reaction. *Scientific American* 262(4):56-61 and 64-65.

[26] Saiki R.K., et al. 1985. Enzymatic amplification of beta-globin genomic sequences and restriction site analysis for diagnosis of sickle cell anemia. *Science* 230(4732):1350-1354; Saiki R.K., et al. 1986. Analysis of enzymatically amplified beta-globin and HLA-DO alpha DNA with allele-specific oligonucleotide probes. *Nature* 324(6093):163-166.

[27] Castilla, J. et al. 2005. In vitro generation of infectious scrapie prions. *Cell* 121(2):195-206.

[28] Carlson, R. 2003. The pace and proliferation of biological techniques. *Biosecurity and Bioterrorism: Biodefense Strategy, Practice, and Science* 1(3):1-12.

[29] Eldredge, N. and S.J. Gould. 1972. Punctuated equilibria: An alternative to phyletic gradualism. *Models in Paleobiology*. Throughout most of the last century, researchers developing the synthetic theory of evolution primarily focused on microevolution, which is slight genetic change over a few generations in a population. Beginning in the early 1970s, this model was challenged by Stephen J. Gould, Niles Eldredge, and other leading paleontologists. They asserted that there is sufficient fossil evidence to show that some species remained essentially the same for millions of years and then underwent short periods of very rapid, major change. Gould suggested that a more accurate model in such species lines would be punctuated equilibrium.

[30] Carlson, R. 2003. The pace and proliferation of biological techniques. *Biosecurity and Bioterrorism: Biodefense Strategy, Practice, and Science* 1(3):1-12.

[31] Craig Venter, briefing to the NSABB, Bethesda, MD on July 1, 2005. Reviewer "J" states that "According to John Mulligan, CEO of Blue Heron Biotechnology, his company has already received a proposal to synthesize a complete bacterial genome. This is technologically feasible at Blue Heron today, which means that the year 2010 may be far too conservative."

[32] Tian, J. et al. 2004. Accurate multiplex gene synthesis from programmable DNA microchips. *Nature* 432(7020):1050-1054. Available online at www.nature.com/nature/journal/v432/n7020/pdf/nature03151.pdf. [accessed January 4, 2006].

[33] www.blueheronbio.com/ [accessed January 14, 2005].

[34] In the next 5 years the net price for long fragments of chemically synthesized DNA seems *very* unlikely to (i.e., will not) drop below $0.10 per base pair. The

$0.01 per base pair number might become possible for the synthesis process itself, but the synthesis number does not include ancillary costs for essential things like handling of intermediate and final materials, sequence verification, and so on.

[35] Lohmann, V. et al. 1999. Replication of subgenomic hepatitis C virus RNAs in a hepatoma cell line. *Science* 285(5424):110-113. Available online at www.sciencemag.org/cgi/reprint/285/5424/110.pdf [accessed January 4, 2006].

[36] Tian, J. et al. 2004. Accurate multiplex gene synthesis from programmable DNA microchips. *Nature* 432(7020):1050-1054. Available online at www.nature.com/nature/journal/v432/n7020/pdf/nature03151.pdf [accessed January 4, 2006].

[37] Couzin, J. 2002. Breakthrough of the year: small RNAs make big splash. *Science* 298(5602):2296-2297. Available online at http://www. sciencemag.org/cgi/reprint/298/5602/2296.pdf [accessed January 4, 2006]; 10 emerging technologies that will change your world," *Technology Review* (February 2004). Available online at www.lib.demokritos.gr/InTheNews/emerging0204.htm [accessed January 4, 2006].

[38] Chen, Y. and C. Mao. 2004. Putting a brake on an autonomous DNA nanomotor. *Journal of the American Chemical Society* 126:8626-8627; Institute of Medicine/National Research Council. 2005. *An International Perspective on Advancing Technologies and Strategies for Managing Dual-Use Risks.* Washington, DC: The National Academies Press: 49-52.

[39] Institute of Medicine/National Research Council. 2005. *An International Perspective on Advancing Technologies and Strategies for Managing Dual-Use Risks.* Washington, DC: The National Academies Press.

[40] Normile, D., G. Vogel, and C. Holden. 2005. Stem cells: Cloning researcher says work is flawed but claims results stand." *Science* 310(5756):1886-1887. Available online at www.sciencemag.org/cgi/reprint/310/5756/1886.pdf [accessed January 4, 2006].

[41] Institute of Medicine/National Research Council. 2005. *An International Perspective on Advancing Technologies and Strategies for Managing Dual-Use Risks.* Washington, DC: The National Academies Press.

[42] Berg, C. et al. 2002. The evolution of biotech. *Nature Reviews. Drug Discovery* 1(1):845-846.

[43] www.bio.org

[44] Wheelis, M. 2002. Biotechnology and biochemical weapons. *The Nonproliferation Review* 9(1):48-53. Available online at cns.miis.edu/pubs/npr/vol09/91/91whee.htm [accessed January 4, 2006].

[45] Of course, there might be an equivalent concern raised by companies holding this information.

[46] Carlson, S. 2000. The Amateur Scientist: PCR at Home. *Scientific American* (July).

[47] The definition of a bioweapon, while meant to be inclusive, does not extend to nuclear weapons or devices.

[48] It should be noted that in a precedent that is quite informative for attempts described later in the report to introduce codes of ethics and conduct for biological scientists, computer "hacking" is coming to stand for activities that are remotely done to computers and networks without the consent of those who own and oper-

ate those machines and networks, regardless of motivation. The Information Technology community is working to develop an ethic that this is not acceptable, even if there is no malice involved. Some such activity may be playful, or pranks, or done without malice; other hacking causes massive damage without any real intent to do so, and still other such activity is intended to, and succeeds at, causing real damage. But all are illegitimate.

[49] "Human security" means to protect the vital core of all human lives in ways that enhance human freedoms and human fulfillment . . . [by] creating political, social, environmental, economic, military and cultural systems that together give people the building blocks of survival, livelihood and dignity. From United Nations Commission on Human Security. 2003. Human Security—Now. Available at www.humansecurity-chs.org/finalreport/English/FinalReport.pdf [accessed February 27, 2006].

[50] One of the earliest recorded instances of biological warfare occurred in 600 BC, when the Athenian leader Solon used the noxious roots of the *Helleborus* plant to poison the water supply in the city of Kirrha. Later, the Greeks and Romans may have used human and animal corpses to poison drinking water wells. And Alexander the Great is thought to have catapulted dead bodies over the walls of besieged cities, possibly as a means of spreading disease and inciting terror among the urban inhabitants. A related technique, used in the Middle Ages, was to deliberately leave dead human or animal corpses behind, in areas that would be occupied shortly by invading troops; catapults were used as well. For further details about these and other later examples of germ-based warfare, including allegations that U.S. government agents deliberately infected the Plains Indians in the 1800s by trading with the Indians smallpox-laden blankets, see National Research Council. 2004. *Biotechnology Research in an Age of Terrorism.* Washington, DC: The National Academies Press: 34-35.

[51] Stockholm International Peace Research Institute (SIPRI). 1971. *The Rise of CB Weapons.* Vol 1. In: *The Problem of Chemical and Biological Warfare.* New York: Humanities Press.

[52] Wheelis, M. 1999. Biological sabotage in World War I. in Geissler E. and J.E. Van Courtland Moon, eds. 1999. *Biological and Toxin Weapons: Research, Development and Use from the Middle Ages to 1945.* SIPRI Chemical and Biological Warfare Studies 18 London: Oxford University Press; 52.

[53] Redmond, C. et al. 1998. Deadly relic of the great war. *Nature* 393(6687):747-748.

[54] Geissler, E., and J.E. Van Courtland Moon, eds. 1999. *Biological and Toxin Weapons: Research, Development and Use from the Middle Ages to 1945.* SIPRI Chemical and Biological Warfare Studies 18 London: Oxford University Press.

[55] Bernstein, B. 1988. America's biological warfare program in the Second World War. *Journal of Strategic Studies* 11(September): 292-317, especially 304 and 308-310. In addition to *Bacillus anthracis* and *Clostridium botulinum*, pathogens studied at Camp Detrick included the causative agents of: glanders; brucellosis; tularemia; melioidosis; plague; smallpox; psittacosis; coccidiomycosis; a variety of plant pathogens including the causative agents for rice blast; rice brown spot disease; late blight of potato; and cereal stem rust. Animal and avian pathogens studied included rinderpest virus, Newcastle disease virus, and fowl plague virus. *The*

Problem of Chemical and Biological Warfare, SIPRI I, London: Oxford University Press, 1971: 122. See also Cochrane, R.C. 1947. Biological Warfare Research in the United States. In *History of the Chemical Warfare Service in World War II* (1 July 1940-15 August 1945), Vol. II (declassified). Historical Section, Office of Chief, Chemical Corps.

[56] U.S. Department of the Army. 1977. *U.S. Army Activity in the U.S. Biological Warfare Programs I*. (unclassified) February 23:1-3.

[57] See Williams, P. and D. Wallace. 1989. *Unit 731: The Japanese Army's Secret of Secrets*. London: Hodder and Stoughton: 280-281; and Harris, S.H. 1994. *Factories of Death: Japanese Biological Warfare, 1932-45, and the American Cover-Up*. London: Routledge.

[58] Ibid.

[59] At least 3,000 people, including Chinese civilians, Russians, Mongolians and Koreans, died in the experiments between 1939 and 1945, Chinese state media have said. Outside the site, more than 200,000 Chinese were killed by biological weapons produced by Unit 731, they said. (Reuters, July 18, 2005)

[60] Guilleman, J. 2001. *Anthrax: The Investigation of a Deadly Outbreak*. Berkeley and Los Angeles, CA: University of California Press.

[61] Meselson, M. et al. 1994. The Sverdlovsk anthrax outbreak of 1979. *Science* 266(5188):1202-1208; Meselson, M. 2001. Note regarding source strength. *The ASA Newsletter* 87: 1, 10-12.

[62] Dimitri Vladimir Pasechnik was a Soviet microbiologist whose defection to Britain in 1989 disclosed the fact that Moscow's germ warfare programme was 10 times greater than previously feared in the West. See portal.telegraph.co.uk/news/main.jhtml?view=DETAILS&grid=&targetRule=5&xml=/news/2001/11/29/db2903.xml.

[63] Kelly, D.C. The trilateral agreement: Lessons for biological weapons verification. In Finlay, T., and O. Meier, eds. 2002. *Verification Yearbook 2002*. London: VERTIC: 93-109; Domaradskij, I.V. and W. Orent. 2003. *Biowarrior: Inside the Soviet/Russian Biological War Machine*. Amherst, NY: Prometheus Books.

[64] Alibek, K. and S. Handelman. 1999. *Biohazard: The Chilling True Story of the Largest Covert Biological Weapons Program in the World—Told from the Inside by the Man Who Ran It*. New York: Random House.

[65] For personnel numbers, see Leitenberg, M. 1993. The Conversion of Biological Warfare research and Development Facilities to Peaceful Uses. in *Control of Dual-Use Threat Agents: The Vaccines for Peace Programme*, SIPRI Chemical and Biological Warfare Series, 15 London, Oxford University Press. For the environmental impacts associated with biological weapons field testing, see Choffnes, E. 2001. Germs on the Loose. *The Bulletin of the Atomic Scientists* 57(March/April):57-61; Alibek, K. and S. Handelsman. 1999. *Biohazard: The Chilling True Story of the Largest Covert Biological Weapons Program in the World—Told from the Inside by the Man Who Ran It*. New York: Random House.

The Soviet military had tested smallpox. Although Moscow has denied that it ever conducted open-air testing of smallpox, a detailed report prepared by the Monterey Institute of International Studies Center for Nonproliferation Studies asserts that the former Soviet Union did conduct such tests on Vozrozhdeniye Island. For more on this program, see Bozheyeva, G., Y. Kunakbayev, and D.

Yeleukenov. 1999. *Former Soviet Biological Facilities in Kazakhstan: Past, Present and Future*. Center for Nonproliferation Studies, Monterey Institute of International Studies: 6.

[66] Alibek, K. and S. Handelsman. 1999. *Biohazard: The Chilling True Story of the Largest Covert Biological Weapons Program in the World—Told from the Inside by the Man Who Ran It.* New York: Random House.

[67] Gould, D. and P. Folb. 2002. *Project Coast: Apartheid's Chemical and Biological Warfare Programme*. Geneva: UNIDR. See also Burgess S. and H. Purkitt. 2001. *The Rollback of South Africa's Chemical and Biological Warfare Program*. USAF Counter Proliferation Center. Maxwell Air Force Base, AL: Air War College.

[68] Institute of Medicine/National Research Council. 2005. *An International Perspective on Advancing Technologies and Strategies for Managing Dual-Use Risks*. Washington, DC: The National Academies Press: 42-43.

[69] Ibid.

[70] Institute of Medicine. 2003. *Microbial Threats to Health: Emergence, Detection, and Response*. Washington, DC: The National Academies Press.

[71] Institute of Medicine. 2003. *Microbial Threats to Health: Emergence, Detection, and Response*. Washington, DC: The National Academies Press.

[72] Specter, M. 2005. Nature's bioterrorist. *The New Yorker* (February 28):50-61.

[73] For detailed discussions of antigenic drift and shift in influenza A virus, see Krug, R.M. 2003. The potential use of influenza virus as an agent for bioterrorism. *Antiviral Research* 57(1-2):147-150; and Wright, P.F. and R.G. Webster. 2001. Orthomyxoviruses. In: D.M. Knipe and P.M. Holwey eds, *Field's Virology* 4th Ed. Philadelphia: Lippincott Williams & Wilkins: 1533-1579.

[74] Influenza viruses are defined by two protein components on the virus surface: haemagglutinin (H) and neuraminidase (N).

[75] Institute of Medicine. 2005. *The Threat of Pandemic Influenza: Are We Ready?* Washington, DC: The National Academies Press.

[76] Ibid.

[77] See www.who.int/csr/disease/avian_influenza/country/cases_table_2006_04_21/en/index.html, [accessed April 25, 2006].

[78]Chen, H. et al. 2004. The evolution of H5N1 influenza viruses in ducks in southern China. *Proceedings of the National Academy of Sciences* 101(28):10452-10457; Institute of Medicine. 2005. *The Threat of Pandemic Influenza: Are We Ready?* Washington, DC: The National Academies Press; Keawcharoen, J. et al. 2004. Avian influenza H5N1 in tigers and leopards. *Emerging Infectious Diseases* 10(12):2189-2191. Available online at www.cdc.gov/ncidod/EID/vol10no12/04-0759.htm [accessed March 17, 2005].

[79] Finlay, B.B. and S. Falkow. 1997. Common themes in microbial pathogenicity revisited. *Microbiology and Molecular Biology Reviews* 61(2):136–169.

[80] Casadevall, A. and L-A. Pirofski. 1999. Host-pathogen interactions: redefining the basic concepts virulence and pathogenicity. *Infection and Immunity* 67(8):3703-3713.

[81] Casadevall, A. and L-A. Pirofski. 1999. Host-pathogen interactions: redefining the basic concepts virulence and pathogenicity. *Infection and Immunity.* 67(8): 3703-3713; Falkow, S. 1997. What is a pathogen? *ASM News* 63:359–365; Finlay,

B.B. and S. Falkow. 1997. Common themes in microbial pathogenicity revisited. *Microbiology and Molecular Biology Reviews* 61(2):136–169.

[82] Savage, D.C. 1977. Microbial ecology of the gastrointestinal tract. *Annual Reviews of Microbiology* 31:107-133, as cited in Hooper, L.V. et al. 1998. Host-microbial symbiosis in the mammalian intestine: exploring an internal ecosystem. *BioEssays* 20(4):336-343; Bäckhed, F. et al. 2005. Host-bacterial mutualism in the human intestine. *Science* 307(5717):1915-1920; Buchanan, M. 2004. A billion bacteria brains are better than one. *New Scientist* (2474):34.

[83] Bäckhed, F. et al. 2005. Host-bacterial mutualism in the human intestine. *Science* 307(5717):1915-1920. Also, Rakoff-Nahoum, S. et al. 2004. Recognition of commensal microflora by toll-like receptors is required for intestinal homeostasis. *Cell* 118(2):229-241. The protective advantage of *Lactobacillus spp.* is being exploited in probiotic therapy—the administration of live, benign microbes, including genetically-engineered bacteria, that benefit the host and aid in the treatment of disease. Institute of Medicine. 2003. *Microbial Threats to Health: Emergence, Detection and Response.* Washington, DC: The National Academies Press; Hooper, L.V. and J.I. Gordon. 2001. Commensal host-bacterial relationships in the gut. *Science* 292(5519):115-1118; Gionchetti, P. et al. 2000. Oral bacteriotherapy as maintenance treatment in patients with chronic pouchitis: a double-blind, placebo-controlled trial. *Gastroenterology* 119(2): 305-309; Cunningham-Rundles, S. and M. Nesin. 2000. Bacterial infections in the immunocompromised host. In Nataro, J., Blaser, M., Cunningham-Rundles, S., eds. *Persistent Bacterial Infections.* Washington, DC: ASM Press: 145-163. See also, Rao et al. 2005. Toward a live microbial microbicide for HIV: Commensal bacteria secreting an HIV fusion inhibitor peptide. *Proceedings of the National Academy of Sciences* 102(34):11993-11998.

[84] Blaser, M. 1997. Ecology of *Helicobacter pylori* in the human stomach." *Journal of Clinical Investigation* 100(4):759–762; Merrell, D.S. and S. Falkow. 2004. Frontal and stealth attack strategies in microbial pathogenesis. *Nature* 430(6996):250-256.

[85] Casadevall, A. and Pirofski, L-A. 2000. Host-pathogen interactions: Basic concepts of microbial commensalism, colonization, infection, and disease. *Infection and Immunity* 68(12):6511–6518; Pirofski, L-A. and Casadevall, A. 2002. The meaning of microbial exposure, infection, colonisation, and disease in clinical practice. *Lancet Infectious Diseases* 2(10):628-35; Casadevall, A. and Pirofski, L-A. 2003. The damage-response framework of microbial pathogenesis. *Nature Reviews Microbiology* 1(1):17-24.

[86] Blaser, M. 1997. Ecology of *Helicobacter pylori* in the human stomach. *Journal of Clinical Investigation* 100(4):759–762. It may also be possible to produce more dangerous pathogens by intentionally or inadvertently disrupting this dynamic equilibrium.

[87] Merrell, D.S. and S. Falkow. 2004. Frontal and stealth attack strategies in microbial pathogenesis. *Nature* 430(6996):250-256.

[88] It should, however, be noted that severe disease or mortality enhances the transmissibility of some pathogens—eg., intestinal pathogens (*Vibrio cholera, Bacillus anthracis*) and host mortality may provide food for others.

[89] Merrell, D.S. and S. Falkow. 2004. Frontal and stealth attack strategies in microbial pathogenesis. *Nature* 430(6996):250-256; Mascie-Taylor, C.G. and E. Karim. 2003. The burden of chronic disease. *Science* 302(5652):1921-1922.

[90] Blaser, M. 1997. Ecology of *Helicobacter pylori* in the human stomach. *Journal of Clinical Investigation* 100(4):759–762.

[91] Staskawicz, B. et al. 2001. Common and contrasting themes of plant and animal diseases. *Science* 292(22):2285-2289.

[92] Plotnikova, J.M. et al. 2000. Pathogenesis of the human opportunistic pathogen *Pseudomonas aeruginosa* PA14 in arabidopsis. *Plant Physiology* 124(4):1776-1774; Woolhouse, M.E. et al. 2001. Population biology of multihost pathogens. *Science* 292(5519):1109-1112.

[93] Reeve, J.N. 1999. Archaebacteria then...Archaes now (are there really no archaeal pathogens?). *Journal of Bacteriology* 181(12):3613-3617; Eckburg, P.B. et al. 2003. Archaea and their potential role in human disease. *Infection and Immunity* 71(2):591–596.

[94] Lepp, P.W. et al. 2004. Methanogenic *Archaea* and human periodontal disease. *Proceedings of the National Academy of Sciences* 101(16):6176-6181.

[95] Worobey, M. 2000. Extensive homologous recombination among widely divergent TT viruses. *Journal of Virology* 74(16):7666-7670. Available online at www.cdc.gov/ncidod/EID/vol10no12/04-0759.htm [accessed January 4, 2006].

[96] Shimono, N. et al. 2003. Hypervirulent mutant of *Mycobacterium tuberculosis* resulting from disruption of the mce1 operon. *Proceedings of the National Academy of Sciences* 100(26):15918; Foreman-Wykert, A. and Miller, J.F. 2003. Hypervirulence and pathogen fitness. *Trends in Microbiology* 11(3):105-108.

[97] Mouslim, C. et al. 2002. Conflicting needs for a *Salmonella* hypervirulence gene in host and non-host environments. *Molecular Microbiology* 45(4):1019-1027.

[98] Lorange, E.A., et al. 2005. Poor vector competence of fleas and the evolution of hypervirulence in *Yersinia pestis. Journal of Infectious Diseases* 191(11):1907-1912.

[99] The Committee recognizes that virulence can evolve to increase or decrease in a pathogen, in response to specific circumstances, such as how the pathogen is transmitted from person to person.

[100] Kagan, E. 2001. Bioregulators as instruments of terror. *Clinics in Laboratory Medicine* 21(3):607-618. See also, Wheelis, M. 2004. Will the new biology lead to new weapons? *Arms Control Today* 34(6):6-13.

[101] Casadevall, A. and L-A. Pirofski, 1999. Host-pathogen interactions: redefining the basic concepts of virulence and pathogenicity. *Infection and Immunity* 67(8): 3703-3713; Ingham, H.R. and P.R. Sisson. 1984. Pathogenic synergism. *Microbiol. Sci.* 1(8):206-208; Janeway, C.A., C.C. Goodnow and R. Medzhitov. 1996. Immunological tolerance: danger—pathogen on the premises! *Current Biology* 6:519-522. For an easy to read guide on Polly Matzinger's work on molecular "danger signals" see en.wikipedia.org/wiki/Polly_Matzinger.

[102] Kagnoff, M.F. and Eckmann L. 1997. Epithelial cells as sensors for microbial infection. *Journal of Clinical Investigation* 100(1):6-10.

[103] Bäckhed, F. et al 2005. Host-bacterial mutualism in the human intestine. *Science* 307(5717):1915-1920; Rakoff-Nahoum, S. et al. 2004. Recognition of commensal microflora by Toll-like receptors is required for intestinal homeostasis. *Cell* 118(2):229-241.

[104] Kobayashi, K.S. et al. 2005. Nod2-dependent regulation of innate and adaptive immunity in the intestinal tract. *Science* 301(5710):731-734; Maeda, S., et al. 2005. Nod2 mutation in Crohn's disease potentiates NF-kappaB activity and IL-

1beta processing. *Science* 307(5710):734-738; Girardin, S.E. et al. 2003. Lessons from Nod2 studies: towards a link between Crohn's disease and bacterial sensing. *Trends in Immunology* 24(12):652-658; Girardin, S.E. et al. 2003. Nod1 detects a unique muroopeptide from gram-negative bacterial peptidoglycan. *Science* 300(5625):1584-1587; Girardin, S.E. et al. 2003. Nod2 is a general sensor of peptidoglycan through muramyl dipeptide (MDP) detection. *Journal of Biological Chemistry* 278(11): 8869-8872; Fiocchi, C. 1998. Inflammatory Bowel Disease: Etiology and Pathogenesis. *Gastroenterology* 115(1):182-205.

[105] Nixdorff, K. and W. Bender. 2002. Ethics of university research, biotechnology and potential military spin-off. *Minerva* 40(1):15-35.

[106] *Proliferation: Threat and Response.* Available online at www.defenselink.mil/pubs/prolif97/annex.html#technical [accessed February 24, 2005].

[107] Block, S. 1999. Living nightmares: Biological threats enabled by molecular biology. In Drell, S.D., A.D. Sofaer, and G.D. Wilson. 1999. *The New Terror: Facing the Threat of Biological and Chemical Weapons.* Stanford, CA: Hoover Institution Press: 39-75.

[108] Krug, R.M. 2003. The potential use of influenza virus as an agent for bioterrorism. *Antiviral Research* 57(1-2):147-150.

[109] Kobasa, D. et al. 2004. Enhanced virulence of influenza A viruses with the haemagglutinin of the 1918 pandemic virus. *Nature* 431:703-707.

[110] Tumpey, T.M. et al. 2005. Characterization of the reconstructed 1918 Spanish influenza pandemic virus. *Science* 310(5745):77-80; Taubenberger, J.K. et al. 2005. Characterization of the 1918 influenza virus polymerase genes. *Nature* 437(7060):889-893. Available online at www.nature.com/nature/journal/v437/n7060/full/nature04230.html [accessed January 4, 2006].

[111] Ibid.

[112] Institute of Medicine/National Research Council. 2005. *An International Perspective on Advancing Technologies and Strategies for Managing Dual-Use Risks.* Washington, DC: The National Academies Press; also Bokan, S. et al. 2002. An evaluation of bioregulators as terrorism and warfare agents. *ASA Newsletter* 02-3(90):1. Available online at www.asanltr.com/newsletter/02-3/articles/023c.htm [accessed January 4, 2006]; Kagan, E. 2001. Bioregulators as instruments of terror. *Clinics in Laboratory Medicine* 21(3):607-618.

[113] In determining whether to list a biological agent, the Secretary of HHS, in consultation with scientific experts representing appropriate professional groups, is required to consider the agent's effect on human health, its degree of contagiousness and methods by which the agent is transferred to humans, and the availability of immunizations and treatments for illnesses that may result from infection by the agent. The list was initiated in 1997, when the Antiterrorism and Effective Death Penalty Act of 1996 required the Secretary of HHS to establish and enforce safety procedures for the transfer of listed biological agents (select agents), including measures to ensure proper training and appropriate skills to handle such agents, and proper laboratory facilities to contain and dispose of such agents. An expanded list of pathogens and toxins went into effect on February 11, 2003. Agricultural plant and animal pathogens are now also included; other changes reflect taxonomic changes and a few reassessments of what constitutes the most dangerous biothreat agents.

[114] See Table 2-2 in National Research Council. 2004. *Biotechnology Research in an Age of Terrorism*. Washington, DC: The National Academies Press; 54-57.

[115] Our current biosafety system and select agents lists are mostly concerned with full systems or whole organisms. But as we start to construct new things via the combination of many functions in novel ways, the current scheme will not scale. Although beyond the scope of this study, governments and regulatory bodies may need to consider whether or not a biosafety system that is based at the "parts" level might be more useful.

[116] Meselson, M. 2000. Averting the hostile exploitation of biotechnology. *The CBW Conventions Bulletin*.48(June): 16-19.

[117] Information about current biological weapons capabilities summarized in Squassoni, S. 2004. Nuclear, biological, and chemical weapons and missiles: Status and trends. CRS Report for Congress, July 2 (RL30699); National Research Council. 2004. *Biotechnology Research in an Age of Terrorism*. Washington, DC: The National Academies Press.

[118] See www.opbw.org/ [accessed October 28, 2004].

[119] Theodore Kaczynski, the Unabomber, was charged in 1998 with making and delivering four bombs that killed two men and maimed two scientists. In all, Mr. Kaczynski was alleged to have killed three people and injured 29, in 16 attacks between 1978 and 1995. See www.cnn.com/US/9805/04/kaczynski.sentencing/index.html [accessed January 4, 2006].

[120] Hacking the genome. 2003/2004; 2600 *The Hacker Quarterly* 20(4), Winter 2003/2004.

[121] "National security experts and even . . . biologists themselves are concerned that rogue scientists could create new biological weapons—like deadly viruses that lack natural foes. They also worry about innocent mistakes: organisms that could potentially create havoc if allowed to reproduce outside the lab.". . . [W]e live in an age that many tools and technologies can be turned into weaponry," said Laurie Zoloth, a bioethicist at Northwestern University. "You always have the problem of dual use in every new technology." See Elias, P. 2005. Light-sensitive bacteria used to create pictures:UCSF Scientists Make Living Film. Associated Press, November 24. Available online at www.montereyherald.com/mld/montereyherald/business/technology/13251114.htm?template= contentModules/printstory.jsp, [accessed January 4, 2006].

[122] Much of this discussion draws from Chyba, C.F. 2002. Toward biological security. *Foreign Affairs* 81(3):122-136; and Chyba, C.F. and A.L. Greninger. 2004. Biotechnology and bioterrorism: An unprecedented world. *Survival* 46(2):143-162.

[123] Weapons of Mass Effect (i.e., truck bombs or hijacked airliners) are used, as *Time* magazine says, "to cause great loss of life and spread chaos and despair" among the populace. See www.worldnetdaily.com/news/article.asp? ARTICLE_ID=24804 [accessed June 14, 2005].

[124] Weiss, L. 2003. Atoms for peace. *Bulletin of the Atomic Scientists* November/December:34-44.

[125] And because these nuclear materials advertise their presence by emitting various distinctive signatures as radioactive emissions from the source.

[126] Many states attached reservations to their instruments of ratification that

had the effect of making this protocol an agreement to *only* ban first use, not retaliation.

[127] Chyba, C.F. and A.L. Greninger. 2004. Biotechnology and bioterrorism: An unprecedented world. *Survival* 46(2):143-162. For a more in-depth discussion of this point see, National Research Council. 2004. *Biotechnology Research in an Age of Terrorism*. Washington, DC: The National Academies Press.

[128] Meselson proposal to make use of biological weapon a crime against humanity; A Draft Convention to Prohibit Biological and Chemical Weapons Under International Criminal Law. The Draft Convention and a discussion about the need for such a convention may be found at www.sussex.ac.uk/Units/spru/hsp/CRIMpreambleFeb04.htm [accessed January 4, 2006].

[129] It should be noted that a biological "arms race" is between protective measures and malevolent applications of potentially benevolent technologies, rather than between protective measures and offensive weapons programs. The protective technologies that are developed in such a competition are very unlikely to be classified (for all the reasons described) and hence may enable malicious applications of that same technology. This means that it is difficult for defensive applications to win, and bears on the question (which should be discussed to a greater extent) of whether defense can win an offense-defense competition. For a discussion of "can defenses run faster than offenses," see the section with that name, 17-19 of Epstein, G.L. 2005. *Global Evolution of Dual-Use Biotechnology: A Report of the Project on Technology Futures and Global Power, Wealth, and Conflict*. Center for Strategic and International Studies.

[130] Some of the implications of creating a regime of "sensitive" information are discussed in Epstein, G.L. 2001. Controlling biological warfare threats: Resolving potential tensions among the research community, industry, and the national security community. *Critical Reviews in Microbiology*, 27(4):321-354, especially pp. 347-348. This analysis was extended in a presentation given to the Committee on June 23, 2004 titled "Sensitive Information in the Life Sciences." A presentation very similar to that one, and available online, was delivered at the International Forum on Biosecurity in Lake Como on March 21, 2005 and can be found online at www7.nationalacademies.org/biso/Biosecurity_Epstein_2.0.ppt [accessed January 4, 2006].

[131] Institute of Medicine/National Research Council. 2005. *An International Perspective on Advancing Technologies and Strategies for Managing Dual-Use Risk*. Washington, DC: The National Academies Press.

[132] Daar, A.S. et al. 2002. Top 10 biotechnologies for improving health in developing countries. *Nature Genetics* 32:229-232.

[133] We are mindful, however, that crops are indeed monocultures and thus exquisitely sensitive to epidemics of the next "new" fungus or virus; they usually require a lot of water (increasingly scarce in our world) and fertilizer (increasingly expensive and polluting); they are often disruptive of local social structures. As an exercise in practical GM crops, consider the lessons of Lansing, J. S. 1991. *Priests and Programmers: Technologies of Power in the Engineered Landscape of Bali*. Princeton, NJ: Princeton Univ Press.

[134] Hoyt, K. and S.G. Brooks. 2003/2004. A double-edged sword. *International Security* 28(Winter):123-148.

[135] Jackson, R.J. et al. 2001. Expression of mouse interleukin-4 by a recombinant ectromelia virus suppresses cytolytic lymphocyte responses and overcomes genetic resistance to mousepox. *Journal of Virology* 75(3):1205-1210.

[136] Cello, J. et al. 2002. Chemical synthesis of poliovirus cDNA: generation of infectious virus in the absence of natural template. *Science* 297(5583):1016-1018.

[137] Rosengard, A.M. et al. 2002. Variola virus immune evasion design: Expression of a highly efficient inhibitor of human complement. *Proceedings of the National Academy of Sciences* 99(13):8808-8813.

[138] Fraser, C.M. and D.R. Dando. 2001. Genomics and future biological weapons: The need for preventive action by the biomedical community. *Nature Genetics* 29(3):253-256. See, also, National Research Council. 2004. *Seeking Security: Pathogens, Open Access, and Genome Databases*. Washington, DC: The National Academies Press.

[139] Arora, N. and S.H. Leppa. 1994. Fusions of anthrax toxin lethal factor with shiga toxin and diptheria toxin enzymatic domains are toxic to mammalian cells. *Infection and Immunity* 62(11):4955-4961.

[140] Broad, W.J. 1998. Gene-engineered anthrax: Is it a weapon? *New York Times* (February 14): A4; Wade, N. 1998. Tests with anthrax raise fears that American vaccine can be defeated. *New York Times* (March 26).

[141] Aldous, P. 2001. Biologists urged to address risk of data aiding bioweapon design. *Nature* 414(6861): 237-238 as cited in Zilinskas, R.A. and J.B. Tucker. 2002. Limiting the contribution of the open scientific literature to the biological weapons threat. *Journal of Homeland Security*. Available online at www.homelandsecurity.org/journal/Articles/tucker.html [accessed December 30, 2005].

[142] Hatta, M., et al. 2001. Molecular basis for high virulence of Hong Kong H5N1 influenza A virus. *Science* 293(5536):1840-1842.

[143] Gibbs et al. 2001. Recombination in the Hemagglutinin Gene of the 1918 'Spanish Flu.' *Science* 293(5536):1842-1845.

[144] Tumpey, T.M. et al. 2005. Characterization of the Reconstructed 1918 Spanish Influenza Pandemic Virus. *Science* 310(5745):77-80; Taubenberger, J.K. et al. 2005. Characterization of the 1918 influenza virus polymerase genes. *Nature* 437:889-893.

[145] Boyce, N. 2002. Flu's Worst Season. *US News and World Report* 133(6):50.

[146] Wang, X.G. et al. 2001. Purified Cholera Toxin B Subunit from Transgenic Tobacco Plants Possesses Authentic Antigenicity. *Biotechnology and Bioengineering* 72(4):490-494.

[147] Boyce, N. 2002. Should Scientists Publish Work that Could be Misused? *US News and World Report* 132(22): 60.

[148] Wein, L.M. and Y. Liu. 2005. Analyzing a bioterror attack on the food supply: The case of botulinum toxin in milk. *Proceedings of the National Academy of Sciences* 102(28):9984-9989.

[149] Berg, P. et al. 1975. Summary statement of the Asilomar Conference on recombinant DNA molecules.*Proceedings of the National Academy of Sciences* 72(6): 1981-1984.

2

Global Drivers and Trajectories of Advanced Life Sciences Technologies

Advances in science and technology with biological dual-use potential are materializing worldwide at a very rapid pace. Over the next five to ten years, the United States, followed by the European Union and Japan, will likely remain the most powerful global players in the life sciences. Yet, many other nations and regions are developing new and strong scientific and technological infrastructures and capabilities, and some states are emerging as regional and global leaders in their respective fields of specialization. Brazil, China, India, and Russia are among those expected to become stronger economic, political, scientific, and technological global players in the future.

A multitude of complex and interacting economic, social, and political forces drive innovation in life sciences-related technologies and the rapid global dispersion of these technologies (e.g., the technologies described in Chapter 3). These forces, or drivers, include:

- **economic forces** (i.e., labor costs,[1] national investment in research and development, and shifting geographic trends in consumerism and purchasing power, as detailed in this chapter);
- **social forces** (e.g., efforts by developing countries to utilize health and agricultural biotechnology and nanotechnology to improve the well-being of their populations, as well as efforts to make agricultural and other practices more environmentally "friendly"); and
- **political forces,** such as the Canadian government's commitment to devote at least five percent of its research and development investment

to a knowledge-based approach to develop assistance for less fortunate countries,[2] the Mexican national agenda to become a regional leader in genomic medicine,[3] Singapore's plan to make biotechnology the "fourth pillar" of its economy (the other three being electronics, chemicals, and engineering),[4] and the U.S. government's current investment in biodefense.

These drivers operate globally but at varying levels of intensity, depending on national priorities and the strength of local and regional economies. This variability is particularly true of the social and political forces that drive this development. Moreover, the relative importance or strength of the different social, economic, and political drivers changes over time. Within the United States, for example, this country's response to the anthrax mailings following the 9/11 terrorist attacks has emerged only recently as an economic driver. While biodefense spending is still tiny in comparison with the pharmaceutical market forces, it is currently contributing to the shaping of national priorities related to life sciences research. The U.S. focus on 9/11 and biodefense research has also resulted in new immigration and other policies that impact international collaborative scientific research and technological exchange and thus could have a broader impact on science and technology in this country (as discussed in Chapter 4).

In Mexico, a relatively recent national aspiration to become a regional leader in genomic medicine is driving a strongly supported effort to bolster the scientific and technological capacity to do so.[5] In addition to the public health and social benefits expected of personalized health care, the Mexican government perceives the issue as one of national security and sovereignty. A Mexican-specific genomic medicine platform would minimize the country's dependence on foreign technological aid in the future. Meanwhile, in Singapore, where similar efforts are focused on building a national genomic medicine platform, the value of genomic medicine lies in its economic potential. The country is investing billions of dollars in biotechnology, much of the money coming from the Ministry of Trade and Industry, rather than the Ministry of Health.[6]

Inseparable from the diverse economic, social, and political drivers described thus far, another driver—or "mega driver"—of the rapid growth and global dispersion of advanced technologies is globalization itself. In the National Intelligence Council's most recent report on future global trends, globalization is referred to as a "mega-trend . . . a force so ubiquitous that it will substantially shape all the other major trends in the world of 2020."[7] Globalization encompasses the expanding international flow of:

- **capital and goods,** as reflected by the growing number of multinational business collaborations and global firms in the life sciences industry, global trends in biotechnology-related patents, and the globalization of consumerism and purchasing power;
- **knowledge,** as reflected by the changing higher education landscape, the intercontinental movement of students, researchers and technology experts, the growing number of scientific publications authored by researchers outside of the United States, and trends in biotechnology-related patents; and
- **people,** again reflected by the changing nature of the intercontinental movement of students, researchers, and life science professionals.

The following discussion is based on these three broad categories of drivers, or mega drivers, rather than on whether a driver is classified as economic, social, or political. Accordingly, the first half of this chapter summarizes evidence and patterns that reflect the increasingly important roles of the global expansion of capital and goods, knowledge, and people in shaping the global technology landscape. In particular, we survey the pharmaceutical, biotechnology, nanobiotechnology, agricultural, and industrial sectors of the global life sciences industry (which reflect the expanding global flow in capital and goods, knowledge, and people); summarize global scientific productivity, in terms of publication and citations in international journals and other indicators and recent biotechnology patent activity (both of which reflect the expanding global flow in knowledge and people); and highlight foreign student enrollment in U.S. graduate science and technology programs (which reflects the expanding global flow of knowledge and people).

The second half of this chapter includes a snapshot of the rapidly evolving global landscape for the creation, adoption, and adaptation of the advanced technologies discussed herein. This section is not intended to be comprehensive, but to illustrate the extent to which advanced technologies are being developed and disseminated worldwide, well beyond the borders of the G8[8] (i.e., Canada, France, Germany, Italy, Japan, United Kingdom, United States, and the Russian Federation). Highlighted regions and countries were selected on the basis of recent known investments in life science research and applied technologies, obvious indications that the countries are expanding their science and technology foundations, and publicized country efforts to become regional centers of excellence in technologies of interest to this study.

THE GLOBAL MARKETPLACE

One of the most significant factors fueling the global dispersion of advancing technologies is the quest for profit and the desire to enter and succeed in the international marketplace. Over the next five to ten years, all sectors of the life sciences industry—most notably health care and agriculture but also food production, the industrial and environmental sector, and homeland defense and national security—are expected to continue to benefit from and thus drive the rapid growth of new biological knowledge and advanced technologies (Table 2-1). The predictions in Table 2-1 are not comprehensive but are illustrative of the wide range of future market-driven applications, or trends, and key technologies that will enable these applications. Of note, information technology stands out as being common to all sectors, trends, and goals. The dual role of information technology as an advanced technology, in and of itself, and as a driver of other advanced technologies is discussed later in this chapter.

Although North America, Europe, and Japan currently dominate the global marketplace, several other countries are poised to become regional or global leaders in the near future. Not only have new globalization strategies emerged over the past few decades, encouraging increased international collaboration and resulting in a greater number of firms operating in the global arena, but a growing number of new businesses are originating in countries outside North America, Western Europe, and Japan. The latter is evident by current trends in the number of biotech companies in Australia, Brazil, Israel, and South Korea, as detailed below. With regard to increased international collaboration, the number of technological cooperation agreements in biotechnology rapidly grew from near zero in 1970 to almost 700 in 1985-1989 (more recent data are not yet available).[9] Technological cooperation agreements between firms in different countries, focusing on either production or research and development (sometimes both), provide the benefits of collaboration without the contentious issues associated with changes in long-term ownership. Although most of those agreements were between U.S. firms (34 percent), nearly as many U.S.-Japanese (10 percent) or U.S.-Western Europe (19 percent) agreements were formed during this same time period. Other agreements were between Western European and Japanese firms (3 percent), between Western European firms (24 percent), and between Japanese firms (5 percent).

International contracting among biotech and pharmaceutical firms has also increased in recent years.[10] These contracts extend across national borders between firms for the production of components, supplies, and products, made possible by advances in transportation and communications technologies. Following its accession to the World Trade Organization (WTO), the national strengths possessed by India in process engineering

TABLE 2-1 Current and Near-Future Applications of Advancing Technologies

Sectors	Trend	Goal	Key Enabling Technologies
Pharmaceuticals	Development of designer drugs ("personalized medicine"); genotype profiling	Individual and genome-specific drugs	Gene and protein chip (i.e., microarray), biomedical databases (i.e., information technology), computing
	Improved drug delivery	Alternative routes for drug administration	Nanotechnology, aerosol technology, microencapsulation, transdermal delivery technologies
Medicine	Improved diagnosis	Automated genomic tests	Databases, gene and protein chips
	Better treatments for infectious disease	Provide cures for difficult-to-treat or untreatable infections	Biomedical and genome databases, high-throughput screening of compound structural libraries, nanotechnology
	Gene therapy	Identify and treat defective genes	Databases, gene chip, high-performance computing
	Xenotrans-plantation	Develop rejection-free tissues and organs for transplantation	Databases, animal models, recombinant methods
Agriculture	Transgenic crops	Development of disease, pest, and environmental insult-resistant crops; manufacture of biological products	Genome sequencing methods, databases
Biomaterials	Artificial tissue and organs	Develop tissue, stem cell, and other engineering methods	Databases, transgenic crops/animals, nanotechnology
	Biopolymers	New materials for biological and industrial applications	Databases, computing, transgenic crops/animals, nanotechnology
Biodefense	Strengthening biodefense capabilities	Improvement and production of vaccines and prophylactics, rapid diagnostics, pathogen detectors, and forensics	Gene chips, databases, nanotechnology, detector hardware
Computing	Performance improvement	Faster computing for intensive analysis and filtering; convergence of technologies	Grid computing and supercomputers
	Expansion of biotech-specific applications	Develop and strengthen biotech-specific software	Advanced software and search algorithms

SOURCE: Adapted from presentation by Terence Taylor, Cuernavaca workshop.

and information technology have made it a potentially very powerful partner for collaborative and outsourcing drug development and other biotechnology applications. So too is China, following its recent accession to the WTO. According to one account, over the last five years, more than 100 global companies have established research and development centers in India.[11] An industry analysis by the business consulting firm Frost & Sullivan, estimated India's pharmaceutical market to be $5.1 billion in 2004, ranking it 13th globally by value and 4th by volume.[12]

Industry, government, and science news reports point to recent activities throughout Asia, particularly China's rapid entry into the global economy, as some of the strongest evidence of the global expansion of biotechnology and related businesses. According to a recent Intercontinental Marketing Services (IMS) Health report, pharmaceutical sales in China reportedly increased 28 percent to reach $9.5 billion annually. Although that figure is relatively small compared to the global pharmaceutical market of $400-450 billion, industry analysts predict that China's large population size and flourishing economy will push the figure even higher in the future.[13] Asia also boasts the emergence of several major stem cell research centers—in China, Singapore, South Korea, and Taiwan—promising not only exciting opportunities for expatriate students and scientists, but also future commercial success. At Taiwan's Academia Sinica, most of the Ph.D.-level stem cell researchers are Taiwanese or Chinese scientists who have returned home from the United States, United Kingdom, or Australia.[14] ES Cell International (Singapore), a regenerative medicine company, is banking on developing a method for transforming stem cells into insulin-producing cells for transplantation into patients with diabetes.

The Pharmaceutical Industry

Worth approximately $400-450 billion and with an annual growth of about nine percent, the pharmaceutical industry dominates the global life sciences landscape and plays a major driving role in technological development.[15] (Compare this figure to those presented in Table 2-6 for the telecommunications industry, where the total telecommunications market revenue for services and equipment was estimated at U.S. $1,370 billion in 2003.) Although North America and the European Union occupy about 75 percent of the current global pharmaceutical market and enjoy annual growth rates of approximately 12 and 8 percent, respectively, the Asian, African, and Australian markets together are worth about $32 billion and enjoy an annual growth rate of 11 percent.[16] According to a pharmaceutical industry overview by Frost & Sullivan, in the next 5 to 10 years

TABLE 2-2 Analysis of the Global Pharmaceutical Market

Region	Annual Worth	Market Share	Annual Growth
North America	$204 billion	51%	12%
Europe	$102 billion	25%	8%
Japan	$47 billion	12%	1%
Asia, Africa, Australia	$32 billion	8%	11%
Latin America	$17 billion	4%	−10%[a]

[a]This figure reflects past trends. According to a Frost & Sullivan report, the Latin America market is expected to grow significantly in the next 5 to 10 years.
SOURCE: Terence Taylor, Cuernavaca workshop, September 21, 2004.

the Asia-Pacific and Latin American markets should grow significantly and increase their presence in the global marketplace.[17]

The majority of the global market is targeted toward chronic diseases among the elderly (i.e., people over the age of 65). The best-selling pharmaceuticals (and their annual market value in parentheses[18]) are anti-ulcerants ($22 billion), cholesterol reducers ($22 billion), antidepressants ($27 billion), antirheumatics ($12 billion), calcium antagonists ($10 billion), antipsychotics ($10 billion), and oral antidiabetics ($8 billion).

The figures in Table 2-2 represent worldwide trends and include purchases in both developed and developing countries. The developing world market for these best-selling pharmaceuticals is expected to expand in the future, even as resource-poor countries continue to face serious public health problems associated with emerging infectious diseases. Over the next 20 years, the aging population in northwestern Europe is expected to increase by 50 to 60 percent.[19] In the developing world, the same demographic is expected to increase 200 percent over the same time period.

Two likely future major pharmaceutical market trends are the use of genome-specific "designer drugs" (i.e., as part of "personalized" health care) and the use of new and improved modes of drug delivery. These trends will depend on and drive the development and global dissemination of a range of technologies, including gene and protein chip technologies, biomedical databases, computing, nanobiology, aerosol drug delivery applications, and other technologies.

Global Growth of the Biotechnology Industry

Biotechnology companies are enterprises that use a variety of tools and technologies—recombinant DNA, molecular biology and genomics, live organisms, cells, or biological agents—to produce goods and services. In contrast to "large pharma," the biotech industry is dominated by small to medium-sized companies. According to the Biotechnology Industry Organization (BIO), the principal U.S. trade organization for this sector, there are currently 1,473 U.S. biotech companies, of which 314 are publicly held. Corporate membership in BIO is currently over 1,000, compared to 502 in 1993. In contrast, the World Nuclear Association, a global industry organization promoting the peaceful use of nuclear energy, has about 125 members, mostly companies.

Canada ranks second in terms of the number of biotech companies and third, behind the United States and United Kingdom, in terms of generating biotech revenue, according to BIOTECanada. Although California and Massachusetts host the two largest biotechnology industries among all U.S. states and Canadian provinces, Quebec and Ontario follow with 158 and 137 companies in each province. The next largest biotech industries are in North Carolina (88), Maryland (84), British Columbia (78), and New Jersey (77).[20]

The number of European biotech companies grew from 720 to 1,570 between 1997 and 2001.[21] EuropaBio, the principal European trade organization for bioindustry, currently represents about 1,500 small and medium-sized businesses involved in research and development, testing, manufacturing, and distribution of biotechnology products. According to the BioIndustry Association (BIA), the principal trade association for the U.K. biotech sector, the United Kingdom has about 550 biotech, or bioscience, companies, employing over 40,000 people. There are about 350 BIA members.

Growth in the biotechnology sector outside the United States, Canada, and the European Union is equally remarkable. For example:

- the number of biotech companies in Brazil grew from 76 in 1993 to 354 in 2001[22];
- the number of biotech companies in Israel increased from about 30 in 1990 to about 160 in 2000[23];
- the number of publicly listed South Korean biotechnology firms grew from one in 2000 to 23 by 2002[24];
- the Japan Bioindustry Association has about 300 corporate members, 100 public organization members, and 1,300 individual members (from universities)[25];
- AusBiotech, the industry body representing the Australian biotechnology sector, boasts nearly 2,400 individual members; and,
- 59 countries were represented at the BIO 2005 annual conference, which drew nearly 19,000 attendees to Philadelphia in June 2005.

According to the most recent BIO report on the industry, the total value of publicly traded biotech companies (U.S.) at market prices was $311 billion as of early April 2005.[26] Total U.S. revenues for the biotech industry at large increased from $8 billion in 1994 to $46 billion in 2004 (Table 2-3); the number of U.S. biotechnology patents granted per year increased from 2,160 in 1989 to 7,763 in 2002; and the number of biotech drugs and vaccine approvals per year increased from two in 1982 to 37 in 2003.[27] Currently, there are 370 biotech drug products and vaccines in clinical trials in the United States.[28]

The Fledgling Nanobiotechnology Industry

Nanotechnology—which includes, but is not limited to, biotechnological applications—is expected to become a $750 billion market by 2015.[29] Nanotechnology has been defined in many ways, including the science involving matter that is smaller than 100 nanometers,[30] anything dealing with "human-built structures measuring 100 nanometers or less,[31] arranging molecules (atoms) as precisely as possible so as to perform a designated function,[32] and doing with real molecules what computer graphics does with molecular models.[33]

For the purposes of this discussion, "nanotechnology involves the manipulation of molecules less than about 100 nanometers in size. (One nanometer is one-billionth of a meter; a hydrogen atom is about 0.1 nanometers wide.)"[34] Semantics aside, an intriguing feature of nanotechnology is that it operates on the scale upon which biological systems build their structural components, like microtubules, microfilaments, and chromatin.[35] In other words, biochemistry, genomics, and cell biology are nanoscale phenomena. Even more intriguing, a key property of these biological structural components is self-assembly. The most successful biological self-assembler is, of course, the DNA double helix. In their quest to emulate these biological phenomena, scientists have created the field of DNA nanotechnology, or nanobiotechnology,[36] as well as the closely related field of DNA-based computation by algorithmic self-assembly.[37]

Although nanotechnology remains a fledgling field, according to a 2005 report published by NanoBiotech News, 61 nanotech-based drugs and drug delivery systems and 92 nano-based medical devices or diagnostics have already entered preclinical, clinical, or commercial development.[38] For example, in January 2005 the Food and Drug Administration (FDA) approved the use of the nanoparticle-based Abraxane, a solvent-free form of the breast cancer drug Taxol (paclitaxel).[39] The reformulated drug consists only of albumin-bound paclitaxel nanoparticles (i.e., made possible by American Bioscience's proprietary nanoparticle albumin-bound nab™ technology) and is thus free of the toxic solvents that cause

TABLE 2-3 U.S. Biotech Industry Statistics, 1994-2004[a]

Year	1994	1995	1996	1997	1998	1999	2000	2001	2002	2003	2004
Sales[a]	7.7	9.3	10.8	13	14.5	16.1	19.3	21.4	24.3	28.4	33.3
Revenues	11.2	12.7	14.6	17.4	20.2	22.3	26.7	29.6	29.6	39.2	46.0
R&D Expenses	7.0	7.7	7.9	9.0	10.6	10.7	14.2	15.7	20.5	17.9	19.8
Net loss	3.6	4.1	4.6	4.5	4.1	4.4	5.6	4.6	9.4	5.4	6.4
No. of public companies	265	260	294	317	316	300	339	324	318	314	330
No. of companies	1,311	1,308	1,287	1,274	1,311	1,273	1,379	1,457	1,466	1,473	1,444
Employees	103,000	108,000	118,000	141,000	155,000	162,000	174,000	191,000	194,600	177,000	187,500

[a]Amounts are U.S. dollars in billions.

SOURCES: Ernst & Young, LLP, annual biotechnology industry reports, 1993-2005. Financial data based primarily on fiscal-year financial statements of publicly traded companies.

certain side effects associated with Taxol. As another example, in February 2005, Angstrom Medica, Inc. (Woburn, MA), received FDA clearance for its nanoengineered synthetic bone material, NanOss™ Bone Void Filler, which can be used in the treatment of bone fractures or as an alternative to the use of donor bone and metallic medical implants.[40]

Outside the biomedical arena, nanobiotechnology advances are being used to improve cosmetic and sunscreen products, among others. For example, Microniser Pty Ltd (Victoria, Australia) has used nanobiotechnology to develop its proprietary nano-sized zinc oxide powders and other products. Zinc oxide, a common ingredient in many cosmetic products, normally has a white appearance. Microniser's nano-sized zinc oxide (Nanosun™) is transparent.[41]

Many developing countries are making efforts to harness the potential of nanotechnology, and several have launched nanotechnology initiatives. The Indian government plans to invest $20 million over the next five years (2004-2009) in the country's Nanomaterials Science and Technology Initiative[42]; researchers at the University of Delhi are commercializing two U.S.-patented nanoparticle drug delivery systems; scientists at Panacea Biotec, in New Delhi, are conducting novel drug delivery research using mucoadhesive nanoparticles; and Dabur Research Foundation, located in Ghaziabad, is participating in Phase-I clinical trials of nanoparticle delivery of the anticancer drug paclitaxel.[43] In China, researchers have tested a nanotechnology bone scaffold (with the ability to repair damaged skeletal tissue caused by injury resulting from car accidents) in patients.[44] The number of nanotechnology patent applications from China ranks third in the world behind the United States and Japan.[45] It is estimated that China's central and local governments will invest the equivalent of $600 million in nanotechnology and nanoscience between 2003 and 2007.[46] Strikingly, scientists in China published more papers in these fields in international peer-reviewed journals than American scientists during 2004.[47] In Brazil, the projected 2004-2007 budget for nanotechnology is the equivalent of $25 million; and three institutes, four networks, and about 300 scientists are working in the field. In South Africa, investigators and institutions active in the field of nanotechnology banded together to form the South African Nanotechnology Initiative (www.sani.org.za), with the goal of establishing a critical mass in nanotechnology research and development to improve industry-university links, increase nanotech R&D spending, develop projects that benefit South Africa, and generally strengthen South Africa's position as a regional and global player in what is predicted to become the next great wave of technological innovation (i.e., nanotechnology). Thailand, the Philippines, Chile, Argentina, and Mexico are also pursuing nanotechnology initiatives.[48]

A 2005 study in *PLoS Medicine* identified the top 10 potential beneficial applications of nanotechnology for developing countries, illustrating the wide range of social issues that, together with economic forces and political motivations, drive not just nanotechnology but all technological growth:[49]

1.   energy storage, production, and conversion (e.g., novel hydrogen storage systems based on carbon nanotubes and other lightweight nanomaterials);

2.   agricultural productivity enhancement (e.g., nanoporous zeolites for slow release and effcient dosage of fertilizers and of nutrients and drugs for livestock);

3.   water treatment and remediation (e.g., nanomembranes for water purification, desalination, and detoxification);

4.   disease diagnosis and screening (e.g., "lab-on-a-chip" nanoliter systems);

5.   drug delivery systems (e.g., nanocapsules, liposomes, dendrimers, buckyballs, nanobiomagnets, and attapulgite clays for slow and sustained drug release systems);

6.   food processing and storage (e.g., nanocomposites for plastic film coatings in food packaging);

7.   air pollution and remediation (e.g., TiO_2 nanoparticle-based photo-catalytic degradation of air pollutants in self-cleaning systems);

8.   construction (e.g., nanomuscular structures to make asphalt and concrete more robust to water seepage);

9.   health monitoring (e.g., nanotubes for glucose sensors and for in situ monitoring of homeostasis); and,

10.   vector and pest detection and control (e.g., nanosensors for pest detection).

Developing countries recognize the potential of novel technologies. Nowhere is this more evident than with nanotechnology.

AGRICULTURAL BIOTECHNOLOGY[50]

The expansion of transgenic crops is expected to be one of the most important future agricultural trends associated with or resulting from advances in biotechnology. Potential benefits of transgenic agriculture include the development of more disease-resistant crops (which obviate the need for environmentally hazardous pesticides) to the production of better-tasting foods. Environmental and societal benefits notwithstanding, ultimately, as with the pharmaceutical industry, economics is the bottom line. Any technology that results in lower production costs and higher profit margins will likely progress more rapidly than other, lower-yield ventures. About 45 percent of the world's crops are lost to disease, insects, drought, and so forth, annually. In the United States alone, $20 billion worth of crops are lost annually (i.e., one-tenth of production), which represents a large margin that could potentially be affected by advances in transgenic technology. The situation per hectare is worse in other parts of the world. For example, while the United States produces about 6 tons of rice per hectare, Europe produces about 5 tons per hectare and Latin America 2.3 tons per hectare. With corn (maize), the United States produces 7 tons per hectare; Europe, 6 tons per hectare; Latin America, 2.1; and Africa, only 1.7. [51]

The recent rapid growth and global dispersion of commercialized genetically modified (GM) or transgenic crops, also known as biotech crops, suggests that efforts to improve and maximize agricultural productivity already serve as yet another powerful driver of advanced technologies. Transgenic food crops have already entered and flourished in the global marketplace. According to a report, issued in January 2005 by the International Service for the Acquisition of Agri-Biotech Applications (ISAAA), in 2004 there were 14 biotech "megacountries," that is, countries that grow more than 50,000 hectares of biotech crops.[52] These countries were, in order of hectarage, the United States (59 percent of the global total), Argentina (20 percent), Canada (6 percent), Brazil (6 percent), China (5 percent), Paraguay (2 percent), India (1 percent), South Africa (1 percent), Uruguay, Australia, Romania, Mexico, Spain, and the Philippines. To put these figures into perspective, the ISAAA report described the accumulated biotech acreage between 1996 and 2004 as equivalent to 40 percent of the land area of the United States or China and 15 times the total land area of the United Kingdom (Figures 2-1 and 2-2).

Although China ranks fifth in terms of commercialized GM crop hectarage, it is expected to become the world's largest GM crop producer in the next 10 to 20 years. With one-quarter of the world's population and only seven percent of the world's arable land, China has made a strong commitment to using transgenic technology and has spent in the past three years the equivalent of $120 million on developing transgenic rice technology alone. Between 2001 and 2005, China's investment in transgenic technology development was 400 percent greater than between 1996 and 2000.[53]

The global area of biotech crop plantings grew for the ninth consecutive year in 2004, at a rate of 20 percent (up from 15 percent growth in 2003), to 81.0 million hectares (equivalent to 200 million acres), compared to 67.7 million hectares (167 million acres) in 2003 and seven million acres in 1997, when biotech crops were first commercially grown (Figure 2-3).[54] Also in 2004, biotech crops were grown by approximately 8.25 million farmers in 17 countries, compared to 7 million farmers in 18 countries in

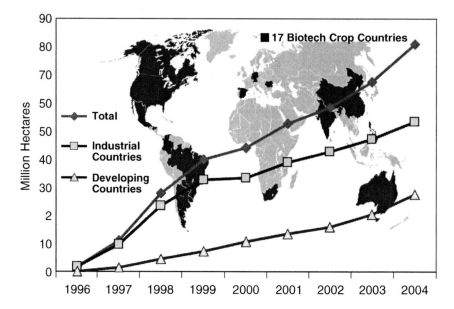

FIGURE 2-1 Global area of biotech crops in million hectares (1996-2004). Increase of 20 percent, 13.3 million hectares or 32.9 million acres between 2003 and 2004. SOURCE: www.isaaa.org/kc/CBTNews/press_release/briefs32/figures/global_area.jpg [accessed January 4, 2006].

2003. Ninety percent of these farmers were in resource-poor countries. In fact, the absolute growth in biotech crop area between 2003 and 2004 was higher in developing countries (7.2 million hectares) than in industrialized countries (6.1 million) for the first time. Brazil and India are expected to become larger sectors of the production market in the near future (see Table 2-4).[55] Other developing countries with small but growing shares of the market include Indonesia, Mexico, Uruguay, Colombia, Honduras, and the Philippines.[56] By the end of the decade, an estimated 15 million farmers are predicted to be growing biotech crops on some 150 million hectares in up to 30 countries.

Based on data from Cropnosis, a crop protection market research firm, and provided by ISAAA, the global market value of biotech crops for 2004 was an estimated $4.7 billion and is expected to grow higher than $5 billion in 2005. Its cumulative global value for the nine-year period between 1996 (when biotech crops were first commercialized) and 2004 was $24 billion. The two most common genetically engineered crop traits are herbicide tolerance (72 percent of global biotech hectares in 2004) and insect resistance (15.6 percent of global biotech hectares in 2004).[57] Major

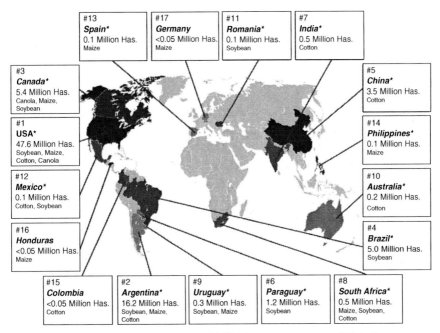

| #13
Spain*
0.1 Million Has.
Maize | #17
Germany
<0.05 Million Has.
Maize | #11
Romania*
0.1 Million Has.
Soybean | #7
India*
0.5 Million Has.
Cotton |

| #3
Canada*
5.4 Million Has.
Canola, Maize,
Soybean | #5
China*
3.5 Million Has.
Cotton |

| #1
USA*
47.6 Million Has.
Soybean, Maize,
Cotton, Canola | #14
Philippines*
0.1 Million Has.
Maize |

| #12
Mexico*
0.1 Million Has.
Cotton, Soybean | #10
Australia*
0.2 Million Has.
Cotton |

| #16
Honduras
<0.05 Million Has.
Maize | #4
Brazil*
5.0 Million Has.
Soybean |

| #15
Colombia
<0.05 Million Has.
Cotton | #2
Argentina*
16.2 Million Has.
Soybean, Maize,
Cotton | #9
Uruguay*
0.3 Million Has.
Soybean, Maize | #6
Paraguay*
1.2 Million Has.
Soybean | #8
South Africa*
0.5 Million Has.
Maize, Soybean,
Cotton |

* 14 biotech mega-countries growing 50,000 hectares, or more, of biotech crops.

FIGURE 2-2 Biotech crop countries and mega-countries, 2004.
SOURCE: www.isaaa.org/kc/CBTNews/press_release/briefs32/figures/crop_
countries.jpg [accessed January 4 , 2006].

transgenic crops include soja (i.e., *Glycien soja*, wild soybean; 61 percent of global market), maize (23 percent), cotton (11 percent), and colza (i.e., canola oil, 5 percent). In 2004 the European Commission approved two biotech maize imports, signaling the end of the 1998 moratorium, and 17 biotech maize varieties for planting in the European Union.

Plant biotechnology is widely recognized throughout Asia as a key tactic for achieving food security and sustainable agriculture.[58] In addition, making recombinant plants is an attractive approach for improving yield. Increases in food production between 1970 and 1995 (i.e., following improvements in agricultural production initiated by the Green Revolution), even as the population grew by one billion, were due largely to the cultivation of new high-yielding varieties of rice and wheat, which were developed by introducing genes that made the plants more responsive to fertilizers and less likely to fall over when fertilized or irrigated. Other factors that contributed to increased yields included expansion of irrigated

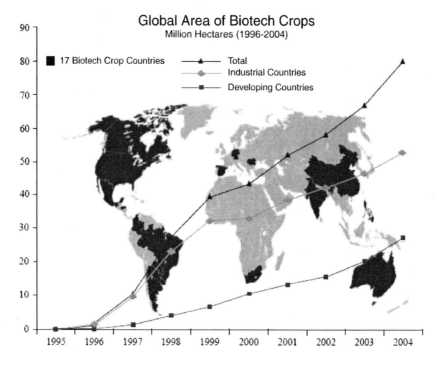

FIGURE 2-3 Global area of biotech crops, 1996-2004. Increase of 20 percent, 13.3 million hectares or 32.9 million acres between 2003 and 2004.
SOURCE: James, C. 2004. Global Status of Commercialized Biotech/GM Products: 2004. International Service for the Acquisition of Agri-Biotech Applications, Executive Summary. Available online at: www.isaaa.org/kc/CBTNews/ press_release/briefs32/ESummary/Executive%20Summary%20(English).pdf [accessed January 4, 2006].

areas, increases in fertilizer and pesticide use, and greater availability of credit.[59]

Plants as Manufacturing Platforms

Transgenic crops are not the only agricultural application of advancing life sciences knowledge. Similar technological advances are being applied to "biopharming," the production of vaccine antigens and other biologically active proteins by transgenic plants. Plant-based manufacturing platforms are considered potentially low-cost, highly efficient, alternatives to other production methods and may be especially suitable for use in developing countries.[60] However, the future of biopharming is unclear. Academic researchers have been investigating the potential for

TABLE 2-4 Top Biotech Crop Countries and Mega-countries, 2004

Country	Hectares (millions)	Key Crops
United States	47.6	soybean maize cotton canola
Argentina	16.2	soybean maize cotton
Canada	5.4	canola maize soybean
Brazil	5.0	soybean
China	3.7	cotton
Paraguay	1.2	soybean
India	0.5	cotton
S. Africa	0.5	maize soybean cotton
Uruguay	0.3	soybean, maize
Australia	0.2	soybean
Romania	0.1	soybean
Mexico	0.1	cotton soybean
Spain	0.1	maize
Philippines	0.1	maize
Colombia	<0.05	cotton
Honduras	<0.05	maize
Germany	<0.05	maize

SOURCE: www.isaaa.org/kc/CBTNews/press_release/briefs32/ESummary/Executive%20 Summary%20(English).pdf; a mega-country is a country that grows 50,000 more hectares of biotech, or transgenic, crops

plant-produced vaccines for over a decade but, despite the promise of the technology, have attracted little venture capital or captured the interest of conventional vaccine manufacturers. Similar efforts are underway with respect to the use of transgenic animals for production of therapeutic proteins. While having considerable potential, these efforts have been slowed by concerns about the potential for gene transfer from transgenic plants to wild type.

Nanotechnological Applications in Agriculture

As mentioned in the previous section, agricultural productivity enhancement has been identified as one of 10 future beneficial applications of nanotechnology in the developing world. Nanotechnology could be used to improve agricultural productivity through the use of nanoporous materials for the slow release and efficient utilization of water and fertilizers for plants (and of nutrients and drugs for livestock), nanocapsules for herbicide delivery, nanosensors for soil quality and plant health monitoring, and nanomagnets for soil contaminant removal.

Industrial Biotechnology

Industrial biotechnology—the application of scientific and engineering principles to the processing of materials by biological agents—has been dubbed the "third wave" of biotechnology, after health and agricultural biotechnology.[61] Chemicals, auto parts, plastics, textiles, and paper are just a few of the many products and industrial sectors that stand to benefit from biological processing, which is generally less expensive, environmentally safer, and more sustainable than petroleum- or chemistry-based manufacturing. Viewed simply from an economic perspective, biotechnology will supplant traditional production technologies as the costs of biotechnology development and production reagents (e.g., glucose) drop below the costs of traditional production reagents (e.g., petroleum). Biological processing can also yield better products. Back in the 1970s, for example, when laundry detergent manufacturers replaced phosphates with cell-derived enzymes, they created a product that produced less waste, cost less to package and transport, and removed stains better than other products then on the market.[62]

Today, cell-, enzyme-, and plant-based processing technologies are being incorporated into a wide range of manufacturing and energy conversion applications. For example, in 2001, Cargill Dow opened a biorefinery in Blair, Nebraska, for the conversion of corn sugar into a polylactide polymer that can be used to produce packaging materials, clothing, and bedding products.[63] April 2004 marked the first commercial shipment of bioethanol—ethanol made from, in this case, wheat straw—by a Canadian biotech company, Iogen. Iogen expects to begin construction of a 50-million-gallon-a-year manufacturing plant in 2006.[64]

Currently, only about five percent of industrial chemicals are of biological origin (e.g., alcohols, amino acids, vitamins, pharmaceuticals). The figure is expected to increase to 10 percent or higher by 2010, depending on such factors as consumer acceptance, governmental policies and support, and the regulatory environment. Also by 2010, biologically-produced ethanol is expected to constitute as much as 6 percent of all transportation fuel used in Europe.[65]

Biodefense

U.S. biodefense spending[66] has increased dramatically over the past few years. Combined Department of Health and Human Services (HHS) and Department of Homeland Security (DHS) biodefense preparedness spending has increased as follows:

- FY 2001—$294 million (HHS budget),
- FY 2002— $3 billion (HHS budget),
- FY 2003—$4.4 billion (combined HHS and DHS budgets), and
- FY 2004—$5.2 billion (combined HHS and DHS budgets).

The National Institutes of Health (NIH) biodefense research funding increased from $53 million in FY 2001 to nearly $1.7 billion in FY 2005 (and a requested $1.8 billion for FY 2006). The FY 2005 budget was used for basic research ($574 million), diagnostics ($149 million), healthcare facilities construction ($51 million), vaccines ($625 million), and antibiotics and antivirals ($259 million).[67] According to data from CRISP, NIH's grant database, the number of NIH grants referencing bacterial bioweapons agents (i.e., agents that cause tularemia, anthrax, plague, glanders, melioidosis, or brucellosis) increased 15-fold between 1996-2001 (33 grants) and 2001-January 2005 (497 grants).

As with other sectors of the life sciences industry, international collaboration and technology exchanges—as reflected by the growing number of co-owned and foreign-owned patents, in addition to the growing number of international subcontracting and technological cooperation agreements—are vital to the success of the U.S. biodefense industry. This is particularly true with respect to vaccine research and development.[68] Vaccines are considered a key component of U.S. biodefense, yet there are few incentives for the pharmaceutical and biotechnology industries to develop new biodefense vaccines. The FDA has licensed vaccines to protect against only a handful of the nearly 50 biological threat agents identified by HHS (i.e., anthrax, cholera, plague, smallpox). Until recently, manufacturers had ceased producing all but one of these FDA-approved vaccines (i.e., anthrax). Recognizing the urgent need for new biodefense vaccines, the BioShield initiative—$5.6 billion in federal funding for the purchase of vaccines and other medical countermeasures over a period of 10 years—was launched in an effort to create incentives for the private sector to develop and produce new vaccines. However, although nationally funded biodefense-related research and development programs have achieved high visibility, the perceived need for biodefense products has thus far received little attention from well-established pharmaceutical companies. This is likely to remain the case as long as the government is perceived to represent the sole market for such products or until significant new incentives are adopted.

In addition to a lack of incentives, scientific, technological, and regulatory advances make it difficult for all but the largest vaccine manufacturers to house the range of expertise and capabilities required to take a vaccine from concept to commercialization. Consequently, as with other sectors of the life sciences industry, smaller commercial vaccine developers have increasingly relied on outsourcing and technological cooperation agreements (e.g., between pharmaceutical companies and biotech startups). As of 2004, of the top six class A biological threat agents identified by the United States, vaccines for all but one (Botulinum toxin) are being developed in cooperation with international biotechnology companies, including firms in Austria, Belgium, Denmark, France, Japan, the Netherlands, and the United Kingdom.[69]

GLOBAL DISPERSION OF KNOWLEDGE

Articles in international peer-reviewed journals and citations of those articles are commonly used as one of a variety of metrics to assess a country's scientific output, which, in turn, reflects a country's ability to generate new knowledge and adapt and benefit from research conducted globally. Likewise, patents are commonly used as an indicator of a country's technological capacity and output. This section summarizes recent trends with respect to these two major categories of indicators while demonstrating that scientific and technological knowledge in general—and life sciences knowledge and technology specifically—is spreading globally at a very rapid pace.

Global Scientific Productivity

In a recent analysis of the number of published research papers and reviews, and their citations, based on data provided by Thomson ISI, which indexes more than 8,000 scientific journals in 36 languages, Professor Sir David A. King, chief scientific officer of the United Kingdom and head of the Office of Science and Technology, London, compared scientific productivity across 31 countries.[70] The selected countries comprised more than 98 percent of the world's "highly" cited papers, which are defined as the one percent most frequently cited by field and year of publication. In terms of the number of publications, number of citations, and share of the top one percent of cited papers, the United States clearly leads. South Africa is the only African country on the list, and Iran is the only Islamic country represented.

With respect to the number of publications between 1997 and 2001, in terms of a percentage of the world's total, the United States (34.86 percent) is followed by, in decreasing order, the United Kingdom (9.43), Japan (9.28), Germany (4.58), France (6.39), Canada (4.58), Italy (4.05), Russia (3.4), China (3.18), Spain (2.85), Australia (2.84), the Netherlands (2.55), India (2.13), Switzerland (1.84), South Korea (1.53), Belgium (1.32), Taiwan (1.25), Brazil (1.21), Poland (1.18), Denmark (1.02), Finland (0.96), Austria (0.93), Greece (0.62), South Africa (0.5), Singapore (0.42), Portugal (0.37), Ireland (0.35), Iran (0.13), and Luxembourg (0.01). See Figure 2-4 for a graphic representation of these numbers.

Compared to the number of publications between 1993 and 1997, there has been a notable decrease in the percentage of papers authored by U.S. scientists (37.47 in 1993-1997, compared to 34.86 in 1997-2001), compared to increases for many other countries (including the United Kingdom, Japan, Germany, France, and particularly China) over the same time periods. With respect to the number of citations and share of the top one percent most frequently cited papers, again the gap between the United States and other countries narrowed slightly. From 1993-1997, U.S.-authored papers comprised 52.3 percent of the world total and 65.6 percent of the top one percent of frequently cited papers. Those figures fell to 49.43 percent and 62.76 percent, respectively, in 1997-2001.

Since the United States is a relatively large nation in terms of its population and gross domestic product (GDP), it is instructive to also look at such data on a per capita basis. Citations are considered a measure of the impact of a nation's publication output. In the graphical comparison of the "citation intensity" (citations per gross national product) and "wealth intensity" (GDP per person) in Figure 2-4, Israel, the Scandinavian countries, Switzerland, and the Netherlands are all above the norm (i.e., they have higher-than-average citation numbers: wealth indices). The United States, Japan, Taiwan, Ireland, and Luxembourg all fall below the norm.

Among the G8 nations and in terms of a national disciplinary "footprint" (i.e., a country's impact on international science based on citation share), notable features include Russia's relative weakness in the life sciences (compared to its relative strength in the physical sciences and engineering). The United States has the highest impact and the United Kingdom the second-largest footprint in the life sciences.

Importantly, as King notes, ranking countries by citation share may hide important recent trends, such as the very rapid growth that China is currently experiencing with respect to establishing a strong science and technology base and the initial steps that many other countries are taking toward strengthening their scientific and technological capabilities. Indeed, in a similar analysis that appeared recently in *Science*, the authors found that, although the gap in scientific output between the world's rich-

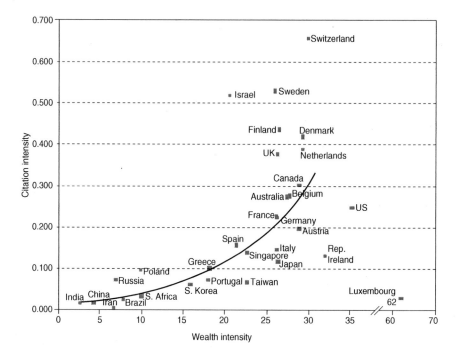

FIGURE 2-4 Comparing economic and scientific wealth. National science citation intensity, measured as the ratio of the citations to all papers to the national GDP, shown as the function of the national wealth intensity, or GDP per person, for the 31 nations in the comparator group. GDP and wealth intensity are given in thousands of U.S. dollars at 1995 purchasing-power parity. Source: Thomson ISI, OECD and the World Bank.
SOURCE: King, D.A. 2004. The scientific impact of nations. *Nature* 430(6997):311-316. Reprinted by permission from Macmillan Publishers Ltd: *Nature* 430(6997), copyright 2004.

est and poorest countries may be widening—the output from the world's 63 poorest countries dropped by about a tenth to just 0.3 percent of all healthcare publications (in more than 4,000 journals) between 1992 and 2001—scientists from middle-income countries, such as China and Turkey, have increased their output by about 20 to 30 percent.[71]

As presented in Table 2-5, scientific productivity does not necessarily immediately or directly translate into commercial development or economic gain.[72] Technological innovation and development involve a highly interactive and dynamic process with multiple influences (e.g., laws, health regulations, cultural norms, social rules, technical standards) and

TABLE 2-5 Comparisons of Private-Sector R&D Spending and the Output of Ph.D.s and Researchers

Country	BERD	BERD as % of GDP	Ph.D.s	Ph.D.s per capita	Full-time researchers	Full-time researchers per 1,000 employed
Japan	65,726	2.12	10,962	0.08	644,208	9.59
U.S.	169,228	1.97	44,955	0.17	1,148,271	8.17
Germany	31,013	1.66	24,940	0.30	238,944	5.93
France	18,186	1.38	10,056	0.17	156,004	5.99
U.K.	15,048	1.22	11,253	0.19	147,035	5.02
European Union	95,733	1.19	6,323	0.18	784,066	5.6
Canada	8,343	1.06	3,871	0.13	90,245	5.88
Russia	6,577	0.72	—	—	—	—
Italy	6,569	0.53	3,494	0.06	69,621	3.09

SOURCE: Adapted from Table 3 in King, D.A. 2004. The scientific impact of nations. *Nature* 430 (6997):311-316 Feature.

institutions (e.g., private-sector firms, governments, research and educational institutions, financial institutions, international linkages) providing constraints and incentives.[73] As such, King also analyzed indicators of business research and development activity among G8 nations. Japan took the lead, and the United Kingdom fell to fifth place, ranked according to BERD (business enterprise research and development, in millions of US$ at 1995 prices, adjusted for purchasing power) per GDP.

Global Growth in Biotech Patent Activity[74]

Patent data reflect the inventive performance of countries and regions and, along with other science and technology indicators, such as those above, can help paint a clearer picture of international advances in biotechnology. According to data provided by the Organisation for Economic Cooperation and Development (OECD), there has been a dramatic increase in patent activity worldwide over the past decade.[75] For example, between 1985 and 2000, the total number of triadic patent families[76] grew by 4.4 percent per year, from about 5,000 in 1985 to close to 44,000 in 2000. Most triadic patenting activity occurs in the United States (about 34 percent), Japan (about 27 percent), Germany (about 13 percent), France (about 5 percent), and the United Kingdom (about 4 percent). These five countries accounted for about 84 percent of triadic patent families in 2000. But several other countries—namely, Brazil, China, and India—showed remarkable growth in patent activity during the 1990s (although their share of the total triadic patent families is still very small, at about 0.1 to 0.2 percent for each country).

There has also been a recent increase in the number of patent rights sought by filing a single international application with a single patent office (in accordance with the Patent Cooperation Treaty, PCT). In 2001 there were an estimated 106,948 PCT applications, compared to only 24,126 in 1991. The United States and the European Union accounted for 74 percent of the applications in 2001, followed by Japan, which accounted for about 12.8 percent. Within the European Union, Germany accounted for the greatest amount of PCT patent application activity in 2001 (12.9 percent of global total), followed by the United Kingdom (5.3 percent), France (4.7 percent), the Netherlands (3.5 percent), Italy (2.8 percent), and Sweden (2.5 percent). Other countries comprising a notable share of PCT applications in 2001 included Korea (2.2 percent of global total), Canada (2.1 percent), Switzerland (1.9 percent), Australia (1.6 percent), Israel (1.2 percent), China (0.8 percent), and Russia (0.6 percent). The number of PCT applications originating from developing countries has increased rapidly, although they still account for only a very small proportion of the total number of applications. For example, the number of PCT applications originating from China, India, and South Africa combined was on par with the number of applications from Australia.

By the late 1990s, an average of 14.5 percent of patents in any OECD country were owned or co-owned by foreign residents, compared to 10.7 percent in the early 1990s, indicating an increasingly global and internationally linked inventive performance. Smaller countries and large non-OECD member countries tend to have higher percentages of patents with foreign coinventors. For example, in 1999-2000, Luxembourg had the greatest share of EPO patent applications with foreign coinventors (56 percent), followed by Singapore and Russia (both 43 percent). The United States, Germany, Japan, and other countries with large numbers of patents tend to have a lower share of patents with foreign coinventors; the United States, Germany, and Japan ranked 28th, 29th, and 33rd, respectively, in terms of the percentage of EPO patent applications in 1999-2000 with at least one foreign coinventor.

The number of patent applications filed at the national patent office of Brazil (INPI) and the State Intellectual Property Office of the People's Republic of China (SIPI) has increased rapidly over the past decade. In Brazil the total number of INPI applications filed in 2000 increased to 16,700, up from an estimated 7,000 in 1991. Most of those applications were filed by inventors from the United States (30.5 percent), European Union (34.8 percent), Brazil (17.8 percent), and Japan (5.7 percent). In China the total number of patent applications filed at SIPI increased from about 12,000 in 1985 to nearly 60,000 in 2000. Again, most of the patent activity in China is from foreign investors (Japan, 20.6 percent of total patent activity; European Union, 16.8 percent; United States, 14.9 percent; and Korea, 3.6 percent), although domestic applications (i.e., from Chinese inventors) have shown a dramatic 15 percent annual growth rate. By 2000, nearly 40 percent of all SIPI patent applications were domestic.

Although all technology fields have experienced patent growth over the past 10 years, biotechnology and information and computer technology (ICT) have grown most rapidly. For example, between 1991 and 2000, biotechnology and ICT patent applications to the EPO increased 10.9 percent and 9.8 percent, respectively, compared to 6.9 percent growth overall. The United States showed particularly rapid growth in biotechnology patent activity, with 9.6 percent of its EPO patents in the field of biotechnology, compared to only 4.2 percent of the European Union's EPO patents and 3.5 percent for Japan.

The United States (45.1 percent), European Union (33.4 percent), and Japan (11.3 percent) have the greatest shares of biotechnology EPO patents. Within the European Union, Germany holds the most EPO biotechnology patents (12.4 percent), followed by the United Kingdom (5.8 percent), France (4.9 percent), Netherlands (3.0 percent), Denmark (1.7 percent), and

Belgium (1.4). Outside the United States, the European Union, and Japan, the countries holding the most biotechnology EPO patents are Canada (2.4 percent), Switzerland (1.4 percent), Australia (1.3 percent), Israel (1.1 percent), and Korea (0.9 percent).

Hungary, Norway, and New Zealand also have shown particularly rapid recent growth in terms of the percentage of their patents that are in biotechnology. Although Singapore, India, and Denmark each have a higher ratio of biotechnology EPO patents than either the United States, the European Union, or Japan (i.e., about one in eight patents issued to Singapore, India, and Denmark are in the biotechnology field), their overall contribution to the total number of biotechnology patents filed at the EPO is quite small.

Information Technology

The expanding global flow of capital, goods, technology, information, and people is made possible, in part, by advances in information technologies, including the Internet and communications. Future breakthroughs in materials science and nanotechnology are predicted by some to lead to the development of next-generation information and communications devices and tools with unforeseen capabilities, which, in turn, will continue to accelerate progress in information technologies and drive globalization.[77] Of note, many poorer nations are successfully gaining access to newer technologies (e.g., mobile telecom services) rather than developing the infrastructure required for older technologies (e.g., telephone landlines).

Some salient statistics that reflect the global spread of advanced information technologies are presented here (see also Table 2-6):

• According to a 2003 report by the U.S. Department of Commerce, the fastest-growing biotechnology-related technical occupation in the United States is R&D-focused computer specialist, which grew 21.8 percent annually between 2000 and 2002.[78]

• All of the world's major economies deploy high-end broadband communications connections. According to 2004 data compiled by the International Telecommunications Union, South Korea leads the world in terms of broadband[79] penetration (24.9 percent total broadband penetration rate, including DSL, cable modems, and other), followed by Hong Kong (20.9 percent), the Netherlands (19.4 percent), Denmark (19.3 percent), Canada (17.6 percent), Switzerland (17.0 percent), Taiwan (16.3 percent), Belgium (16.9 percent), Iceland (15.5 percent), Sweden (15.1 percent), Norway (15.0 percent), Israel (14.3 percent), Japan (14.1 percent), Finland (12.3 percent), Singapore (11.6 percent), United States (11.4 percent), France (11.2 percent), the United Kingdom (10.3 percent), and Austria (10.1 percent).[80]

TABLE 2-6 Key Global Telecom Indicators for the World Telecommunications Service Sector

	1991	1995	1999	2000	2001	2002	2003
Telecom market revenue							
(US$ Billions)							
Services	403	596	854	920	968	1,020	1,070
Equipment	120	183	269	290	264	275	300
Total	523	779	1,123	1,210	1,232	1,295	1,370
Main telephone lines (millions)	546	689	905	983	1,053	1,129	1,210
Cellular subscribers (millions)	16	91	490	740	955	1,155	1,329
International telephone traffic minutes (billions)	38	63	100	118	127	135	140
Personal computers (millions)	130	235	435	500	555	615	650
Internet users (millions)	4.4	40	277	399	502	580	665

SOURCE: International Telecommunications Union, available online at www.itu.int/ITU-D/ ict/statistics /at_glance/Key Telecom99.html.

- Averaged across 182 countries and according to data compiled by the International Telecommunications Union, there were 10.13 personal computers (PCs) per 100 inhabitants in 2003. The top 20 countries on the list (also the only countries with more than 40 PCs per 100 inhabitants) are Switzerland (70.87 PCs per 100 inhabitants), the United States (65.98), Singapore (62.2), Sweden (62.13), Luxembourg (62.02), Australia (60.18), Denmark (57.68), South Korea (55.8), Norway (52.83), Canada (48.7), Germany (48.47), Taiwan (47.14), Netherlands (46.66), Iceland (45.14), Finland (44.17), Estonia (44.04), Hong Kong (42.2), Ireland (42.08), New Zealand (41.38), and United Kingdom (40.57).[81] Averaged across 51 countries, there are only 1.44 PCs per 100 inhabitants in Africa, with the highest concentrations in Seychelles (15.53) and Mauritius (14.87), then dropping down to 9.93 in Namibia, 7.77 in Cape Verde and 7.23 in South Africa.

- Averaged across 182 countries and according to data compiled by the International Telecommunications Union, 55.1 percent of all telephone subscribers worldwide use cell phones (i.e., they have cellular mobile subscriptions). Five of the six countries where more than 90 percent of all telephone subscribers use cell phone technologies are in Africa: D.R. Congo (99.0 percent), Congo (97.9 percent), Uganda (92.7 percent), Cameroon (90.7 percent), and Mauritania (90.2 percent). The sixth country is Cambodia, with 93.2 percent. By region, Africa has the greatest percentage of cell phone subscribers (67.5 percent), followed by Europe (58.9 percent), Oceania (57.2 percent), Asia (54.0 percent), and the Americas (50.3 percent). In the United States, 46.7 percent of all telephone subscriptions are for cell phones.[82]

- In terms of absolute numbers, the top five countries with the largest number of cell phone subscribers are China (269 million in 2003, representing a dramatic increase from 23.8 million in 1998); the United States (158 million in 2003, up from 69.2 million in 1998); Japan (86.6 million in 2003, up from 47.3 million in 1998); Germany (64.8 million in 2003, up from 13.9 million in 1998), and Italy (55.9 million in 2003, up from 20.4 million in 1998). [83]

- China Mobile is the largest cellular operator worldwide, with about 200 million subscribers as of December 2004.

Global Dispersion of People

While the previous sections addressed the global distribution of scientific and technological knowledge as represented by the use and development of advanced technologies in the life sciences industry and changing trends in relevant patents and publications, this section focuses on another vehicle for the global dispersion of knowledge: people. Global travel and migration of scientists, whether for a weekend conference, several years of study, or permanent relocation, is vital to scientific and technological progress—in both the basic research arena and the commercial development of tools and technologies into commercial applications. For example, the United States has maintained its overall leadership in science and engineering in part because it has been able to recruit the most talented people worldwide for positions in academe, industry, and government.[84] The proportion of foreign-born U.S. scientists and engineers has grown rapidly over the past three decades. For example, in 1966, 23 percent of science and engineering doctorates were foreign born, compared to 39 percent in 2000; the percentage of science and engineering postdoctoral scholars in the United States who are considered temporary residents increased from 37 percent in 1982 to 59 percent in 2002; the percentage of doctoral-level employees in science and engineering occupations who are foreign born increased from 24 percent in 1990 to 38 percent in 2000; and more than one-third of all U.S. Nobel laureates are foreign born (Figure 2-5).[85]

Several emerging trends suggest that fewer of the most talented foreign born scientists and engineers are studying or working in the United States, either on a temporary or permanent basis. These trends include the growing two-way flow of scientific and advanced technology brain power among high-, middle-, and low-income countries; increasing global competition for the best science and engineering students and scholars; and new visa and immigration policies, brought about in the aftermath of 9/11. The following section details some of these trends among U.S. doctoral degree recipients.

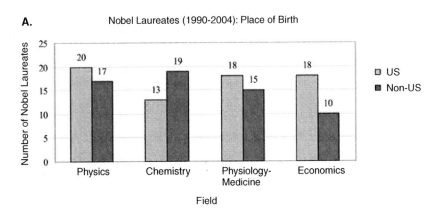

A. Nobel Laureates (1990-2004): Place of Birth

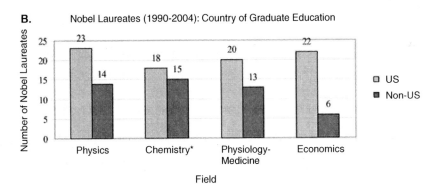

B. Nobel Laureates (1990-2004): Country of Graduate Education

FIGURE 2-5 Exceptional contributions: U.S. Nobel laureates' place of birth and country of graduate school education.

*Data from "Chronology of Nobel Prize Winners in Physics, Chemistry, and Physiology or Medicine." *Nobel e-museum—the official Website of the Nobel Foundation.* Available online at www.nobel.se/index.html. Note that one laureate in chemistry had two Ph.Ds.
SOURCE: National Research Council. 2005. *Policy Implications of International Graduate Students and Postdoctoral Scholars in the United States.* Washington, DC:The National Academies Press: 55.

Trends in Higher Education

As a dominant world force in science and technology, the United States has long attracted and trained students in science and engineering from around the world. The international mobility of foreign recipients of U.S. doctoral degrees leads to a globally dispersed, highly-skilled labor force. Even among the majority of foreign recipients of U.S. doctoral degrees who remain in the United States, continued communication and exchange with their countries of origin fosters and reinforces international ties. Not only do these links play an important role in developing the "bright side" of technology worldwide, they are vital to maintaining the strength of the international collaborations on which U.S. science and technology research and development depends. It is also clear that foreign students and scientists have added historically to the science and technology capacity of the United States and continue to do so today.

Yet foreign interest in U.S. graduate school education in science and technology seems to be waning. According to a recent Council of Graduate Schools survey, from 2003 to 2004, the number of foreign students applying to U.S. graduate programs in the life sciences fell by 24 percent. Applications from Chinese students fell 45 percent, those from Indian students by 28 percent, and those from Korean students by 14 percent. Overall enrollment by foreign students in U.S. graduate school programs fell by 6 percent over the same time period.[86]

The drop in foreign applications may be partly due to the greater restrictive environment that has emerged since 9/11, including a tightening of U.S. visa policies and the tracking of non-U.S. citizens who study or work in the United States for either short-term or extended stays.[87] Importantly, however, it may also be due to the reality that, as other countries develop and strengthen their science and technology bases (including educational and training capacities), there is less reason to study in the United States. The fact that a similar pattern is being seen in the United Kingdom (one survey that reported a 50 percent decline from 2003 to 2004 in university enrollments by Chinese students) suggests that increasing domestic opportunities are creating a competitive global environment.[88]

The trend is particularly striking given the fact that, according to 2004 data from the National Science Foundation (NSF), non-citizens accounted for most of the growth in U.S. science and engineering doctorates from the late 1980s through 2001.[89] During that time, the number of doctorates awarded to non-U.S. citizens rose from 5,100 (26 percent of total) to 9,600 (35 percent); it peaked in 1996, leveled off and declined until 1999, and then rose again. Between 1985 and 2001, a total of approximately 148,000 U.S. doctoral degrees in science and engineering were awarded to foreign students. Foreign students studying in U.S. institutions earn a larger proportion of degrees at the doctoral level than at any other degree level (the proportion varies, depending on the field of study).

Country-specific data from a 2001 NSF report indicate that the largest pool of foreign doctoral degree awardees in the United States in science and engineering fields originated from China (2,405 doctoral degrees awarded in 2001), followed by South Korea (862), India (808), Taiwan

(538), Canada (305), Turkey (304), Thailand (233), Germany (220), and Mexico (205).[90, 91] Based on more recent data from the 2004 NSF Survey of Earned Doctorates, which covers the period between 1985 and 2000, students from 11 major foreign countries/economies and three regional groupings together accounted for nearly 70 percent of all foreign recipients of U.S. science and engineering doctorates.[92]

The major Asian countries/economies sending doctoral students to the United States between 1985 and 2000 were China, Taiwan, India, and South Korea, in that order. Altogether, students from these countries earned more than 50 percent of science and engineering doctoral degrees awarded to foreign students (68,500 out of 138,000), four times more than students from Europe (16,000, most of whom were from Germany, Greece, the United Kingdom, Italy, and France). Chinese students earned, cumulatively, more than 26,500 science and engineering U.S. doctoral degrees, mainly in engineering and the biological and physical sciences. In 1985 only 138 science and engineering doctoral degrees were awarded to Chinese citizens. That number jumped to almost 3,000 by 1996 (see Box 2-1).

Over that same time period, students from Taiwan earned, cumulatively, far fewer degrees than students from China (about 15,500), again mostly in engineering and the biological and physical sciences. Interestingly, in 1985, students from Taiwan earned more U.S. science and engineering doctoral degrees than students from India and China combined (746). As Taiwanese universities increased their capacity to provide advanced science and engineering education in the 1990s, the number of students sent abroad declined. Although Indian students in U.S. institutions earn their doctoral degrees mainly in engineering and the biological and physical sciences, they also comprise the largest number of doctoral degrees awarded to any foreign group in computer and information sciences. U.S. doctorates awarded to South Koreans are mainly in engineering, physical sciences, psychology, and the social sciences. U.S. doctorates awarded to students from Western Europe are mainly in psychology, the social sciences, and engineering; degrees awarded to eastern European students are mainly in the physical sciences, engineering, and mathematics. Eighty-three percent of all doctoral degrees earned by Mexican students are in science and engineering, mainly engineering, psychology/social sciences, biological sciences, and agricultural sciences.

Stay Rates

According to a 1998 NSF report, the majority of foreign students who earned science and engineering doctorates from U.S. institutions between 1988 and 1996 planned to stay in the United States; nearly 40 percent had received firm offers of postdoctoral appointments or employment with

BOX 2-1
Numbers of Engineering Undergraduates in China and India

The Gathering Storm report provided the following numbers:

"In 2004, China graduated over 600,000 engineers, India 350,000 and America about 70,000."

—Geoffrey Colvin, "America isn't ready."
Fortune Magazine, July 25, 2005.

Numbers of Indian Engineering Graduates[a]
The numbers for 1990 (the most recent available from the Indian government) indicate that some 4.9 percent of undergraduates were enrolled in engineering and technology degree programs,[b] yielding 29,000 graduates in 1990.[c] The current data (~2004) indicate that there are 6.7 million students enrolled in state universities and colleges; if the 1990 proportion still holds, that would yield about 300,000 students currently *enrolled* in engineering programs. According to India's National Association of Software and Services Companies (NASSCOM), 341,649 students were enrolled in engineering undergraduate programs in 2004, and that same year 184,347 students graduated.[d]

Numbers of Chinese Engineering Graduates[e]
Data from the Ministry of Science and Technology (MOST) of the People's Republic of China indicate that 1,877,500 undergraduate degrees

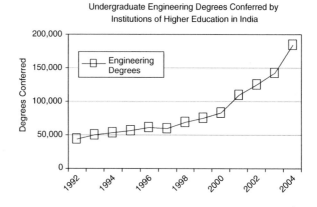

Undergraduate Engineering Degrees Conferred by Institutions of Higher Education in India

CHART SOURCE: NASSCOM. 2005. *Knowledge Professionals.* National Association of Software and Service Companies, India.

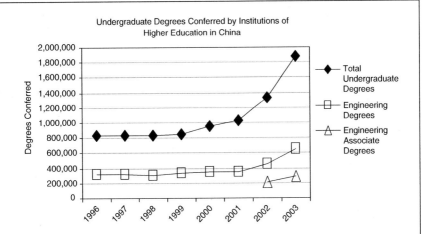

were conferred in 2003, of which 644,106 were in engineering (34.3 percent). Of the engineering degrees, 292,569 (45 percent) are three-year associate degrees.[f] A 2005 McKinsey Global Research Institute study lists China as having had 1.7 million college graduates in 2003, 33 percent of whom were in engineering, for a total of 550,000 engineering degrees.[g]

---

[a]NASSCOM. 2005. *Knowledge Professionals.* National Association of Software and Service Companies, India.

[b]Jayaram, N. 2004. Higher Education in India: Massification and Change. In: *Asian Universities, Historical Perspectives and Contemporary Challenges,* Eds. P.G. Altbach and T. Umakoshi. Baltimore, MD: The Johns Hopkins University Press.

[c]National Science Board. 2004. *Science and Engineering Indicators 2004.* National Science Foundation, Arlington, VA, Appx Table 2-33.

[d]NASSCOM. 2005. *Knowledge Professionals.* National Association of Software and Service Companies, India. Available online at www.nasscom.org/articleprint.asp?art_id=1260 [accessed January 4, 2006]. A subsequent NASSCOM report indicated that about 36% of these graduates have sufficient skills to qualify for interviews with leading companies, see NASSCOM. 2005. *Engineering Graduate Talent Pool in India.* Available online at: www.nasscom.org/download/Engineering_Talent_Pool_Reseach_Highlights1.0.pdf [accessed January 4, 2006].

[e]NASSCOM. 2005. *China S&T Statistics Data Book,* MOST, People's Republic of China.

[f]Ministry of Science and Technology. 2004. *China Statistical Yearbook 2004.* People's Republic of China. Chapter 21, Section 21-11. Available online at www.stats.gov.cn/english/statisticaldata/yearlydata/yb2004-e/indexeh.htm [accessed January 4, 2006].

[g]Farrell, D. and A.J. Grant. 2005. The Emerging Global Labor Market. McKinsey Global Research Institute, New York. The report states "few of China's vast number of university graduates are capable of working successfully in the services export sector, and the fast-growing domestic economy absorbs most of those who could."

industry or elsewhere.[93] According to 2004 report, between 1998 and 2001, 76 percent of foreign doctoral degree recipients in science and engineering fields with known plans intended to stay in the United States, and 54 percent accepted firm offers to do so.

Stay rates for foreign students are not static and are influenced by a variety of factors, including U.S. immigration policies, the number and quality of job opportunities in the home countries of the students, and political change. They also vary by place of origin. For example, in the 1990s both the number of science and engineering students from South Korea and Taiwan and the number who intended to stay in the United States after receipt of their doctoral degree dropped. Both countries have expanded and improved their advanced science and engineering programs and created research institutions that offer more attractive careers for their expatriate scientists and engineers. Between 1985 and 2000, only 26 percent of South Koreans and 31 percent of Taiwanese doctorate recipients reported accepting offers of employment to remain in the United States.

According to a 2003 article in *The Economist*, China's Ministry of Personnel estimated that some 580,000 Chinese students had studied overseas since the late 1970s, with only about 160,000 returning.[94] For example, in 2001, 70 percent of science and engineering doctoral degree recipients from China reported accepting firm offers for employment or postdoctoral research in the United States. But the trend may be changing, as greater numbers of expatriates return home every year. Although only about 9,000 Chinese returned home in 2000 after completing their doctoral studies in the United States, the number had doubled to 18,000 by 2002. Over the past decade, China has made major efforts to lure well-trained expatriates back home to work in academia or start-up companies. For example, in 1994 the Chinese Academy of Sciences launched the Hundred Scholars Project by offering returning young scientists lucrative salary and laboratory set-up packages.[95]

A SNAPSHOT OF THE GLOBAL TECHNOLOGY LANDSCAPE

The section below highlights, on a per-country basis, global advances in life sciences technology. This section is by no means intended to be comprehensive. Rather than an exhaustive analysis, a snapshot is provided of the current global technology landscape, the forces that drive it, and the features that may emerge with respect to the dual-use nature of advancing technologies. Many regions and countries, including the United States, the European Union, and Japan, have already been profiled quantitatively in the previous sections of this chapter. This section highlights and qualitatively profiles other regions and countries that may not be considered global leaders currently but that nonetheless represent focal points for life science-related technological growth. They were selected on the basis of recent known investments in life sciences research and technology, indications that the countries are expanding their science and technology foundations (e.g., a recent series of papers in *Nature Biotechnology* highlighted six so-called innovating developing countries,[96]— developing countries that have demonstrated success in utilizing advanced technologies for the purposes of improving the well-being of their populations) and are making well-publicized national efforts to become regional centers of excellence.

East Asia and the Pacific[97]

The East Asia and Pacific region has made significant economic progress in recent years, with the proportion of the population living in extreme poverty falling from 29.6 percent in 1990 to 14.9 percent in 2001. The region leads the developing world in terms of high-technology exports, including not only pharmaceuticals but also scientific instruments, computers, and aerospace products. However, economic and technological progress in this region is highly uneven, with some countries, such as China, having accomplished much more than others. Indeed, as already indicated and as elaborated below, China is poised to become a future global leader in life-sciences-related technologies.

China

Although China's scientific capacity quickly fell further behind that of the developed world during the Cultural Revolution, which began in 1966, extensive government reforms in the late 1970s and 1980s identified science as central to the country's process of modernization and economic development.[98] In its effort to "catch up" with the rest of the world, China has made dramatic progress in recent years. Indeed, in the most recent Global Trends report by the National Intelligence Council,[99] the likely emergence of China, along with India, as new major global players is compared to the rise of Germany in the 19th century and the United States in the early 20th century. Several factors will fuel this rapid rise in economic and political power, including the active promotion of advanced technologies and the purchasing powers afforded by such large populations (China's population is projected by the U.S. Census Bureau to reach 1.4 billion and India's 1.3 billion by 2020). China is already the third-largest producer of manufactured goods, its share having risen from less than 5 percent in 1980 to about 12 percent today. It is expected to surpass Japan with respect to manufacturing share and exports in the next few years.

Competition from Chinese-manufactured products already restrains manufacturer prices worldwide.

China has made remarkable strides in agricultural biotechnology by accelerating its investments in this area and focusing on commodities that have been largely ignored by commercial interests in other nations.[100] Employing more than 70,000 scientists, China has one of the most successful agricultural research systems in the developing world and is said to be experimenting with more than 120 functional genes among 50 different crop species.[101] Between 1996 and 2000, China's Office of Genetic Engineering Safety Administration approved more than 250 genetically modified (GM) plants, animals, and recombinant microorganisms for field trials, environmental releases, or commercial use. The Chinese government funds almost all plant biotechnology research and in 2001 announced plans to raise the biotechnology research budget by 400 percent over the next 5 years. This is in contrast to most of the industrialized world, where private companies carry out most agricultural biotechnology research.[102]

China's most successful venture with GM crops has been with Bt (*Bacillus thuringiensis*) cotton;[103] from only 730 sown hectares in 1997, by the year 2000, 700,000 hectares had been sown with Bt cotton by 2000. According to one report, Bt cotton is the world's most widespread transgenic crop sown by small farmers, reducing the cost for farmers by U.S. $762 per hectare per season (i.e., largely due to reduced pesticide use).[104] According to the ISAAA report referenced previously, Bt cotton is ranked third in terms of millions of hectares grown in 2004 by all farmers, small or large; the number one biotech crop is herbicide-tolerant soybean, followed by Bt maize.

To date, biotech crops worldwide are primarily used for non food-related purposes (e.g., for fiber, animal feed). This is true despite heavy investments in biotech food crop research. Two of four GM rice varieties, both insect resistant, are already in farm-level preproduction trials (i.e., the last step before commercialization).[105]

Notable achievements in China's health biotechnology sector include the country's participation in the Human Genome Project (China was the only developing country that participated); the 2002 sequencing by Chinese scientists of the rice genome;[106] the approval for market of several Chinese-produced vaccines, diagnostics, and therapeutics (with more than 150 health biotechnology products in clinical trials); the 2003 announcement that a Chinese firm had obtained the world's first drug license for a recombinant gene therapy; and China's liberal environment and access to human embryos for biotechnology research.[107] Additionally, Chinese-authored health biotech publications appearing in the Institute for Scientific Information (ISI)-tracked journals increased from less than 50 in 1991 to more than 300 in 2002.[108] Similarly, there was a modest increase in the number of U.S. Patent and Trademark Office (USPTO) patents in health biotechnology over the same time period (from near zero in 1991 to between 10 and 15 in 2001 and 2002).[109]

Today, there are about 500 biotechnology firms throughout China, employing more than 50,000 people. These include well-funded state-owned enterprises and smaller but more innovative private companies, the latter often established by returned expatriates. The societal role of Chinese universities has changed extensively over the past few decades, with significant implications for biotechnology.[110] Traditionally, higher education concentrated on teaching and training human resources. Now, with an increasing focus on research and industrialization, universities are becoming strong producers of biotechnology knowledge.[111] Additionally, China's large population base and market potential have attracted multinational and other foreign companies, several of which have established joint ventures with domestic companies.[112]

Although the large proportion of Chinese scientists who study abroad and remain to work may be limiting the growth of China's domestic biotechnology sector (e.g., Chinese scientists comprise the largest segment of U.S. science and engineering doctorates awarded to foreign citizens), an increasing number of former expatriates are returning to China to form startups and otherwise engage in biotech research and development. Limited local collaboration is frequently cited as another obstacle to biotechnology growth in China. For example, although scientists from the Beijing Genomics Institute made great efforts to find cooperative partners during the SARS outbreak in China, they were unable to obtain virus samples for testing[113] until after a Canadian group had already posted the entire genome sequence on the Internet. There are signs that this too is changing. For example, while only 13.6 percent of all articles published by Chinese scientists in the international peer-reviewed literature in 1991 included authors from more than one institution, the figure rose to 30 percent by 2002.[114]

Of note, in 2004, China's President Hu Jintao reportedly mentioned brain and cognitive science as one of China's next scientific research frontiers.[115] Not only does this represent an important scientific policy change for China, it is of interest because of the greater dual-use potential that brain and cognitive science research is expected to pose in the future (Chapter 1).

On the nanotech front, as indicated above, the number of nanotechnology patent applications from China ranks third in the world behind the United States and Japan, and Chinese papers on nanoscience and nanotechnology in peer-reviewed international journals now outnumber those from the United States. China is also experiencing one of the fastest rates of increase of Internet and cell phone users in the world and is the leading market for broadband communication.[116]

Singapore

Recent biotechnological growth in Singapore promises to push the country to the forefront as a regional and global biotechnology hub. At least that is its vision: to create infrastructure and industry pipelines that will serve both basic research and the health delivery system. The primary driver is economics. As discussed previously, Singapore wishes to establish biotechnology as the "fourth pillar" of it economy. Strengthening its biotechnological capacity is expected to slow or stop the outsourcing of "high-tech" jobs to India and China.[117]

In the late 1990s, Eli Lilly and Company opened its only clinical pharmacology unit outside the United States, in Singapore, and is recruiting talent from around the globe.[118] In 2001, the company entered into an agreement with the Singapore Economic Development Board to establish an R&D center in Singapore to focus on systems biology.[119] Adding further to the international investment in biotechnology in Singapore, in January 2003, Novartis opened the Novartis Institute for Tropical Diseases in Singapore.[120]

In its efforts to become a global genomic hub with strong ties to the international community, the Singaporean government took a major step forward when it established Biopolis, which is already considered a world-class biomedical research and development hub. Comprising five different research institutes, Biopolis serves as a site for both public and corporate R&D (e.g., including Novartis). Remarkably, the facilities evolved from initial groundbreaking to official opening in a single year. In November 2004, Biopolis hosted the 5th Human Genome Organization (HUGO) Pacific Meeting and the 6th Asia-Pacific Conference on Human Genetics.[121]

In partnership with the Centers for Disease Control and Prevention (CDC), NIH, and FDA, the Singaporean government recently opened the Regional Emerging Diseases Intervention Center (REDI) to conduct research on new viruses and bioterrorist threats and to establish public health policies for emerging infectious diseases. REDI is already beginning to serve as a regional reference center for molecular diagnostics.[122]

South Korea

In 1994, the South Korean government announced intentions to make South Korea one of the world's top seven biotechnology-producing countries by 2010. In 2002, South Korea won the bid to be the permanent host of the International Vaccine Institute, claiming for itself an important platform in the global biotechnology arena.[123] South Korea has demonstrated that it has the technological potential to transform itself rapidly into one of the world's major economies.[124] Known more for its strengths in consumer electronics, heavy industry, and information technology, its biotechnology sector has attracted worldwide attention in recent years when its scientists reported significant advances in the efficient production of patient-specific embryonic stem cell cultures, a critical step toward successful human therapeutic cloning.[125]

Between 2000 and 2007, the government will have invested over 5.2 trillion South Korean Won (about U.S. $4.4 billion) in the biotechnology field. Additionally, the government has initiated financing mechanisms for technology transfer from academia to the private sector, and has made progress in developing a legal framework to encourage the growth of Internet Protocol (IP)-dependent biotech enterprises. Currently, an estimated 450 to 600 Korean companies use biotechnology in their business. As mentioned previously, in 2000, there was only one publically listed South Korean biotech company; by 2002, the number had risen to 23. There are over 40 South Korean pharmaceutical firms with approximately 130 new drugs in phase I or II clinical trials.

Basic scientific productivity in South Korea has increased markedly over the past decade. The number of health biotechnology-related publications (in international, peer-reviewed journals) by South Korean researchers increased from less than 50 in 1991 to almost 350 in 2002. Of note, about one-third of these articles were co-authored by international collaborators. Moreover, the number of U.S. patents increased from practically zero in 1991 to between 20 and 40 per year from 1999 to 2002. The focus in the biotechnology sector is on medical treatments for chronic diseases. In addition to meeting domestic demand, biotechnology exports are expected to increase 10-fold over the next 10 years.[126]

Taiwan

Like South Korea, Taiwan has emerged from a less well-developed economic position in the 1960s to become a powerful technology player with a strong research infrastructure; it ranks 19th in the world in the Scientific Citation Index, up from 35th in 1986.[127] Despite overall budget cuts, the government has been investing an extra 8 to 10 percent annually in R&D, and Taiwan's leading research institution, Academia Sinica, which consists of 25 institutes and more than 800 researchers, has initiated recruitment campaigns and other efforts to keep Taiwanese students from emigrating and to attract foreign talent to its international graduate school. Academia Sinica also now houses a technology transfer office. Although the academy failed to secure a single U.S. patent during its first 70 years of existence, since 1998 it has filed 80 applications, 10 of which have been approved.[128]

Although Taiwan's major national technological strength is in the semiconductor and electronics industry, the Taiwanese government envisions a new economy with a strong biotechnology component. The future is unclear, however, as the biotech sector has yet to approach the success of the country's electronics industry. Despite efforts to recruit high-level researchers, the shortage of talent and personnel, particularly postdoctoral researchers, has been cited as one of the key obstacles to realizing the biotech vision.[129] Fewer Taiwanese students are going abroad, and many of those who do study abroad favor business over science and engineering, resulting in fewer expatriates to recruit; as opportunities increase in Beijing and Shanghai, Taiwan attracts fewer postdocs from mainland China, which in the past was a demographic that fueled growth of the electronics field. Rather than pursuing a research career, those who stay home to study science and engineering often take R&D jobs in one of Taiwan's industrial science parks upon graduation.[130]

Eastern Europe and Central Asia[131]

This region in general is experiencing positive economic growth, leading to a reduction in poverty in some areas, but it is also still contending with serious health and social problems, such as the world's fastest growing HIV/AIDS epidemic, an aging population, and a shrinking workforce. In terms of advanced technologies, notable regional trends include the widespread use of personal computers and the Internet. Between 1995 and 2002, PC availability in European and Central Asian developing countries grew about 22 percent per year. In 2002, the region experienced the highest average rate of PC access of all developing countries (73 per 1,000 people). The Czech Republic, Croatia, Estonia, and the Slovak Republic have the highest availability of PCs in the region. Likewise, Internet use has skyrocketed. The number of Internet users per 1,000 people increased 40-fold in 8 years, from 4 per 1,000 people in 1995 to 160 per 1,000 in 2003. The most rapid growth occurred in Estonia (444 Internet users per 1,000 people), Latvia (404 per 1,000), the Czech Republic (308 per 1,000), and the Slovak Republic (256 per 1,000).

In terms of the global flow of capital, after falling to $37.7 billion in 2001, private capital flows into the region increased by 80 percent to $67.1 billion in 2003, exceeding the flow of private capital into East Asia and the Pacific.[132] Foreign direct investment, on the other hand, still lags significantly behind that provided to East Asia and Pacific. In 2003, foreign direct investment into Europe and Central Asia was about $35.6 billion, compared to nearly $60 billion into East Asia and the Pacific. In 2003 the largest Europe and Central Asia recipients of foreign direct investment were Russia, Poland, Azerbaijan, and the Czech Republic.

Russia

Russia currently faces a severe demographic challenge due to low birth rates, high emigration rates, and a high death rate from emerging and re-emerging infectious diseases—including multidrug-resistant tuberculosis, hepatitis C, and HIV/AIDS)—all of which are expected to continue to contribute to a shrinking working-age population. These and other problems, including poor governance, border conflicts, poor funding for basic science, threaten the ability of this nation, with its longstanding scientific tradition, to participate as a global player in the life sciences.[133] However, Russia's energy resources—particularly its oil and gas exporting potential—give it a leverage that may, over time, boost its economic growth and scientific and technological capacity.

Already, according to some industry analysts, Russia is undergoing an unexpected economic recovery in some areas. In the life sciences industry, according to an Epsicom Business Intelligence industry, the Russian pharmaceutical market is growing at a rate of about nine percent per year.[134] According to the U.S. Department of Commerce, the biologically-active food supplements market is one of the fastest-growing sectors of the Russian biotech industry, with estimates of its total market value ranging from U.S. $1.5 billion to $2 billion.[135] The food additives market is growing faster than the pharmaceutical market presumably because compared to drugs, food supplements are relatively easy to develop, produce, and register. There are 2,000 registered biologically-active food additives in Russia and 556 local manufacturers of such products, mostly small companies.

Despite poor domestic funding for basic life sciences research (e.g., in 2001 the relative percentage of life sciences funding was less than 22 percent of all research funding, compared to 50 percent in the United States), Russia's scientific community has benefited from international relationships and collaborations. For example, the State Research Center of Virology and Biotechnology Vector (SCR VB Vector), and other state laboratories play a leading role in Russia's participation in international efforts to prevent and control the emergence and reemergence of infectious diseases, including smallpox. This particular center has also developed a wide range of diagnostic test kits, some of which are currently being produced by shareholder companies.

Latin America and the Caribbean[136]

The Latin American and Caribbean region has the highest per capita income and highest life expectancy at birth of all developing regions. However, the region as a whole still faces many significant health, education, and other social challenges. With respect to entering the life sciences industry marketplace, one notable challenge is the cumbersome procedure associated with starting a new business in an "informal" economy, in which businesses pay few taxes, workers generally lack health insurance, products are not subject to quality assurance, and businesses have difficulty accessing courts to resolve disputes. On average, Latin American and Caribbean economies require more days (71) to start a business than do economies in other developing regions (56 days is the global average). The time required to complete the procedures necessary to legally operate a business is considered a gauge of the ease of doing business in a country. The ease of doing business reflects expanding opportunities to become involved in life-sciences-related or other industrial sectors. Countries with the shortest time associated with starting a new business are Panama (19 days), Chile (27 days), Jamaica (31 days), Argentina (32 days), and Guatemala (39 days). Three countries in this region—Brazil, Cuba, and Mexico—are highlighted below. Each demonstrates unique potential to develop and strengthen its life-sciences-related technological capacity.

Brazil

Following China and India, the National Intelligence Council recently identified Brazil, along with Indonesia, Russia, and South Africa, as a "rising star" with respect to economic and political growth and power over the next couple of decades.[137] Brazil has a vibrant democracy, a diversified economy, an entrepreneurial population, and solid democratic institutions. Already Brazil enjoys a critical mass of very well-trained scientists and strong public-sector support for research coupled with unparalleled biodiversity, offering the potential for the development of unique plant-based medicines and treatments.[138]

Brazil's healthcare biotechnology success is exemplified by Sao Paulo-based Biobras's development and patenting of a recombinant human insulin in the 1990s—one of only four companies worldwide to have done so at the time; the 2000 sequencing of the plant pathogen *Xylella fastidiosa*, which has encouraged other health-related genomics projects country-wide;[139] the steadily increasing number of Brazilian-authored publications in international peer-reviewed journals; and, as mentioned previously, the rapid expansion of the private biotechnology sector, from 76 Brazilian firms in 1993 to 354 by 2001.[140] Although the healthcare biotechnology sector has not been as successful as it could be in terms of translating basic scientific knowledge into useful commercial products, given the generally excellent conditions for doing so, there are signs that this is changing.[141] For example, in 2004 a bill to encourage private sector participation of university professors was under discussion by the Senate.[142] In March 2005, the president of Brazil signed a biosafety law that will legalize human embryonic stem cell research and establish a clear process for the approval of genetically modified crops, facilitating research and commercialization in both stem cell and agricultural biotechnology.[143]

Brazil has also served as a model for how a well-functioning information system can enable large public initiatives, in this case antiretroviral therapy (ART) scale-up for patients with HIV/AIDS through expanding access to antiretroviral health care for people with HIV/AIVS to all in need.[144] There was an awareness that, in a country the size of Brazil, the logistical demands of ART scale-up would require the ready availability and usability of valuable, accurate information pertaining to all aspects of drug delivery, from procurement to patient compliance. In response, two national computerized systems were created and deployed: SICLOM (Sistema de Controle Logistico de Medicamentos, or System of Logistical Control of ARV), to register and track the distribution of antiretrovirals;[145] and SISCEL (Sistema de Controle de Exames Laboratoriais, or Systems for Control of Laboratory Exams), to track CD4 and viral load laboratory test results.[146]

Cuba

Health-related Cuban biotechnology products are exported to more than 50 countries.[147] Its vaccine industry has generated attention from the international community. Cuba was one of the first countries to have developed a vaccine against the group B meningococcus, although its efficacy may be restricted to a limited number of strains. More recently, Cuban scientists played a leading role in developing the world's first human vaccine with a synthetic carbohydrate antigen, for use in protecting against *Haemophilus influenzae* Type b (Hib) disease.[148]

Cuba has built international collaborations to promote innovation within its biotechnology sector, particularly health biotechnology. This has occurred despite limited financial resources and the U.S. trade embargo, which was imposed in 1966. For example, in July 2004, a joint venture between the Center for Molecular Immunology (Havana) and YMBiosciences (Canada) made an agreement with a U.S. firm in Carlsbad, CA, to undertake development and licensing of two Cuban anticancer therapeutics.[149] More recently, in April 2005 it was announced that Cuban Defense Minister Raul Castro and Malaysian Deputy Prime Minister Najib

Razak had established a bilateral trade and cooperation agreement that will allow Cuban researchers to test experimental vaccines and drug products on the very ethnically-diverse Malaysian population.

Like many developing countries, Cuba suffers from a "brain drain," because many of its current leaders in biotechnology have been influenced by studying abroad in the United States and Europe, and Cuban firms are involved in international collaborations with companies worldwide. The main driving force behind the growth of Cuba's health biotechnology sector has been the desire to improve the health of its citizens, as evidenced by strong governmental support.[150] Cubans have one of the longest life expectancies in the Americas (76.7 years), universal access to health care, and an integrated research/healthcare system enterprise that encourages the creation of innovative products.

Mexico

In 1999 a group of Mexican biomedical experts met to analyze the potential impact that genomic medicine could have in Mexico and designed a plan with three 5-year periods to establish and develop a genomic medicine platform—the first in Latin America and one that is expected to serve as a regional model for other countries. The plan has three components: (1) development of a new Institute of Genomic Medicine (Instituto de Medicina Genomica);[151] (2) an intramural program that will provide expertise, research, and technological capabilities; and (3) a strong extramural program that will foster domestic and international collaborations. Some early accomplishments include the identification of more than 100 candidates for the intramural faculty, most of whom are Mexicans working in Mexico or abroad, with expertise in genomics, bioinformatics, and medical research; sponsorship of dozens of lectures on topics related to genomic medicine, many of which are available on the Internet; and the establishment of several new graduate courses in genomic medicine.[152]

Given the country's ethnically diverse population composition (i.e., more than 65 ethnic groups) and health demands (including both chronic and emerging infectious diseases), Mexican officials assert that a genome-based medicine in Mexico should be based on applications that have been developed specifically for the Mexican population—as opposed to importing products that have been developed for use in some other population. Advancing genomic technology and developing and owning a Mexican-specific genomic database is viewed as an important step, toward improving not only public health and economic development but also national security.[153]

Middle East and North Africa[154]

The Middle East and North African region has a well-developed natural resource infrastructure. Over 85 percent of its population has access to clean water, despite critical shortages of freshwater resources. The region has only 761 cubic meters of internal freshwater resources per person, compared to an average 6,441 cubic meters per person across developing regions worldwide. Sixty-four percent of its roads are paved. Yet the region has made little progress with respect to reducing the number of people living on less than U.S. $1 to $2 a day. Although energy production has increased substantially—low- and middle-income countries in the region produced 26 percent more energy in 2002 than in 1990—energy use per capita has been increasing at an even more rapid rate (by 36 percent over the same time period). The average annual increase in energy use has been the greatest of any developing region at 4.5 percent/year.

In terms of life sciences research and biotechnology development, the countries highlighted here were selected based on information recently published in the scientific literature.

Egypt

Egypt has emerged as a scientific leader among Arab states, particularly in agricultural biotechnology but also in the health biotechnology sector, as evidenced by its ability to rapidly respond to local health crises.[155] For example, in response to an acute insulin shortage in 2002, an internationally partnered emergency plan led to local production of recombinant insulin, which had previously been largely (90 percent) imported at a cost of U.S. $35 million annually. Now, the country continues to rely on local manufacture of insulin for its estimated 3 million to 5 million diabetics who can be treated using the recombinant hormone.[156] Similarly, Egypt has developed several diagnostic and therapeutic products for hepatitis C, rates of which are higher in Egypt than in neighboring countries and other countries with comparable socioeconomic conditions.[157] Also in cooperation with U.S. partners, the Schistosomiasis Research Vaccine Development Project is developing two vaccine candidates for use against what has emerged as the leading parasitic disease in rural Egypt and the number one cause of death among men aged 22 to 44 years.[158] Egypt's government is actively promoting local health care biotechnology with the aim of reducing dependency on importation—for example, by channeling funds toward the building of multipurpose biotechnology pilot manufacturing plants (which are located in the Mubarak City for Scientific Research and Technology Applications, the National Research Centre (NRC), and El Monoufiya University) and by introducing biotechnology educational programs into Egypt's higher education system. The National Strategy for Genetic Engineering and Biotechnology, which was developed in the mid-1990s, includes both short- and long-term plans for the production and marketing of a range of vaccines and diagnostic and other products.[159]

Egypt's health biotechnology sector still relies on the strength of its international linkages, however, in terms of education and training and the actual technology. The country currently exploits knowledge in the public domain more than it does novel contributions by its own research community, and most Egyptian biotechnology companies rely on international contacts rather than local academic research. This reliance on the international community has led to the creation of a global network of alliances among foreign experts and Egyptian scientists living both abroad and at home.[160]

Israel

Israel has seen significant growth in the life sciences and biotechnology development over the past two decades. Currently, almost 60 percent of Israeli-authored scientific publications are in the life sciences, including medicine and the agricultural sciences. As an example of the type of innovative, cutting-edge life sciences research emerging from Israel, in March 2005, scientists from the Institute of Catalysis Science and Technology (Technion, Israel) reported that they had developed a biological computer composed entirely of DNA molecules and enzymes.[161] If borne out, this type of molecular computer could potentially be used in the future for any of a variety of practical applications, including the encryption of information.

According to the Israel National Biotechnology Committee, the number of biotech companies increased from only a handful in 1980 to about 160 by 2000, and the number of people employed in the industry rose tenfold from about 400 in 1988 to some 4,000 in 2000.[162] According to the Ministry of Industry and Trade, therapeutic pharmaceuticals comprise about 67 percent of Israeli biotechnology sales, agricultural and veterinary products about 23 percent (not only genetically engineered hybrid seeds but also poultry and farm animal vaccines, etc.), and diagnostics another 4 percent.[163]

The Office of the Chief Scientist provides $400 million in grant money annually to life sciences companies and has created a network of 24 technology "incubators" for promoting technology transfer from academic institutions to industry. The U.S.-Israel Science and Technology Foundation—a bilateral joint venture between the Israeli and U.S. governments, is promoting entrepreneurship in Israel through distance training of Israeli companies (e.g., through the Larta Institute, Los Angeles, CA).

Libya[164]

Although Libyans enjoy the highest per capita income in Africa, their 12 years of isolation while the country was under an international sanctions regime effectively halted their scientific and technological progress. The last sanctions were lifted in 2004. The newly constructed $100 million Center for Infectious Disease Control in Africa (Tripoli) is one of several recent developments with the aim of making Libya a regional, and eventually a global, center for scientific and technical collaboration. Other initiatives include an exchange program between Italian and Libyan disease researchers, and there are plans to build an on-site factory for generic drug production (for HIV/AIDS, malaria, and tuberculosis). However, a recent highly politicized trial of western healthcare workers on charges that they had allegedly promoted the dissemination of HIV may have dampened current enthusiasm for collaborations with Libya.

Saudi Arabia

In December 2004, Abdul Latif Jameel Company, Ltd.[165] (Jeddah, Saudi Arabia) announced that it would be making a $1 million annual donation to the Arab Science and Technology Foundation (ASTF) to launch and support a new research fund that will provide merit-based support for research projects in nanotechnology, biotechnology, pharmaceuticals, and science. Modeled after the U.S. National Science Foundation, scientists from the 22 Arab nations will be eligible to compete for the grant money. Although this is a drop in the bucket compared to what U.S. and other foundations, even small ones, provide, it represents the first pan-Arab science fund.[166]

South Asia

South Asia has experienced rapid economic growth over the past 15 years, averaging 5.3 percent annually. Since 1990, India has reduced its poverty rate by 5 to 10 percent and, indeed, is forecast by some to become the world's fastest-growing economy in the future (see below). The region has also experienced tremendous growth in modern information and communications technologies. Bangalore, India—the "Silicon Valley" of India—has emerged as a global hot spot for the information technologies industry, with nearly 1,200 companies, including more than 100 multinational companies, operating there. The first company to enter Bangalore, for offshore development, was Texas Instruments in 1984. India's biotechnology achievements are discussed below.

India

Although India currently lags behind China with respect to most economic measures, it is expected to sustain high levels of economic growth and could eventually overtake China as the fastest-growing economy in the world (not because its economic growth rate will match those previously achieved by China over the past decade, but because China's ability to sustain its current rapid growth rate is at risk). Like China, India's rise to economic prominence will have a regional impact, including throughout Southeast and Central Asia and in Iran and other Middle East countries, with whom India will likely pursue strategic partnerships in many sectors, including the life sciences and its associated industries.

India's strength has been in bulk and generic manufacturing and low-cost processing, but there are signs that this may be changing.[167] In the basic research arena, the new Science Advisory Council to the Prime Minister has recommended a National Science and Engineering Research Foundation. Modeled after the U.S. National Science Foundation, the new foundation is widely viewed as being an important step toward strengthening the country's scientific establishment.[168]

According to a 2003 Ernst & Young report,[169] over 328 companies and 241 institutions in India use some form of biotechnology in agricultural, medical, or environmental applications. There are 96 Indian biotechnology companies, giving India the third-largest biotech sector in the Asian region (behind Australia, with 228 enterprises, and China/Hong Kong, with 136).[170] This includes both small and medium-sized enterprises, most of which focus on biopharmaceuticals.

As a result of its emphasis on manufacturing processes rather than on developing novel products, India has developed a strong generic and bulk pharmaceutical manufacturing base and considerable expertise in manufacturing and process innovations. For example, one of the country's more notable biotechnology successes was the development and production of a recombinant hepatitis B vaccine. Although Shanvac-B, as it was named, was not a novel product, its development relied on novel expression technology, lessened the dependence on imports, and provided vaccine at a very low price: $0.50 per dose. It is now supplied to the United Nations Children's Fund (UNICEF). Currently, India is the third-largest producer and prime exporter of generic drugs in the world.

Using their processing strength to their advantage, Indian entrepreneurs and local companies have cultivated a diverse network of global relationships, which they are using to diversify and expand into the global marketplace. For example, Shantha Biotechnics has forged a joint venture with East West Laboratories (San Diego, CA) to develop novel therapeutic monoclonal antibodies for the treatment of various types of cancer.[171]

In the agricultural arena, India has actively promoted the development and use of genetically modified crops nationwide and throughout Asia. In 2002, the Indian government held a conference in New Delhi, at which 18 Asian countries formed an alliance to deal with issues surrounding the introduction of GMOs.[172] According to a 2002 report in *Nature Biotechnology*, the participating countries planned to help each other build scientific capacity to assess the environmental and food safety of GMOs, establish appropriate administrative and legal frameworks, and provide training and other facilities for strengthening the infrastructure for handling GMOs.

As another example of its regional or, in this case, global service with respect to science and technology generally, New Delhi is the site of one of two headquarters for the International Center for Genetic Engineering and Biotechnology (ICGEB; the other office is in Trieste, Italy). The ICGEB was founded in 1983 as a mechanism for involving developing countries in biotechnology. It is an intergovernmental organization with 69 signature states, 52 member states, and a 35-center network. As summarized in the Cuernavaca workshop report[173] and as detailed on the ICGEB website[174] the center performs several functions, including its current agreement with the United Nations Secretariat to draft a code of conduct for scientists.[175] The operational group tasked with drafting the code of conduct is composed of members of the ICGEB and the National Academies of Sciences of China, Cuba, Italy, Nigeria, and the United States. The draft code of conduct was presented to the Secretary-General of the United Nations in April 2005, and in August 2005, it was transmitted as a working document to the BWC.[176]

Along with Brazil, India is one of several developing countries that has launched a major nanotechnology initiative. The country's Department of Science and Technology will invest U.S. \$20 million over the next five years for its Nanomaterials Science and Technology Initiative.

Sub-Saharan Africa

The recent increased access to cell phone service across sub-Saharan Africa, as previously discussed, reflects a positive development in terms of establishing the infrastructure necessary for a modern economy. Yet sub-Saharan Africa has the largest proportion of people living on less than U.S. \$1 a day, and even though the regional economy is expected to improve over the next 10 years or so, with an average per capita growth of

1.6 percent, the number of poor is expected to continue to rise as well (from 313 million in 2001 to 340 million by 2015). A few countries, such as Uganda and Ghana, have sustained remarkable progress in terms of poverty reduction, despite the many social, economic, and political challenges facing the region.

Below, South Africa's recent success in biotechnology is highlighted.

South Africa

Despite its many social and health challenges—including HIV/AIDS, poverty, and crime—South Africa's economy is expected to grow by about 4 to 5 percent per year over the next 10 years, propelling the country even further ahead than it already is in relation to its sub-Saharan neighbors. By focusing on arms, textiles, and mining, South Africa has developed a strong scientific and technological base over the past several decades, even while remaining relatively isolated from the international community while under the apartheid regime.[177] South Africa's industrial success in these areas led to a confidence that has fostered more recent huge strides in agricultural and health biotechnology.

In terms of health biotechnology, the government has established initiatives to encourage international partnerships in the life sciences industry; biotech start-ups, like Shimoda Biotech (with a focus on cyclodextrin drug delivery) and Bioclones (with a focus on monoclonal antibody technology testing for use in diagnostics and immunohistology), are emerging from universities and preexisting generic product companies; diagnostic testing and clinical trials are expanding; and recent controversy over HIV/AIDS national policy has raised awareness about recombinant vaccine trials.

In addition to developing its own national biotech sector, South Africa is hoping to use regional initiatives—such as the New Partnership for African Development—to export its products to other sub-Saharan countries and to use its biotechnological strength to address HIV/AIDS and other regional public health problems. The University of Cape Town and University of Stellenbosch are currently evaluating six different potential novel HIV/AIDS vaccine candidates; in 2002, two Phase I trials were launched, making South Africa the first country with multiple HIV vaccine trials and the first country to have executed a trial on a preventative vaccine against the HIV-1 C subtype.[178]

Elsewhere in Africa, in January 2005 a group of African scientists, engineers, and educators announced plans for an African Institute for Science and Technology, with the aim of strengthening sub-Saharan Africa's tertiary education and research. Currently, the region has only about 83 scientists or engineers per million residents, which is one-sixth of the ratio for all developing countries. Modeled on the Indian Institutes of Technology, the first institute is expected to open in Tanzania in 2007 and will offer undergraduate and graduate degrees in science, engineering, economics, and management. The aim is to attract as many African Ph.D.s working abroad as possible.[179]

SUMMARY

Although providing no more than a high level survey of current trends in the globalization of advanced technologies in the life sciences, the data provided in this chapter do provide evidence that both basic and cutting-edge life sciences technologies are highly dispersed worldwide, and will continue to become more so in the near-term future. The drivers for this are several and vary by nation and region. Developing countries recognize the potential of novel technologies to boost their economies, promote their development, and enhance their regional standing. Turner T. Isoun, Nigeria's minister of science and technology, has observed that "developing countries will not catch up with developed countries by investing in existing technologies alone. [In order] to compete successfully in global science today, a portion of the science and technology budget of every country must focus on cutting-edge science and technologies."[180] This statement, echoing the aspirations of many lesser developed countries, has important implications for the future dispersion of knowledge in the global life sciences community. The trends are profound and well rooted.

ENDNOTES

[1] For most of the core reagents for DNA synthesis, there are no longer any significant U.S. suppliers. As a result, DNS synthesis technology is being "off-shored" to countries with lower labor costs at least as fast as the technology is being developed. This trend can only be expected to escalate in the coming years.

[2] See www.pm.gc.ca/eng/news.asp?id=277 .

[3] Jimenez-Sanchez, G. 2003. Developing a platform for genomic medicine in Mexico. *Science* 300(5617):295-296.

[4] www.biomed-singapore.com/bms/sg/en_uk/index/newsroom/speeches/2000/minister_for_trade.html.

[5] Institute of Medicine/National Research Council. 2005. *An International Perspective on Advancing Technologies and Strategies for Managing Dual-Use Risks*. Washington, DC: The National Academies Press.

[6] For more detailed discussion of the national genomic medicine initiative in Mexico and Singapore's genomic medicine and other biotechnology initiatives, see Institute of Medicine/National Research Council. 2005. *An International Per-*

spective on Advancing Technologies and Strategies for Managing Dual-Use Risks. Washington, DC: The National Academies Press.

[7] National Intelligence Council. 2004. Mapping the Global Future, Report of the National Intelligence Council's 2020 Project. Available online at www.cia.gov/nic/nic_globaltrend2020.htm#contents [accessed April 26, 2006].

[8] It should be noted that article X of the Biological and Toxin Weapons Convention (BWC), and Article XI of the Chemical Weapons Convention (CWC), mandate peaceful cooperation among nations in biology and chemistry.

[9] Hoyt, K. and S.G. Brooks. 2003/2004. A double-edged sword. *International Security* 28(Winter):123-148.

[10] Dicken, P. 1998. *Global Shift: Transforming the World Economy*, Third Edition. New York: The Guilford Press.

[11] Mashelkar, R.A. 2005. India's R&D: reaching for the top. *Science* 307(5714):1415-1417.

[12] See www.inpharm.com/External/InpH/1,2580,1-3-0-0-inp_intelligence_art-0-307722,00.html [accessed May 9, 2005].

[13] See www.ims-global.com/insight/news_story/0503/news_story_050330.htm [accessed May 9, 2005].

[14] Normile, D. and C.C. Mann. 2005. Asia jockeys for stem cell lead. *Science* 307(5710): 660-664.

[15] Although $400 billion was quoted at the Cuernavaca workshop by Terrence Taylor, a Frost & Sullivan analysis puts the figure at $447.5 billion for 2004. See www.frost.com/prod/servlet/vp-further-info.pag?mode=open&sid=2850225 [accessed May 5, 2005].

[16] From Terence Taylor's presentation at the Cuernavaca workshop, September 21, 2004. National Research Council/Institute of Medicine. 2005. *An International Perspective on Advancing Technologies and Strategies for Managing Dual-Use Risks.* Washington, DC: The National Academies Press; Table 3-2, pg. 38.

[17] See www.frost.com/prod/serv/vp-further-info.pag?mode=open&sid=2850225 [accessed May 9, 2005].

[18] Supra, note 16.

[19] Kinsella, K. and V.A. Velkoff. 2001. An Aging World: 2001. U.S. Census Bureau, Series P95/01-1. Washington, DC: Government Printing Office.

[20] See www.bio.org/speeches/pubs/er/statistics.asp [accessed May 6, 2005].

[21] Berg, C. et al. 2002. The evolution of biotech. *Nature Reviews* 1(11):845-846. Although these figures may not seem remarkable at first glance, they are impressive in light of the fact that this time period covered the dot-com crash.

[22] Ferrer, M. et al. 2004. The scientific muscle of Brazil's health biotechnology. *Nature Biotechnology* 22(Suppl.):DC8-DC12.

[23] See www.larta.org/lavox/articlelinks/2004/040510_usisrael.asp [accessed May 9, 2005].

[24] Wong, J. et al. 2004. South Korean biotechnology—a rising industrial and scientific powerhouse. *Nature Biotechnology* 22(Suppl.):DC42-DC47.

[25] See www.jba.or.jp/eng/jba_e/index.html [accessed May 9, 2005].

[26] Biotechnology Industry Facts, 2005, http://www.bio.org/speeches/pubs/er/statistics.asp.

27 Biotechnology Industry Organization (BIO), 2005. Guide to Biotechnology. Available online at www.bio.org/speeches/pubs/er/ [accessed May 5, 2005].

28 Ibid.

29 Cutiss, E.T. 2005. Nanotechnology—Market Opportunities, Market Forecasts, and Market Strategies, 2004 to 2009. Research Report # WG8270, electronics.ca publications, January. Available online at www.electronics.ca/reports/nanotechnology/opportunities.html [accessed January 3, 2006].

30 Blumenstyk, G. 2004. Big bucks for tiny technology. *The Chronicle of Higher Education* 51(3):A26. Available at chronicle.com/free/v51/i03/03a02601.htm [accessed January 4, 2006].

31 Monastersky, R. 2004. The dark side of small. *The Chronicle of Higher Education* 51(3):A12. Available online at chronicle.com/free/v51/i03/03a01201.htm [accessed January 4, 2006].

32 As defined by N. Seeman at the Cuernavaca Workshop; Institute of Medicine/National Research Council, 2005. *An International Perspective on Advancing Technologies and Strategies for Managing Dual-Use Risks.* Washington, DC: The National Academies Press; 50.

33 Ibid.

34 DiJusto, P. 2004. Nanosize me: nebulous naming-nano knack not needed. *Scientific American* (December).

35 Seeman, N.C. and A.M. Belcher. 2002. Emulating biology: building nanostructures from the bottom up. *Proceedings of the National Academy of Sciences* 99(Suppl. 2):6451-6455.

36 Seeman, N.C. 1999. DNA engineering and its application to nanotechnology. *Trends in Biotechnology* 17(11):437-443; Fortina, P. et al. 2005. Nanobiotechnology: the promise and reality of new approaches to molecular recognition. *Trends in Biotechnology* 23(4):168-173.

37 Nanobiotechnology is an emerging area of scientific and technological opportunity. Nanobiotechnology applies the tools and processes of nano/microfabrication to build devices for studying biosystems. Researchers also learn from biology how to create better micro-nanoscale devices. www.nbtc.cornell.edu/.

38 2005 Nanomedicine, Device & Diagnostic Report, available online at www.nhionline.net/products/nddr.htm.

39 www.corporate-ir.net/ireye/ir_site.zhtml?ticker=APPX&script=410&layout=6&item_id=660605 [accessed May 9, 2005].

40 www.angstromedica.com/images/NanOss%20Clearance.htm [accessed May 9, 2005].

41 www.micronisers.com [accessed May 9, 2005].

42 newdelhi.usembassy.gov/wwwhpr0812a.html [accessed June 23, 2005].

43 Bapsy P.P. et al. 2004. DO/NDR/02 a novel polymeric nanoparticle paclitaxel: Results of a phase I dose escalation study. *Journal of Clinical Oncology* 22(14S): 2026; Salamanca-Buentello, F. et al. 2005. Nanotechnology and the developing world. *PloS Medicine* 2(5):383-386.

44 Court, E. et al. 2005. Will Prince Charles et al. diminish the opportunities of developing countries in nanotechnology? Available online at www.nanotechweb.org/articles/society/3/1/1 [accessed February 21, 2005].

[45] Salamanca-Buentello, F. et al. 2005. Nanotechnology and the developing world. *PloS Medicine* 2(5):383-386.

[46] Hassan, M.H.A. 2005. Small things and big changes in the developing world. *Science* 309(5731):65-66.

[47] Ibid.

[48] Salamanca-Buentello, F. et al. 2005. Nanotechnology and the developing world. *PloS Medicine* 2(5):383-386.

[49] Ibid.

[50] It should be noted that with the application of any new technology to the consumer market there is often controversy. This is no less so for "genetically modified" foods. It is beyond the scope of this report to provide an in depth treatment of the debate over the safety and ethical use of GM crops and commodities. For an overview of this issue, please see, Department of Energy, 2005. Genetically Modified Foods and Organisms, on the Human Genome Project Information Website, www.ornl.gov/sci/techresources/Human_Genome/elsi/gmfood.shtml [accessed January 4, 2006].

[51] These production differences are likely due to geographic differences in sunlight, temperature, nutrients, and water.

[52] Global Status of Commercialized Biotech/GM Crops: 2004. Available online at www.isaaa.org/ [accessed February 21, 2005].

[53] Information on China presented by Luis Herrera-Estrella to committee at Cuernavaca; Institute of Medicine/National Research Council. 2005. *An International Perspective on Advancing Technologies and Strategies for Managing Dual-Use Risks*. Washington, DC: The National Academies Press; 21.

[54] Global Status of Commercialized Biotech/GM Crops: 2004. Available online at www.isaaa.org/ [accessed February 21, 2005].

[55] Based on presentation by Luis Herrera-Estrella to committee at Cuernavaca; Institute of Medicine/National Research Council. 2005. *An International Perspective on Advancing Technologies and Strategies for Managing Dual-Use Risks*. Washington, DC: The National Academies Press; 21.

[56] Ibid.

[57] Global Status of Commercialized Biotech/GM Crops: 2004. Available online at www.isaaa.org/ [accessed February 21, 2005].

[58] Asian Development Bank. 2001. *Agricultural biotechnology, poverty reduction, and food security*. Manila, Philippines: Asian Development Bank. Available online at www.adb.org/Documents/Books/Agri_Biotech/default.asp [accessed February 9, 2005].

[59] Asian Development Bank. 2001. *Agricultural biotechnology, poverty reduction, and food security*. Manila, Philippines: Asian Development Bank. Available online at www.adb.org/Documents/Books/Agri_Biotech/default.asp [accessed February 9, 2005].

[60] Arntzen, C.J. and M.A. Gomez-Lim. 2005. BioPharming: plant-derived vaccines to overcome current constraints in global immunization. Institute of Medicine/National Research Council. 2005. *An International Perspective on Advancing Technologies and Strategies for Managing Dual-Use Risks*. Washington, DC: The National Academies Press; 19.

[61] Vinas, T. 2004. Making waves. IndustryWeek.com (August). Available online at www.bio.org/ind/pubs/IndustryWeek_81704.pdf [accessed January 4, 2006].

[62] International Association of Soaps, Detergents, and Maintenance Products, 2002: Poster; An Overview of the major European and international developments, the key association activities, and the main technological innovations of the industry. See www.aise-net.org/PDF/ar_2002_poster.pdf.

[63] See www.natureworksllc.com/corporate/nw_pack_home.asp.

[64] See www.iogen.ca/.

[65] See www.bio.org/ind/background/SummaryProceedings.pdf.

[66] It should be noted that these figures are most likely underestimates of the total expenditures in "biodefense" in the United States since what constitutes biodefense spending has never been consistently defined either within or across government departments and agencies.

[67] Schuler A. 2004. Billions for biodefense: federal agency biodefense funding, FY2001-FY2005. *Biosecurity and Bioterrorism: Biodefense Strategy, Practice, and Science* 2(2):86-96. A more recent article is: Schuler A. 2005. Billions for biodefense: federal agency biodefense budgeting, FY2005-FY2006. *Biosecurity and Bioterrorism: Biodefense Strategy, Practice, and Science* 3(2):94-101; Enserink, M. and J. Kaiser. 2005. Has biodefense gone overboard? *Science* 307(5714):1396-1398.

[68] Hoyt, K. and S.G. Brooks. 2003/2004. A double-edged sword. *International Security* 28(Winter):123-148.

[69] Ibid.

[70] King, D.A. 2004. The scientific impact of nations. *Nature* 430(6997):311-316 Feature.

[71] Paraje G., R. Sadana, and G. Karam. 2005. Public health. Increasing international gaps in health-related publications. *Science* 308(5724):959-960.

[72] OECD, Eurostat. 1997. *The Measurement of Scientific and Technological Activities: Proposed Guidelines for Collecting and Interpreting Technological Innovation Data.* Paris: OSLO Manual.

[73] This sub-section is text that has been adapted from Thorsteinsdottir, H. et al. 2004. Introduction: promoting global health through biotechnology. *Nature Biotechnology* 22(Suppl.):DC3-DC9.

[74] All of the data presented in this section is from the OECD 2004 Compendium of Patent Statistics Report. Available at www.oecd.org/dataoecd/60/24/8208325.pdf [accessed January 4, 2006].

[75] OECD Compendium of Patent Statistics 2004. Available at www.oecd.org/dataoecd/60/24/8208325.pdf [accessed January 4, 2006].

[76] Triadic patent families are sets of patents registered at the world's three largest patent offices: the European Patent Office, EPO, the Japanese Patent Office, JPO, and the U.S. Patent and Trademarks Office, USPTO.

[77] Anton, P.S. et al. 2001. *The Global Technology Revolution: bio/nano/materials trends and their synergies with information technology by 2015.* RAND Corporation. Available online at www.rand.org/pubs/monograph_reports/2005/MR1307.pdf [accessed January 4, 2006].

[78] A Survey of the Use of Biotechnology in U.S. Industry, 2003. Available online at www.technology.gov/reports/Biotechnology/CD120a_0310.pdf [accessed January 4, 2006].

[79] The word "broadband" is a generic term. It refers to the wide bandwidth characteristics of a transmission medium and its ability to carry numerous voice, video or data signals simultaneously. The medium could be coaxial cable, fiber-optic cable, UTP Media Twist or a wireless system. See www.unt.edu/telecom/ Services/broadband.htm [accessed January 4, 2006].

[80] See www.itu.int/ITU-D/ict/statistics/at_glance/top20_broad_2004.html [accessed June 15, 2005].

[81] See www.itu.int/ITU-D/ict/statistics/at_glance/Internet03.pdf [accessed June 15, 2005].

[82] See www.itu.int/ITU-D/ict/statistics/at_glance/cellular03.pdf [accessed June 15, 2005].

[83] Ibid.

[84] National Research Council. 2005. *Policy Implications of International Graduate Students and Postdoctoral Scholars in the United States*. Washington, DC: The National Academies Press.

[85] Ibid.

[86] Kernodle, K. 2005. Combating Continued Drops in Foreign Student Enroll-ment—U.S. Driven to Increase App*eal of Colleges and Universities*. *Frances Kernodle Associates*. Available online at www.fkassociates.com/Combating%20Continued% 20Drops%20in%20Foreign%20Student%20Enrollment.html [accessed January 6, 2006].

[87] National Research Council. 2004. *Biotechnology Research in an Age of Terror-ism*. Washington, DC: The National Academies Press; this may be changing, since the United States announced, in February, 2005, that it had changed its visa rules to make it easier for foreign scientists and students working on "sensitive tech-nologies" to reenter the United States after overseas trips (e.g., to attend confer-ences or visit their home countries).

[88] See www.universitiesuk.ac.uk/international/intlstrategy.pdf [accessed May 10, 2005].

[89] Science and Engineering Indicators—2004. Available online at www.nsf. gov/statistics/seind04/ [accessed January 4, 2006].

[90] National Science Foundation, Division of Science Resources Statistics, *Sci-ence and Engineering Doctorate Awards: 2001*, NSF 03-300, Susan T. Hill, Project Of-ficer (Arlington, VA 2002).

[91] See www.nsf.gov/statistics/nsf03300/pdf/secta.pdf: 53 [accessed January, 2006].

[92] See www.nsf.gov/statistics/seind04/c2/c2s4.htm [accessed January 6, 2006].

[93] From National Science Foundation, Division of Science Resources Studies. 1998. *Statistical Profiles of Foreign Doctoral Recipients in Science and Engineering: Plans to Stay in the United States*. NSF 99-304, Author, Jean M. Johnson (Arlington, VA).

[94] Cited in Zhenzhen, L. et al. 2004. Health biotechnology in China— reawakening a giant. *Nature Biotechnology* 22(Suppl.):DC13-DC18.

[95] Breithaupt, H. 2003. China's leap forward in biotechnology. *EMBO Reports* 4:111-113.

[96] Morel, Carlos M. et al. 2005. Health Innovation Networks to Help Develop-ing Countries Address Neglected Diseases. *Science* 309(5733):401-404. This term

was first proposed by Charles Gardner of the Rockefeller Foundation, based on the 2003 Zuckerman Lecture delivered at the UK Royal Society by R.A. Mashelkar.

[97] Data in the overview of this section is from World Bank data www.worldbank.org/data/databytopic/eap_wdi.pdf

[98] Zhenshen, L. et al. 2004. Health biotechnology in China—reawakening of a giant. *Nature Biotechnology* 22(Suppl.):DC13-DC18.

[99] National Intelligence Council. 2004. Mapping the Global Future, Report of the National Intelligence Council's 2020 Project. Available online at www.cia.gov/nic/nic_globaltrend2020.htm#contents [accessed April 26, 2006].

[100] Huang, J. et al. 2002. Plant biotechnology in China. *Science* 295(5555):674-677.

[101] Ibid.

[102] Huang, J. et al. 2002. Plant biotechnology in China. *Science* 295(5555):674-677.

[103] BT plants carry the gene for an insecticidical toxin produced by the bacteria *Bacillus thuringiensis*, reducing the need for chemical insecticides.

[104] Ibid.

[105] Huang, J. et al. 2005. Insect-resistant GM rice in farmers' fields: assessing productivity and health effects in China. *Science* 308(5722):688-690.

[106] Yu, J. et al. 2002. A draft sequence of the rice genome (*Oryza sativa L. ssp. indica*). *Science* 296(5655):79-92.

[107] Chien, K. and L. Chien. 2004. The new Silk Road. *Nature* 428(6979):208-209.

[108] This figure refers only to health biotech papers, not all Chinese–authored scientific papers in international peer-reviewed journals, nor does it include papers published in local journals not covered by ISI.

[109] This figure does not reflect trends in non-U.S. patents and does not cover all health biotech patents.

[110] Zhenshen, L. et al. 2004. Health biotechnology in China—reawakening of a giant. *Nature Biotechnology* 22(Suppl.):DC13-DC18.

[111] Ibid.

[112] Ibid.

[113] The Beijing Genomics Institute was unable to obtain SARS samples from Guandong, despite efforts, due to safety regulations banning the transfer of viruses.

[114] Ibid.

[115] Harding, A. 2005. The politics of Science. *The Scientist* 19(2):37-40.

[116] National Intelligence Council. 2004. Mapping the Global Future, Report of the National Intelligence Council's 2020 Project. Available online at www.cia.gov/nic/NIC_globaltrend2020_s3.html [accessed January 4, 2006].

[117] Based on materials presented to Committee by Tan Boon Ooi. See Institute of Medicine/National Research Council. 2005. *An International Perspective on Advancing Technologies and Strategies for Managing Dual-Use Risks*. Washington DC: The National Academies Press.

[118] See www.med.nus.edu.sg/lilly/ [accessed October 21, 2004].

[119] See www.lsb.lilly.com.sg/ [accessed October 21, 2004].

[120] See www.nitd.novartis.com [accessed October 21, 2004].

[121] Institute of Medicine/National Research Council. 2005. *An International Perspective on Advancing Technologies and Strategies for Managing Dual-Use Risks*. Washington, DC: The National Academies Press.

[122] Ibid.

[123] Wong, J. et al. 2004. South Korean biotechnology—a rising industrial and scientific powerhouse. *Nature Biotechnology* 22(Suppl.):DC42-47.

[124] Ibid.

[125] Hwang, W.S. et al. 2004. Evidence of a pluripotent human embryonic stem cell line derived from a cloned blastocyst. *Science* 303(5664):1669-1674; Hwang, W.S. et al. 2005. Patient-specific embryonic stem cells derived from human SCNT blastocysts. *Science* 308(5729):1777-1783.

[126] Wong, J. et al. 2004. South Korean biotechnology—a rising industrial and scientific powerhouse. *Nature Biotechnology* 22(Suppl.):DC42-47.

[127] Swinbanks, D. and D. Cyranoski. 2000. Taiwan backs experience in quest for biotech success. *Nature* 407(6802):417-426.

[128] The latest USPTO patent statistics for 2003 reveal that Taiwan's 6,676 patents place it fourth in the world behind the U.S., which posted 98,598 patents, Japan (37,250) and Germany (12,140). See investintaiwan.nat.gov.tw/en/news/200406/2004062501.html [accessed January 4, 2006].

[129] Ibid.; and Cyranoski, D. 2003. Biotech vision Taiwan. *Nature* 421(6923):672-673.

[130] Taiwan aims to become sci-tech island. *Nature* 394(6693),1998:603. Available at www.nature.com/cgi-taf/DynaPage.taf?file=/nature/journal/v394/n6693/full/394603a0_fs.html&content_filetype=pdf [accessed January 4, 2006].

[131] Data in the overview of this section is from World Bank www.worldbank.org/data/databytopic/eca_wdi.pdf.

[132] Private capital flows refer to investments by the private sector into a sector of a country's economy. Foreign direct investments are investments made to acquire a lasting interest by a resident entity in one economy in an enterprise resident in another economy. See www.nscb.gov.ph/fiis/default.asp.

[133] National Intelligence Council. 2004. Mapping the Global Future, Report of the National Intelligence Council's 2020 Project. December. Available online at www.cia.gov/nic/NIC_globaltrend2020.html#contents [accessed May 3, 2005].

[134] See www.inpharm.com/External/InpH/1,2580,1-3-0-0inp_intelligence_art-0-305987,00.html [accessed May 9, 2005].

[135] See The Biologically Active Food Supplement Market in Russia. Available online at www.bisnis.doc.gov/bisnis/bisdoc/0401food.htm [accessed May 10, 2005].

[136] Unless otherwise indicated, data in the overview of this section is from the World Bank www.worldbank.org/data/databytopic/lac_wdi.pdf.

[137] National Intelligence Council. 2004. Mapping the Global Future, Report of the National Intelligence Council's 2020 Project. Available online at www.cia.gov/nic/nic_globaltrend2020.htm#contents [accessed April 26, 2006].

[138] Ferrer, M. et al. 2004. The scientific muscle of Brazil's health biotechnology. *Nature Biotechnology* 22(Suppl.):DC8-DC12.

[139] Simpson, A.J. et al. 2000. The genome sequence of the plant pathogen *Xylella fastidiosa*. *Nature* 406(6792):151.

[140] Ferrer, M. et al. 2004. The scientific muscle of Brazil's health biotechnology. *Nature Biotechnology* 22(Suppl.):DC8-DC12.

[141] Ibid.

142 See www.adunicamp.org.br/noticias/universidade/leideinova%E7%E3o. pdf [accessed February 7, 2005].

143 See www.ctnbio.gov.br/index.php?action=/content/view&cod_objeto= 1296 [accessed May 10, 2005].

144 For more details, see the description in Institute of Medicine. 2005. *Scaling Up Treatment for the Global AIDS Pandemic.* Washington, DC: The National Academies Press; 156-157.

145 Viols, V. et al. 2000. Promoting the rational use of antiretrovirals through a computer aided system for the logistical control of AIDS medications in Brazil. Presentation at the 13th International AIDS Conference in Durban, South Africa.

146 Lima R.M., and Veloso, V. 2000. SICLOM: Fistruicao informatizada de medicamentos para HIV/AIDS. *Acao Anti-AIDS* 43:6-7.

147 Thorsteinsdottir, H. et al. 2004. Cuba—innovation through synergy. *Nature Biotechnology* 22(Suppl.):DC19-DC24.

148 Verez-Bencomo,V. et al. 2004. A synthetic conjugate polysaccharide vaccine against *Haemophilus influenzae* type b. *Science* 305(5683):522-525.

149 See www.ymbiosciences.com/presspop.cfm?newsID=3024 [accessed February 7, 2005].

150 Thorsteinsdottir, H. et al. 2004. Cuba—innovation through synergy. *Nature Biotechnology* 22(Suppl.):DC19-DC24.

151 See www.inmegen.org.mx.

152 Jimenez-Sanchez, G. 2003. Developing a platform for genomic medicine in Mexico. *Science* 300(5617):295-296.

153 National Research Council/Institute of Medicine. 2005. *An International Perspective on Advancing Technologies and Strategies for Managing Dual-Use Risks.* Washington, DC: The National Academies Press.

154 Information in the overview of this section is from World Bank www.worldbank.org/data/databytopic/mna_wdi.pdf .

155 Abdelgafar, B. et al. 2004. The emergence of Egyptian biotechnology from generics. *Nature Biotechnology* 22(Suppl.):DC25-DC30.

156 Ibid.

157 See www.who.int/csr/disease/hepatitis/whocdscsrlyo2003/en/index4.html [accessed May 9, 2005].

158 Ibid.

159 Ibid.

160 Ibid.

161 Soreni, M. et al. 2005. Parallel biomolecular computation on surfaces with advanced finite automata. *Journal of the American Chemical Society* 127(11):3935-3943.

162 See www.larta.org/lavox/articlelinks/2004/040510_usisrael.asp [accessed May 9, 2005].

163 Ibid.

164 The information for this section is from Bohannon, J. 2005. From pariah to science powerhouse? *Science* 308(5719):182-184.

165 The ALJ business was founded by the late Sheikh Abdul Latif Jameel in 1945. In1955 he was granted the sole distributorship for Toyota vehicles in Saudi Arabia which the ALJ Group has maintained ever since. On March 8, 2005 Abdul

Latif Jameel Company Limited commemorated 50 years of successful and fruitful partnership with the Toyota Motor Corporation. www.alj.com/about03.html.

[166] See www.astf.net/site/news/news_dtls.asp?news_id=1015&ogzid=0 [accessed May 10, 2005].

[167] Kumar, N.K. 2004. Indian biotechnology—rapidly evolving and industry led. *Nature Biotechnology* 22(Suppl.):DC31-DC36.

[168] Bagla, P. 2005. Prime minister backs NSF-like funding body. *Science* 307(5715):1544.

[169] Cited in Kumar, N.K. 2003. Biotech Consortium India Ltd. *Directory of Biotechnology Industries & Institutions in India*. New Delhi: BCIL.

[170] Again, as cited in the Kumar paper: Ernst & Young. 2004. *On the threshold. The Asia Pacific Perspective Global Biotechnology Report*. SF.

[171] See www.shanthabiotech.com/shantha-west.asp [accessed February 9, 2005].

[172] Jayaramam, K.S. 2002. India promotes GMOs in Asia. *Nature Biotechnology* 20(7):641-642.

[173] Institute of Medicine/National Research Council. 2005. *An International Perspective on Advancing Technologies and Strategies for Managing Dual-Use Risks*. Washington, DC: The National Academies Press.

[174] www.icgeb.org.

[175] Institute of Medicine/National Research Council. 2005. *An International Perspective on Advancing Technologies and Strategies for Managing Dual-Use Risks*. Washington, DC: The National Academies Press.

[176] Ibid., 98.

[177] Motari, M. et al. 2004. South Africa—blazing a trail for African biotechnology. *Nature Biotechnology* 22(Suppl.):DC37-DC41.

[178] Ibid.

[179] Normile, D. 2005. Fundraising begins for network of four African institutes. *Science* 307(5709):499.

[180] Cited in Hassan, M.H.A. 2005. Small Things and Big Changes in the Developing World. *Science* 309(5731):65-66.

3

Advances in Technologies with Relevance to Biology: The Future Landscape

This chapter provides an overview and a perspective on the breadth and types of technologies that may have an impact on the life sciences enterprise of the future, with the understanding that there are inherent difficulties in anticipating or predicting how any of these technologies alone or in combination will alter the nature of the future threat "landscape."

Rather than attempt to cover the technology landscape in a comprehensive manner, this chapter (1) highlights technologies likely to have obvious or high-impact near-term consequences; (2) illustrates the general principles by which technological growth alters the nature of future biological threats; and, (3) highlights how and why some technologies are complementary or synergistic in bolstering defense against future threats while also enhancing or altering the nature of future threats.

There is immense diversity and rapid evolution of technologies with relevance to (or impact on) the life sciences enterprise. Their impact(s) may be beneficial or detrimental depending on how these tools and technologies are applied. Some may be seen as "coming out of left field"; that is, these technologies may have very different applications from those originally intended, or may be combined in unexpected, nontraditional configurations. The combination of nanotechnology and biotechnology is one such example of a synergistic combination.

Many of the technologies discussed in this chapter create novel opportunities for scientists (and others) to explore aspects of biological and chemical diversity that cannot be accessed through natural mechanisms or processes. Given the unpredictable nature of technological change, it is difficult if not impossible to describe in definite terms what the global technology landscape will look like in 5 to 10 years, both with regard to the emergence of technologies with dual-use applications and the global geography of future breakthroughs. New, unexpected discoveries and technological applications in RNAi and synthetic biology arose even during the course of deliberations by this committee. If this report, with the same charge, were prepared even a year or two in the future, many of the details presented in this chapter would likely be different.

A CLASSIFICATION SCHEME FOR
BIOLOGICAL TECHNOLOGIES

Despite the seemingly disparate and scattered goals of recent advances in life sciences technologies, the committee concluded that there are classes or categories of advances that share important features. These shared characteristics are based on common purposes, common conceptual underpinnings, and common technical enabling platforms. Thus, the technologies outlined in this chapter are categorized according to a classification scheme devised by the committee and organized around four groupings:

1.   **Acquisition of novel biological or molecular diversity.** These are technologies driven by efforts to acquire or synthesize novel biological or molecular diversity, or a greater range of specificity, so that the user can then select what is useful from the large, newly-acquired diversity pool. The goal is to create collections of molecules with greater breadth of diversity than found so far in nature, as well as with types of diversity that may not exist in nature. The kinds of molecules that might be generated include, for example, enzymes with enhanced or altered activities, as well as molecules composed of "unnatural" amino acids. Technologies in this category include those dedicated toward DNA synthesis; the generation of new chemical diversity (i.e., through combinatorial chemistry); those that create novel DNA molecules (from genes to genomes) using directed in vitro molecular evolution (e.g., "DNA shuffling"[1]); and those that amplify or simply collect previously uncharacterized sequences (genomes) directly from nature (i.e., bioprospecting). All of these technologies require a subsequent selection step, such that molecules, macromolecular complexes, or even microbes with the desired properties can be identified and isolated from a large and very diverse pool of possibilities. Toward this end, new high-throughput screening (including the use of robotics and advanced information management systems) have become critical enabling technologies.

2. Directed design. These are technologies that involve deliberate efforts to generate novel but predetermined and specific biological or molecular diversity. The use of these technologies begins with a more defined, preexisting understanding of the desired endproduct and its molecular features. One then synthesizes or re-engineers the desired product or its components. Examples include but are not limited to rational, structure-aided design of small-molecule ligands; the genetic engineering of viruses or microbes; and, the emerging field of "synthetic biology."

3. Understanding and manipulation of biological systems. These are technologies driven by efforts to gain a more complete understanding of complex biological systems and an ability to manipulate such systems. Examples include "systems biology"; gene silencing (e.g., RNA interference); the generation of novel binding (affinity) reagents; technologies focused on developmental programs (e.g., embryonic stem cells); genomics and genomic medicine; the study of modulators of homeostatic systems; bioinformatics; and, advanced network theory.

4. Production, delivery, and "packaging." These are technologies driven by efforts in the pharmaceutical, agriculture, and healthcare sectors to improve capabilities for producing, reengineering, or delivering biological or biology-derived products and miniaturizing these processes. Examples include the use of transgenic plants as production platforms, aerosol technology, microencapsulation, microfluidics/microfabrication; nanotechnology; and, gene therapy technology. [Some of these technologies are related to the manipulation of biological systems—e.g., nanotechnology—and may also be applied to the generation (category 1) or design (category 2) of novel biological diversity or to the analysis of complex biological systems (category 3).]

The classification scheme serves several important purposes. It:

- highlights commonalities among technologies and, by so doing, draws attention to critical enabling features;
- provides insight into some of the technical drivers behind biology-related technology;
- facilitates predictions about future emerging technologies; and,
- lends insight into the basis for complementarities or synergies among technologies and, as such, facilitates the analysis of interactions that lead to either beneficial or potentially malevolent ends.

Limitations of the classification scheme include the fact that it is based on a relatively small number of relevant technologies (i.e., the committee's list of technologies may be biased and is inevitably incomplete) and the acknowledgment that there are many ways to categorize these technologies. As a reflection of the latter dilemma, the committee found that some of the technologies discussed in this chapter could have been classified in more than one category. The category assignment in these cases was guided by the nature of the particular applications that the committee had in mind when considering each of the relevant technologies.

The examples below serve as a finite set of future technologies that represent and illustrate each of the four categories. For each example the following issues are addressed: the purpose of the technology, its current state of the art, and future applications. The coverage of these issues for each of the technologies is not intended to be exhaustive. The technologies covered in this chapter include not only those that open up new possibilities for the creation of novel or enhanced biological agents but also those that expose new vulnerabilities (i.e., targets for biological attack). Details are limited to those necessary for a clear explanation of the plausibility of use.

1. ACQUISITION OF NOVEL BIOLOGICAL OR MOLECULAR DIVERSITY

Given the clear capability of at least some microbes and viruses to evolve quickly, acquire new genes, and alter their behavior, it might seem reasonable that over hundreds of thousands of years all conceivable biological agents have been "built" and "tested" and that the agents seen today are the most "successful" of these. Thus, is there any reason to think that it might be possible to create a more successful biological agent? Possibly not, but it is important to understand that "successful" in this context means the most able to survive within, on, or near human populations over time. "Success" does not necessarily equate with virulence or pathogenicity, the ability to cause disease or injury.

The kinds of basic biological diversity found in nature today, or those that have potentially evolved in the natural world and been tested for fitness over time, may have been (and are still) limited by certain natural constraints, including available building blocks—nucleotides and amino acids; natural mechanisms for generating genetic diversity; and, the strength and nature of selective pressures over time. Nor has there been enough time over the history of the earth for nature to have explored more than a tiny fraction of the diversity that is possible.[2] The technologies described in this section are those that seek to create a much wider and deeper set of diverse biological molecules, many of which may never have been generated or given a fair chance for succeeding in nature (although success may be defined in different ways).[3]

Techniques have been developed to expand both the diversity of nucleotide or amino acid sequences of nucleic acids or proteins, respectively (which in both cases ultimately hold the information specifying the folding and thus the conformation of biologically active molecules), or for creating a diversity of small molecules with different shapes, sizes, and charge characteristics. In addition, some investigators are creating unnatural nucleic acids and amino acids in order to test and explore possible structural constraints on molecules with biological function. All of these approaches result in novel types of genetic or molecular diversity that then require assessment of functional potential. This assessment typically takes the form of a screening process (i.e., deliberate examination of all molecules for a desired feature or function) or a selective process (i.e., one that imposes a selective advantage on those molecules that have a property of interest). While the technological processes of assessing and selecting molecules of interest—high-throughput screening and selection— have some features in common with the next category of technologies (i.e., directed design), they are included in this first category because of their critical enabling role in the exploration of molecular and biological diversity.

DNA Synthesis

Description

DNA synthesis is a technology that enables the de novo generation of genetic sequences that specifically program cells for the expression of a given protein. It is not new, but technical enhancements continue to increase the speed, ease, and accuracy with which larger and larger sequences can be generated chemically. By the early 1970s, scientists had demonstrated that they could engineer synthetic genes.[4] However, it was the automation of de novo DNA synthesis and the development of the polymerase chain reaction (PCR) in the early 1980s that spawned the development of a series of cascading methodologies for the analysis of gene expression, structure, and function. Our ability to synthesize short oligonucleotides (typically 10 to 80 base pairs in length) rapidly and accurately has been an essential enabling technology for a myriad of advances, not the least of which has been the sequencing of the human genome.

The past few years have seen remarkable technological advances in this field, particularly with respect to the de novo synthesis of increasingly longer DNA constructs. The chemical synthesis and ligation of large segments of a DNA template, followed by enzymatic transcription of RNA led to the de novo creation of the poliovirus genome in 2002 (about 7,500 nucleotides in length), from which the infectious, virulent virus was rescued following its transfection into permissive cells.[5] The following year scientists announced the successful assembly of a bacterial virus genome.[6] Parallel efforts in industry and academia led to the synthesis and assembly of large segments of the hepatitis C virus genome, from which replication competent RNA molecules were rescued. These studies raised concerns in the media that larger, more complex organisms, such as the smallpox virus (which is approximately 186,000 base pairs long), might be within reach.[7]

DNA synthesis technology is currently limited by the cost and time involved to create long DNA constructs of high fidelity as well as by its high error rate. Current estimates for generating even simple oligonucleotides are at least $0.10 per base (including synthesis of the oligonucleotides plus error correction).[8] See Figure 3-1.

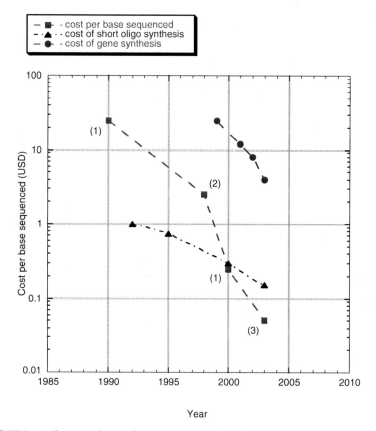

FIGURE 3-1 Cost per base of sequencing and synthesis.
SOURCE: Rob Carlson presentation to the committee, February 2004.

Current State of the Art

Several recent studies have demonstrated important steps toward making gene synthesis readily affordable and accessible to researchers with small budgets, by decreasing its cost and improving its error rate.[9] For example, in December 2004, as this committee deliberated its charge, scientists described a new microchip-based technology for the semiautomated multiplex synthesis of long oligonucleotides.[10] The researchers used the new technology to synthesize all 21 genes that encode proteins of the *E. coli* 30S ribosomal subunit. Almost simultaneously, another research group described a novel approach for reducing errors by more than 15-fold compared to conventional gene synthesis techniques, yielding DNA with one error per 10,000 base pairs.[11]

Future Applications

Developments in DNA synthetic capacity have generated strong interest in the fabrication of increasingly larger constructs, including genetic circuitry,[12] the engineering of entire biochemical pathways,[13] and, as mentioned above, the construction of small genomes.[14] As a specific example of a potential future beneficial application of DNA synthesis, one research group has described a method for synthesizing terpenoid, a natural product used in commercial flavors, fragrances, and antimalarial and anticancer therapeutics, using recombinant DNA constructs.[15] Terpenoids are normally isolated from plant tissue and can only be recovered in small amounts. DNA synthesis technology could be used as an alternative method for producing high-value compounds.

DNA synthesis technology could allow for the efficient, rapid synthesis of viral and other pathogen genomes—either for vaccine or therapeutic research and development, or for malevolent purposes or with unintentional consequences. Given the latter risks, in 2004, George Church (Harvard Medical School, Cambridge, MA) drafted a proposal for decreasing biohazard risks (i.e., creating nearly extinct human viruses, such as polio, or novel pathogens, like IL-4 poxvirus) while minimizing the impact on legitimate research. The proposal focuses on instrument and reagent licensing (e.g., restricting the sale and maintenance of oligonucleotide synthesis machines to licensed entities); regulation for the screening of select agents; establishing a method for testing these newly implemented licensing and, screening systems; criteria for exemption from the whole process; and, strategies for keeping the cost down.[16] The proposal is mentioned here not to endorse it, but rather to highlight the need for a careful analysis and thoughtful discussion of the issues.

DNA Shuffling

Description

Classical genetic breeding has proven itself over and over again throughout human history as a powerful means to improve plant and animal stocks to meet changing societal needs. The late 20th century discovery of restriction endonucleases, enzymes that cut DNA molecules at sites comprising specific short nucleotide sequences, and the subsequent emergence of recombinant DNA technology provided scientists with high-precision tools to insert (or remove) single genes into the genomes of a variety of viruses and organisms, leading, for example, to the introduction of production-enhancing traits into crop plants.[17] Most recently, a powerful mode of directed evolution known as "DNA shuffling"—also known as multigene shuffling, gene shuffling, and directed in vitro molecular evolution—has allowed scientists to greatly improve the efficiency with which a wide diversity of genetic sequences can be derived. A quantum leap in the ability to generate new DNA sequences, DNA shuffling can be used to produce large libraries of DNA that can then be subjected to screening or selection for a range of desired traits, such as improved protein function and/or greater protein production.

"Classical" single-gene breeding starts with a "parental" pool of related sequences (genes, etc.) and then breeds "offspring" molecules, which are subjected to screening and selection for the "best" offspring. The process is repeated for several generations. With DNA shuffling, sequence diversity is generated by fragmenting and then recombining related versions of the same sequence or gene from multiple sources (e.g., related species), resulting in "shuffling" of the DNA molecules. Basically, it allows for the simultaneous mating of many different species. The result is a collection of DNA mosaics. The reassortment that occurs during the shuffling process yields a higher diversity of functional progeny sequences than can be produced by a sequential single-gene approach.

In one of the earliest demonstrations of the technology, which involved shuffling four separately evolved genes (from four different microbial species), the shuffled "hybrids" encoded proteins with 270 to 540 times greater enzymatic activity than the best parental sequence.[18] Even if that same recombined enzyme could have been evolved through single-gene screening, the process would have been dramatically slower. But chances are it never would have evolved. Evidence from at least one study shows that the best parent is not necessarily the one closest in sequence to the best chimeric offspring and thus would probably not represent the best starting point for single-gene evolution (i.e., some other better-looking parental sequence would have been chosen for single-gene directed evolution).[19]

Current State of the Art

The technology has developed quickly, such that scientists are not just shuffling single genes, they are shuffling entire genomes. In 2002, biologists used whole-genome shuffling for the rapid improvement of tylosin production from the bacterium *Streptomyces fradiae*; after only two rounds of shuffling, a bacterial strain was generated that could produce tylosin (an antibiotic) at a rate comparable to strains that had gone through 20 generations of sequential selection.[20] Also in 2002, a portion of the HIV genome was shuffled to create a new strain of HIV that was able to replicate in a monkey cell line that previously had been resistant to viral infection.[21] By 2003 the technique had advanced to the point where many mammalian DNA sequences could be shuffled together in a single bacterial cell line. In one study, scientists shuffled one gene of a cytokine from seven genetically similar mammalian species (including human) to generate an "evolved" cytokine that demonstrated a 10-fold increase in activity compared to the human cytokine alone.[22] It should be emphasized that the power of this technology (and any diversity generating procedure) is only fully realized if the molecules generated with the most enhanced, desired properties can be identified and isolated. Despite continual improvements in the throughput of current screening procedures, the use of conditions that impose strong selective pressures for emergence of molecules with the desired properties is far more efficient in finding the most potent molecule in the pool.

Future Applications

Ultimately, this rapid molecular method of directed evolution will allow biologists to generate novel proteins, viruses, bacteria, and other organisms with desired properties in a fraction of the time required with classical breeding and in a more cost-effective manner. For example, virologists are using DNA shuffling to optimize viruses for gene therapy and vaccine applications.[23] Synthetic biologists are using the technology to speed up their discovery process (see "Synthetic Biology" later in this chapter).

Bioprospecting

Description[24]

Bioprospecting is the search for previously unrecognized, naturally occurring, biological diversity that may serve as a source of material for use in medicine, agriculture, and industry. These materials include genetic blueprints (DNA and RNA sequences), proteins and complex biological compounds, and intact organisms themselves. Humans have been exploiting naturally-derived products for thousands of years. Even as high-throughput technologies like combinatorial chemistry, described above, have practically revolutionized drug discovery, modern therapeutics is still largely dependent on compounds derived from natural products. Excluding biologics (products made from living organisms), 60 percent of drugs approved by the Food and Drug Administration and pre-new drug application candidates between 1989 and 1995 were of natural origin.[25] Between 1983 and 1994, over 60 percent of all approved cancer drugs and cancer drugs at the pre-new drug application stage and 78 percent of all newly approved antibacterial agents were of natural origin.[26] Taxol, the world's first billion-dollar anticancer drug, is derived from the yew tree.[27] Artemisinin, one of the most promising new drugs for the treatment of malaria, was discovered as a natural product of a fernlike weed in China called sweet wormwood. And aspirin—arguably one of the best known and most universally used medicines—is derived from salicin, a glycoside found in many species in the plant genera *Salix* and *Populus*.

Bioprospecting is not limited to plants, nor is drug discovery its only application. Most recently, with the use of molecular detection methods, scientists have uncovered a staggering number of previously unrecognized and uncharacterized microbial life forms.[28] Indeed, microbial genomes represent the largest source of genetic diversity on the planet—diversity that could be exploited for medical, agricultural, and industrial uses. Natural products discovered through bioprospecting microbial endophytes—microorganisms that reside in the tissues of living plants—include antibiotics, antiviral compounds, anticancer agents, antioxidants, antidiabetic agents, immunosuppressive compounds, and insectides. With respect to the last, bioinsecticides are a small but growing component of the insecticide market. Bioprospected compounds exhibiting potent insecticidal properties include nodulisporic compounds for use against blowfly larvae (isolated from a *Nodulisporium spp.* that inhabits the plant *Bontia daphnoides*)[29] and benzofuran compounds for use against spruce budworm (isolated from an unidentified endophytic fungus from wintergreen, *Gaultheria procumbens*).[30] Of note, naphthalene, the ingredient in mothballs, is a major product of an endophytic fungus, *Muscodor vitigenus*, which inhabits a liana, *Paullina paullinioides*.[31]

Prospecting directly for DNA and RNA sequences that encode novel proteins with useful activities has become a potentially important scientific and business enterprise. This approach entails searches based on random expression of thousands or millions of sequences, followed by screening or selection for products with desired activities.[32] Sometimes the search focuses on families of related sequences that are predicted to encode products of interest, which are recovered directly from environments using sequence amplification technology. This kind of approach can synergize with the DNA shuffling technology described above. Recent, early forays into "community genomics," or large-scale random sequencing of the DNA from complex environmental microbial communities, reflect the immense future potential of this approach for the discovery and harnessing of previously unimagined biological activities.[33]

For example, Diversa Corporation (San Diego, CA) utilizes bioprospecting of microbial genomes to develop small molecules and enzymes for the pharmaceutical, agricultural, chemical, and industrial markets.[34] After collecting environmental samples of uncultured microorganisms and extracting the genetic material, researchers search for novel genes and gene pathways for potentially useful products, like enzymes with increased efficiencies and stabilities (e.g., high and low temperature stability, high or low pH tolerance, high or low salt tolerance). The samples are collected from environments ranging from thermal vents to contaminated industrial sites to supercooled sea ice.

Bioprospecting has also been applied to the discovery of microbial agents in efforts to better understand the diversity of microbes in the environment that might serve as human pathogens if provided the opportunity. It has been argued that by deliberately scrutinizing the kinds of vectors and reservoirs that exist in a local environment for previously unrecognized microbes, novel agents might be identified long before they are discovered to be human, animal or plant pathogens, thus providing early warning of potential disease-causing agents.[35] At the least, these surveys could expand our appreciation of microbial diversity and inferred microbial function.[36] For example, in 2002, using a broad-range PCR approach (i.e., using conserved priming sites for a group of related sequence targets, as opposed to specific primers for single unique targets), scientists discovered four novel *Bartonella* DNA sequences in 98 arthropod specimens (fleas, lice, and ticks) from Peru; three of the sequences were significantly different from previously characterized *Bartonella* species.[37] *Bartonella* s are vectorborne bacteria associated with numerous human and animal infections.[38] Rather than having any immediate known clinical implications, this study illustrates the power of this generic approach as well as our incomplete understanding of *Bartonella* diversity.

Current State of the Art

Current methods include recovery of microbes using cultivation-based methods, serologic surveys of potential hosts, extraction/separation/purification of molecules with desired properties, amplification of families of related nucleic acid sequences using broad-range PCR (and similar techniques), shotgun cloning and sequencing of bulk DNA or cDNA from environments of interest, and the use of subtractive hybridization methods[39] to enrich for novel nucleic acid sequences in hosts or environments.

Future Applications

One might consider both molecular and traditional cultivation-based approaches for examining hosts, such as fruit bats and small rodents, which are already known to serve as reservoirs for important human microbial pathogens (Hendra and Nipah viruses, *Borrelia spp.* and other genera, respectively). As described above, the potential benefits associated with the discovery of novel products and microbial genetic diversity are innumerable.

Combinatorial Chemistry: Generating Chemical Diversity

Description

Combinatorial chemistry refers to technologies and processes used for the rapid creation of large numbers of synthetic compounds ("libraries"), typically for the purposes of screening for activity against biological drug targets (see "High-throughput Screening"). Whereas DNA synthesis enables the acquisition of genetic sequence diversity, these techniques allow for the generation of libraries of chemical compounds having a diversity of shapes, sizes, and charge characteristics—all of which may be of interest for their varied abilities to interact with and bind to biologically active proteins or macromolecular complexes, thereby altering the biological properties of these proteins and complexes. Combinatorial chemistry techniques can be used to create a wide range of chemotypes or molecular motifs, ranging from large polycyclic compounds of a peptidic nature to smaller, presumably more druglike, compounds. Initially, it was believed that when used in combination with high-throughput screening technologies, combinatorial techniques would dramatically accelerate the drug discovery process while reducing the associated up-front costs with the drug discovery effort. While this has not yet proven to be the case, most pharmaceutical companies are still heavily invested in combinatorial chemistry and are exploring the development and implementation of novel methods to create additional libraries of compounds. A recent trend noted in the pharmaceutical industry is the move from the development of large, unfocused, general screening libraries to smaller, less diverse libraries for screening against a particular target or family of related targets.

The origins of this new branch of chemistry can be traced back to the early 1960s, when methods were developed for the solid-phase synthesis of peptides.[40] This involved attaching an amino acid to a solid support (i.e., beads of plastic resin) and then adding amino acid residues, one by one in a stepwise fashion through the creation of covalent peptide chemical bonds, until the desired peptide product is created. The final polypeptide is released by chemically breaking its bond with the solid support and washing it free.[41] Subsequent modifications of the solid-phase synthesis process greatly enhanced the ability to generate a large number of peptides with specific amino acid sequences.[42] Individual peptides were synthesized on the ends of "pins" that were spatially oriented in a two-dimensional array designed to match up with the wells of a 96-well microtiter plate. This reduced the scale of the process and greatly facilitated the parallel synthesis of large numbers of peptides. A further modification of the technique enhanced the ability to create a diversity of peptide sequences by incorporating a combinatorial approach.[43] In this case, the solid-phase resin bearing the nascent synthetic peptide was enclosed in a mesh, or "tea bag." Like the pin-based method, the tea-bag process facilitated the numerous washing and drying steps required for peptide synthesis and thus allowed for the parallel synthesis of many different peptides, each in its own tea bag. However, by mixing the resin from different tea bags after each individual stepwise addition of an amino acid residue, combinatorial peptide libraries involving a great diversity of amino acid sequences could be readily generated, in which each resin bead bears an individual peptide with a unique amino acid sequence.[44]

After the compounds are synthesized and a library is constructed, a selection or screening strategy is needed to identify unique compounds of interest to the biological sciences. The most obvious method involves affinity isolation of the peptide of interest on an immobilized target molecule, followed by release of the peptide and analysis utilizing combinations of gas-phase chromatography, high-performance liquid chromatography (HPLC), mass spectrometry, and nuclear magnetic resonance (NMR). It is also possible to determine the structure of compounds still attached to the resin, using "on-bead" analytical techniques such as infrared analysis, gel-phase NMR, matrix-assisted laser desorption ionization time-of-flight mass spectrometry, electrospray mass spectrometry, and HPLC chemiluminescence nitrogen detection.[45]

While direct determination of structure, as described in the previous paragraph, works well for small libraries, these techniques are generally not applicable to large, mixture-based libraries. For libraries, various strategies have been developed that govern the reaction sequence by attaching a readable chemical "tag" to the bead while the molecule is being synthesized. One of the earliest tagging approaches employed the use of oligonucleotides.[46] In this approach, for every amino acid added to the peptide chain, a specific set of oligonucleotides was added to a separate chain that was attached to the same bead. PCR and DNA sequencing techniques were then used to decode the structure of the peptide. Numerous additional tagging techniques and agents have since been developed.[47]

Current State of the Art

Solution-phase parallel synthesis is becoming the combinatorial chemistry technique of choice in the pharmaceutical industry, driven primarily by advances in laboratory automation, instrumentation, and informatics. Compounds can be synthesized either as single discrete compounds per reaction vessel or as mixtures of compounds in a single reaction vessel, so many of the same principles described above for solid-phase (resin-bound) principles are applicable here as well. The primary advantage of solution-phase combinatorial chemistry lies in the increase in the number of chemical reactions/transformations that can be accessed, thereby greatly increasing the range of chemotypes (chemical scaffolds) that can be created.

The earliest reports of solution phase combinatorial chemistry techniques involved the use of a common multicomponent reaction, termed the Ugi reaction, in which an isocyanide, an aldehyde, an amine, and a carboxylic acid are combined in a single-reaction vessel to create a single major product. Using this synthetic approach coupled with advanced data analysis techniques, scientists were able to identify compounds with the desired biological effect after synthesizing only a 400-compound subset of the 160,000 possible products. This represents a 400-fold increase in discovery efficiency over conventional approaches.

The current trend in parallel solution-phase chemistry is leaning toward the development of smaller arrays (12 to 96 compounds) of simple to moderately complex chemical compositions. As the robotics and laboratory instrumentation required for parallel synthesis become more affordable and readily accessible, the technology is being transferred into basic medicinal chemistry laboratories and becoming instrumental in the optimization of lead compounds (i.e., compounds that show potential to be developed into drugs). Such efforts are ideally carried out with knowledge of the structure of the target molecule, usually gained by application of either x-ray crystallography or NMR techniques. Structure-activity relationships are determined as lead compounds, identified initially through the screening of large libraries of compounds, are modified at specific sites, and the impact of the chemical modification on the desired biological properties of the compound is determined.

The purity and identity of combinatorially-produced compounds have been a source of recent great discussion and technological advance since, in order for any meaningful data to be produced from a biological assay, the purity of the compound of interest must be as high as possible.[48] The activity of the compound must also be confirmed by resynthesis of the specific molecule and repeat assays for biological activity.

Future Applications

Combinatorial chemistry techniques are not only useful for drug discovery and development, they are being used in the search for better superconductors, better phosphors for use in video monitors (phosphors are substances that emit light), better materials for use in computer magnetic and other storage devices, and better biosensors for the detection of medically-important molecules and environmental toxins. Combinatorial approaches have been used to develop a "nose chip" sensor capable of detecting and distinguishing among seven common solvents (toluene, chloroform, tetrahydrofuran, acetone, ethyl acetate, ethanol, and methanol).[49]

Using combinatorial and high-throughput methods, the pharmaceutical industry synthesizes and screens several million new potential ligands annually. Although most companies have little use for the tens of thousands of these compounds identified each year as toxic, some might have potential as biochemical weapons (Chapter 1).[50] Although most of the information derived from combinatorial and high-throughput technology is held in proprietary databases, a new public database recently proposed as part of the National Institutes of Health (NIH) Roadmap raises concerns about public access to dual-use information (Chapter 1, Box 1-1). The NIH Roadmap discovery effort is particularly worrisome in this regard, because of plans to optimize lead compounds shown to be capable of targeting specific cellular proteins. The goal is not to develop therapeutic agents but rather to provide a series of reagents, facilitating further exploration of protein function and systems biology.[51] Such compounds may be relatively potent poisons.

While the technologies applied in combinatorial chemistry are not exceedingly complex, a wide variety of laboratory automation and instrumentation is needed to stage an effective combinatorial chemistry campaign.

High-Throughput Screening[52]

Description

High-throughput screening (HTS) refers to the process of examining large numbers of diverse biomolecular or chemical compounds in a rapid and efficient manner for properties of interest. Such technologies are essential to achieving any benefit from the construction of large and diverse libraries of compounds, as they are used to select a particular compound having the desired properties. These properties might include biochemical or enzymatic activities desired of a potential therapeutic agent or toxicity in such an agent that under usual circumstances one would wish to avoid. Advances in miniaturized screening technologies, bioinformatics, robotics, and a variety of other technologies have all contributed to the improved biological assay efficiency that characterizes HTS. In contrast to this paradigm, in which a large library of compounds (i.e., samples) is tested for one specific activity or set of activities, a variation on the HTS theme involves the testing of a single biological sample for a wide variety of activities. The best example of this is the use of DNA or oligonucleotide microarrays—also known as DNA chips. These are routinely used in both basic and applied research to facilitate the large-scale screening and monitoring of gene expression levels, gene function, and genetic variation in biological samples, and to identify novel drug targets.

The process of screening large numbers of compounds against potential disease targets is characterized by a collection of technologies that strive to increase biological assay efficiency through the application of miniaturized screening formats and advanced liquid handling, signal detection, robotics, informatics, and a variety of other technologies. Over the past several years, the industry has witnessed an evolution in screening capabilities, resulting in the ability of a user to screen more than 100,000 compounds per day for potential biological activity. Evaluating upward of 1 million compounds for biological (or various other) properties in a screening campaign is now commonplace in the pharmaceutical industry.

Current State of the Art

Effective HTS relies on robust assays that can detect and then translate biological or other activities into a format that can be readily interpreted. A wide variety of assays are currently in use, including:

- cell-free colorimetric or chemiluminescence assays;
- cell-free fluorescence resonance energy transfer assays;
- cell-based reporter gene assays, usually with an enzymatic read-out;
- cell-based fluorescence imaging assays;
- NMR assays, which involve identifying small molecule ligands for macromolecular receptor targets;
- affinity chromatography assays;
- DNA microarrays (high density arrangements of double-stranded DNA clones (cDNA) or oligonucleotides that serve as identical or complementary probes, respectively, for specific genes, transcripts, or genome sequences); and
- Other types of microarrays, including high-density arrangements of antibodies, nucleic acid or peptide aptamers, antigens (protein or lipid), MHC[53]-peptide antigen complexes, and intact cells.

Future Directions

Future advances in HTS—such as the development of one-step assays and increased miniaturization—will continue to increase the throughput and reduce the cost of HTS assays and may eventually allow the simultaneous monitoring of multiple endpoints (e.g., biological, toxicological) across a wide variety of targets. An analysis of the current HTS technology landscape reveals the following as potential opportunities and future directions:

- further development of one-step (homogeneous) assays;
- development of improved primary screening hardware;
- miniaturization as a means to increase throughput and decrease cost;
- improvements in the capabilities and efficiency of robotic systems in the life sciences;
- application of HTS to lead compound optimization; and,
- novel approaches for identification of biologically-relevant targets.

In short, HTS assays and technologies will permeate new sectors in the life sciences, affecting the productivity and speed of advances and discoveries in these varied sectors. The cost effectiveness of HTS assays and technologies will improve, such that tasks previously believed to be impractical will become quite tractable. Coupled with methods to generate enhanced sequence and structural diversity beyond that seen in nature, these assays and technologies will permit the identification and selection of novel molecules with important biological functions, with ramifications for all of the life sciences.

2. DIRECTED DESIGN

There are other technologies, besides those described in the previous category of technologies, that seek to generate new kinds of genetic or molecular diversity. However, in contrast to the technologies in the first category, these "directed design" approaches are more deliberate, and rely on preexisting knowledge with regard to what needs to be created.

Rational Drug Design

Description

The methods described above, wherein a large library of diverse chemical compounds are screened using HTS methods to identify a smaller number of potential lead compounds with desired activities, are gradually being enhanced by less empirical approaches that are based on a greater understanding of biological systems (i.e., target: ligand interactions), identification of specific target molecules, and determination of the structure of a target molecule whose activity has been shown to be critical for the production of a particular disease or for maintenance of health. Such structural knowledge has grown rapidly over the past decade due to advances in x-ray crystallography, NMR technologies, and associated computational techniques that now allow for rapid determination of the structure of even large proteins or nucleic acid molecules at atomic-level resolution. A quick survey of the Protein Data Bank (PDB),[54] the global resource for all publicly available biological macromolecular structures, reveals that the number of structures deposited on an annual basis witnessed nearly a 10-fold increase between 1994 (3,091) and 2004 (28,992); see Figure 3-2. With such structural knowledge of targets in hand, chemists can rationally pursue the design of novel chemical compounds that either bind to selected sites on the surface of these target molecules or mimic the structure of the target molecule and thereby compete for the binding to a receptor molecule.

157

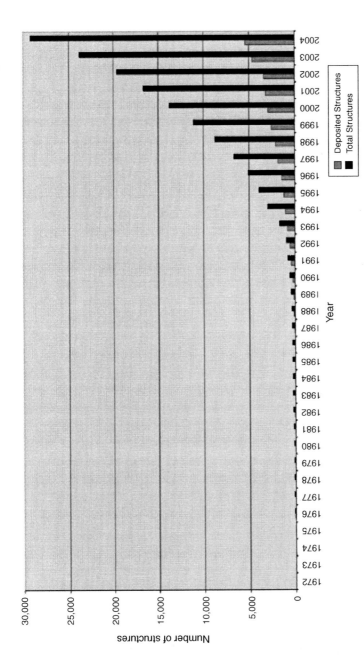

FIGURE 3-2 Growth in the number of structures deposited per year (gray) and total holdings of the PDB (black) from the time the bank was founded.
SOURCE: Reprinted from Dutta, S. and H.M. Berman. 2005. Large macromolecular complexes in the protein data bank: A status report. *Structure* 13(3):382, with permission from Elsevier.

An excellent example of technological convergence exists with the field of *in silico*, or virtual, screening. This methodology capitalizes on the advances described above with respect to the determination of structures for target molecules as well as advances in computer hardware and specialized chemical informatics algorithms, so-called docking and scoring programs. Many thousands of virtual compounds can be rapidly and effectively assessed for potential target molecule complementarity,[55] as a prerequisite for biological activity, prior to any actual chemistry being carried out or biological assays being performed. The product of this computational effort is thus a rationally designed molecule that, once synthesized, can potentially serve as a lead compound in the drug discovery process.

Current State of the Art

Although rational drug design has received a great deal of attention from the pharmaceutical industry and is recognized as having great potential for the future, most efforts today by the drug discovery industry reflect a combination of structure-aided rational design of compounds and the HTS screening of libraries of diverse compounds. Thus, the use of structure, when known for a given molecular target, may come into play once a lead compound has been identified through an HTS process and efforts are made to optimize this lead and improve the biological activity or pharmacological properties of the compound. The field today is such that absence of knowledge of the structure of a targeted molecule is viewed as a critical impediment to the development of a new drug.

In contrast to the rational design of small-molecule therapeutics, the rational design of therapeutic nucleic-acid-based compounds is much easier in that such compounds are synthesized to be complementary to the targeted nucleic acid sequence. While nucleic acid therapeutics based on antisense oligonucleotides or ribozymes, enzymatically-active RNAs that cleave specific RNA target sequences, have been pursued for over a decade, their promise has not yet been realized due to difficulties in delivering stable compounds to desired sites. Significant advances are now occurring, however, in providing desired pharmacological properties to siRNA-based compounds and morpholino antisense oligonucleotides.

Future Applications

As the structure of greater numbers of potential target molecules are identified in the future and as both *in silico* screening and chemical synthesis methods continue to advance, it seems clear that a greater reliance is likely to develop on these types of approaches. Greater application of rational, structure-based design approaches is likely to speed the discovery process significantly. While there are dual-use implications for such technologies, as there are for almost any advancing life sciences technology, the infrastructure required to pursue such structure-based design of novel biologically active compounds is likely to limit its use to the legitimate pharmaceutical industry for a number of years. It should be noted, however, that like the nucleotide sequence databases that are open to the public, rapidly growing numbers of protein structures are being placed in the public domain. This trend is likely to continue and even accelerate, and as the computer hardware and software requirements for viewing and interpreting such structures becomes increasingly simple, these approaches will become increasingly accessible to scientists outside the pharmaceutical industry.

Synthetic Biology

Description

The fledgling 5-year-old-field of synthetic biology—which is attracting engineers and biologists in equal measure—means different things to different researchers. Engineers view it primarily as a way to fabricate useful microbes to do what no current technology can do (i.e., they view it as an engineering discipline). Biologists see it as a powerful new way to learn about underlying principles of cellular function.

Unlike systems biologists (see description later in this chapter), who adopt a big-picture approach to biology by analyzing troves of data on the simultaneous activity of thousands of genes and proteins, synthetic biologists reduce the very same systems to their simplest components. They create models of genetic circuits, build the circuits, see if they work, and adjust them if they do not, learning underlying principles of biology in the process. By examining simple patterns of gene expression and treating pieces of DNA as modules, which, like Legos™, can be spliced together, synthetic biologists construct what are effectively biochemical logic boards that control both intra- and extracellular activity.

Because the molecular nature of many cellular reactions is only partially understood, most synthetic genetic circuits require considerable further empirical refinement after the initial computational work. Some scientists use DNA shuffling to streamline the empirical process. After inserting mutated DNA circuits into cells and selecting for those cells (and the circuits therein) that performed the best, researchers can evolve an effective product in just a couple of generations.[56]

Current State of the Art

One of the goals of the field is to transform bacteria into tiny programmable computers. Like electronic computers, the live bacterial circuits would use both analog and digital logic circuits to perform simple computations. For example, researchers are working to develop modular units, such as sensors and actuators, input and output devices, genetic circuits to control cells, and a microbial chassis in which to assemble these pieces. If they are successful, a "registry of biological parts" will allow researchers to go to the freezer, get a part, and hook it up.[57] The computing power of programmable cells will likely never rival that of their electronic counterparts. Rather, the beauty of synthetic biology lies in what living cells can do.

In 2000, a genetic "circuit" was created in *E.coli* that caused the cells to blink like a lighthouse.[58] The circuit, which was called "the repressilator," was comprised of three repressor genes, one of which turned on a gene for green fluorescent protein (GFP), which, when activated, emits a green glow. Three years later another research group created a genetic circuit by crafting a "toggle switch" that could oscillate the circuit and alter its pattern depending on growth conditions.[59] Using this technique, investigators subsequently developed a procedure to re-engineer a bacterial protein that binds to TNT (an explosive) and that, when bound, activates a gene circuit that produces GFP.[60] This demonstrates an initial effort to engineer organisms that operate as biological sentinels, pinpointing explosives or detecting the presence of biological weapons.

In 2004, researchers in Israel designed a prototype "DNA computer" with the capacity to logically analyze mRNA disease indicators in vitro (i.e., in this case, early signs of prostate and lung cancer) and control the administration of biologically active ssDNA molecules, including drugs.[61] The procedure is relatively innocuous, requiring the injection of a very small amount of fluid containing billions of nanoparticles, each of which operates as a tiny computer by effectively interrogating the cell and detecting the presence of diagnostic DNA markers (e.g., mutated mRNA sequences or underexpressed or overexpressed mRNA). If the markers are present, the nanoparticle sends out a therapeutic short nucleic acid that can affect the level of gene expression.

Future Applications

Synthetic biology technology has many potential applications, including designing bacteria that can detect chemical or biological agent signatures, engineering bacteria that can clean up environmental pollutants, and engineering organisms or compounds that can diagnose disease or fix faulty genes. Although initial efforts are focused on microbial cells, some synthetic biologists imagine a day when they will be able to program adult stem cells for therapeutic purposes (e.g., to patch up a damaged heart).

Engineering ethicist Aarne Vesilind (Bucknell University) is one of many scientists promoting the idea that synthetic biologists and ethicists hold an Asilomar-like conference on synthetic biology—much like that held at the dawn of genetic engineering research in the mid-1970s—to define bioengineers' "responsibilities to society" should these engineered organisms survive outside the laboratory to cause harm to human health or the environment.[62] Several efforts have now been planned to examine the implications of this kind of work, including one foundation-funded study involving three institutions, two of which play a major role in synthetic genomics research.[63] In addition, the National Science Advisory Board for Biosecurity has identified synthetic genomics as a major area of interest. Many of the same issues are raised by the genetic engineering of viruses.

Genetic Engineering of Viruses

Description

As described above, the development of recombinant DNA technology and the ability to manipulate DNA sequences in bacterial species such as *E. coli* has resulted over time in the capacity to insert almost any desired gene into almost any kind of prokaryotic or eukaryotic cell. Placing the DNA inserted under appropriate transcriptional controls, and the protein encoded by it under appropriate translational control, allows that gene to direct the expression of almost any kind of protein: a fluorescent marker (as in the GloFish described in Chapter 1), an enzyme that might function as a reporter, an antibiotic resistance marker, or even a toxin. Using very similar techniques, genes of interest (subject to size constraints) can be introduced into the genomes of many different types of DNA viruses, ranging from adenoviruses to herpesviruses. Such capabilities raise obvious and compelling dual-use concerns.

The introduction of heterologous gene sequences into the genomes of RNA viruses, or other types of modifications to the RNA genomes of these viruses, presents a special set of technical difficulties due to the fact that the genetic material is RNA, which is less stable than DNA and not as amenable to the genetic splicing techniques that have made recombinant DNA technology as versatile. However, this has been accomplished for a growing number of different types of RNA viruses. Moreover, given the small size of these RNA genomes, it has proven possible to synthesize completely de novo all the genetic material needed to recover fully infectious virus particles with near wild-type infectivity, virulence and replication potential.

RNA viruses come in several types, depending on the number of strands of RNA in each molecule of their genome (i.e., single-stranded or double-stranded RNA molecules) and the number of genomic segments (one or more). Genetic engineering of single-stranded RNA viruses in which the RNA is of positive polarity (i.e., the same sense as the messenger RNA that encodes the viral proteins) has proven most straightforward. It has been known for many years that genomic RNA isolated from positive-strand RNA viruses, such as poliovirus, is intrinsically infectious. When transfected (i.e., introduced) into a permissive cell in the absence of any accompanying proteins, such RNA will lead directly to the synthesis of the viral proteins, which will then begin to assemble the necessary replicative machinery to make additional copies of the RNA as well as more viral protein, leading ultimately to the assembly and "rescue" of fully infectious virus, which is then generally released from the cell.

To manipulate the viral RNA genome, scientists in the age of molecular biology have developed efficient enzymatic methods for creating complementary DNA (cDNA) copies of the viral genomic RNA using reverse transcriptase enzymes encoded by retroviruses. This cDNA can be engineered to have "sticky" ends, allowing it then to be molecularly cloned into *E. coli*, in which it can be manipulated by all the modern methods available. This can include the deletion of protein coding sequences, the creation of deletion or point mutations, or even the introduction of completely novel protein-coding sequences. The modified cDNA can then be placed downstream of an appropriate promoter sequence for a DNA-dependent RNA polymerase and a novel, molecularly engineered viral RNA genome efficiently transcribed in an in vitro transcription reaction. The transcribed RNA can then be transfected back into a permissive cell and, if the introduced mutations are compatible with continued viability of the virus, will give rise to novel infectious viruses.

The process by which virologists use this method, involving the conversion of the genetic sequence of the virus from RNA to DNA and back to RNA, generally in order to assess the impact of mutations on the viral life cycle or pathogenic properties, is known as "reverse genetic engineering." This approach is widely used by positive-strand molecular virologists. First carried out in 1980 with poliovirus,[64] infectious cDNA clones have now been constructed for members of many positive-stranded RNA virus families, including brome mosaic virus,[65] yellow fever virus,[66] Sindbis virus,[67] citrus tristeza virus,[68] and equine arteritis virus.[69] In the case of hepatitis C virus, a positive-strand virus in the *Flaviviridae*, virus rescue has generally required injection of the synthetic RNA directly into the liver of a chimpanzee. On the other hand, fully infectious poliovirus, a member of the family *Picornaviridae*, has been recovered in a cell-free reaction carried out in vitro in an optimized cell extract system.

In the past, coronaviruses, which have the largest genomes of all positive-strand RNA viruses (around 30 kilobases long), were difficult to reverse engineer because of the sheer size and instability of their full-length cDNA clones in bacterial vectors.[70] However, recent technological advances have made it possible to reverse engineer even these largest of all known RNA viruses,[71] including the causative agent of severe acute respiratory syndrome (SARS), a previously undescribed coronavirus.[72]

Similarly, the reverse genetic engineering of negative-strand RNA viruses[73] has proven much more difficult, given the fact that the RNA genomes of these viruses do not function directly as messenger RNAs and thus do not give rise to infectious virus progeny following their introduction into permissive cells. These RNAs require the expression of certain viral proteins, in order to make positive-strand copies of the negative-stranded RNA genome and to initiate the replicative cycle. The technology to accomplish this was first developed for influenza A virus in the late 1980s to early 1990s. Like the earlier efforts with positive-strand RNA viruses, these efforts not only have dramatically improved our understanding of how these viruses replicate, but have also created the means for genetically manipulating viral genomes in order to generate new viruses for use as live, attenuated vaccines or vectors.[74]

Initially, reverse engineering of the influenza virus required the use of helper viruses, which provided proteins and RNA segments that the reconstituted in vitro RNPs (i.e., reconstituted ribonucleoprotein complexes containing RNA transcribed from the molecularly cloned cDNA) needed in order to be infectious following transfection into cells. Later, alternative methods for introducing influenza RNPs into cells were developed, including entirely plasmid-driven rescue that did not require the involvement of a helper virus.[75] The latter plasmid-based system allowed for easy engineering of viral genomes with multiple specific mutations. By 2001 at least one laboratory had generated a pathogenic H5N1 virus using reverse engineering.[76]

In addition to influenza A virus, and as summarized in a paper that appeared in the *Journal of Virology* in 1999,[77] in its first decade the technology was used to reverse engineer, or "recover" many other negative-stranded RNA viruses including rabies virus,[78] vesicular stomatitis virus,[79] respiratory syncytial virus,[80] measles virus,[81] Sendai virus,[82] human parainfluenza type 3,[83] rinderpest virus,[84] simian virus,[85] bovine respiratory syncytial virus,[86] Newcastle disease virus,[87] and bunyavirus.[88]

Current State of the Art

Most recently, as mentioned in Chapter 1, reverse engineering has been used to produce infectious influenza A viruses containing the viral haemagglutinin (HA) and neuraminidase (NA) genes of the strain that caused the devastating 1918-1919 "Spanish" influenza pandemic. Scientists demonstrated that the HA of the 1918 virus confers enhanced pathogenicity in mice to recent human viruses that are otherwise nonpathogenic in their murine host. HA is a major surface protein that stimulates the production of neutralizing antibodies in the host, and changes in the genome segment that encodes it may render the virus resistant to preexisting neutralizing antibodies, thus increasing the potential for epidemics or pandemics of disease. Moreover, the reverse engineered viruses expressing 1918 viral HA elicited hallmark symptoms of the illness produced during the original pandemic.[89]

With the complete genetic sequencing of the H1N1 influenza A virus, referred to in Chapter 1, some have questioned whether these studies should have been published[90] in the open literature given concerns that terrorists could, in theory, use the information to reconstruct the 1918 flu virus.[91] It should be noted that in addition to the "normal" scientific peer review, the editors of *Science* required the authors to demonstrate that they had obtained approval to publish their research from the director of the Centers for Disease Control and Prevention, and the director of the National Institute of Allergy and Infectious Diseases.[92] Furthermore, the National Science Advisory Board for Biosecurity (NSABB) was asked to consider these papers prior to publication and determined that the scientific benefit of the future use of this research far outweighed the potential risk of misuse.[93]

Future Applications

Reverse engineering of the causative agent of SARS illustrates the many potential beneficial applications of the technology. In addition to opening up new opportunities for exploring the complexity of the SARS-coronavirus genome, the availability of a full-length cDNA provides a genetic template for manipulating the genome in ways that will allow for rapid and rational development and testing of candidate vaccines and therapeutics.[94] By mutating the many small proteins seemingly expressed by this unique coronavirus, scientists will learn their function in viral replication and/or pathogenesis and potentially identify useful targets for drug discovery efforts.

The influenza A reverse genetic engineering system serves as an excellent example of the potential for this technology to be used with the intent to do harm. As summarized in a 2003 article on the potential use of influenza virus as an agent for bioterrorism, with respect to advances that allowed for helper virus free production of a pathogenic H5N1 virus, virologist Robert M. Krug (University of Texas, Austin) has written:

There is every reason to believe that the same recombinant DNA techniques can be used to render this H5N1 virus transmissible from humans to humans. Furthermore, it should be possible to introduce mutations into such a recombinant virus so that it is resistant to currently available influenza virus antivirals (M2 inhibitors: amantadine and rimantadine; and NA inhibitors: zanamivir and oseltamivir), and so that it possesses an HA antigenic site that is unlike those in recently circulating human viruses. In fact, several viruses with different HA antigenic sites could be generated. The human population would lack immunological protection against such viruses, existing antiviral drugs would not afford any protection, and these viruses could be spread simply by release of an aerosol spray in several crowded areas. [95]

3. UNDERSTANDING AND MANIPULATION OF BIOLOGICAL SYSTEMS

A more holistic understanding of complex biological systems (e.g., the workings of an intact cell, multicellular organism, or complex microbial community) is emerging through a set of technologies that allow for the collection of vast, comprehensive (highly parallel) sets of data for multiple kinds of biological processes, the integration of these data sets, and the identification of critical components or pathways. Critical components can then serve as targets for therapeutic and preventive intervention or manipulation; they can also serve as targets for malevolent manipulation and as the basis for novel kinds of biological attack. Concurrently, technologies that facilitate a better understanding of intracellular, organ, and whole-animal control "circuitry" will enhance the ability of scientists to manipulate these complex systems.

Examples of some technologies that are leading to this type of holistic overview include the emerging field/discipline of "systems biology"[96] and genomic medicine. Examples of the tools that could be used to manipulate complex biological systems include gene silencing, novel binding reagents (e.g., nucleic acid and peptide aptamers, engineered antibodies), and small-molecule modulators of physiological systems. In many ways this category of technologies opens up entirely novel aspects of the future biodefense and biothreat agent landscapes and changes the fundamental paradigm for future discussions on this topic.

RNA Interference

RNA interference—also known as RNAi and RNA silencing—was first observed in plants when it was noted that endogenous and "foreign" genes appeared to be turning each other off by a process initially termed "co-suppression."[97] What was initially thought to be peculiar to petunias was later found in other plants and also animals. The phenomenon is now known as RNA interference, and is recognized to be a common antiviral defense mechanism in plants and a common phenomenon in many other organisms, including mammals. It is also increasingly apparent that RNAi is intimately related to widespread regulation of gene expression by very small endogenously expressed RNA molecules, so-called micro-RNAs (miRNA). This field is exploding with new discoveries almost daily concerning the role of miRNAs in regulating gene expression during development and after. The interaction of endogenous miRNAs with cellular mRNAs encoding specific proteins leads to suppression of protein expression, either by impairing the stability of the mRNA or by suppressing its translation into protein. The fact that small, largely double-stranded RNAs of this type, about 21 nucleotides in length, could play such an apparently broad and fundamental role in development and in the control of cellular homeostasis was not at all appreciated just a few years ago and highlights the sudden, unpredictable paradigm shifts and sharp turns in the way scientists think that are possible in the advance of the life sciences (Figure 3-3).

The basic molecular mechanism of RNAi is as follows. Long, double-stranded RNAs (dsRNAs; typically >200 nucleotides long) silence the expression of target genes upon entering a cellular pathway commonly referred to as the RNAi pathway. First, in the so-called initiation step, the dsRNAs are processed into 20 to 25 nucleotide small interfering RNAs (siRNAs) by an Rnase III-like enzyme called Dicer. The siRNAs then assemble into endoribonuclease-containing complexes known as RNA-induced silencing complexes (RISCs), unwinding in the process. The siRNA strands subsequently guide the RISCs to complementary RNA molecules, where RISC complex cleaves and destroys the cognate RNA (i.e., this is the effector step). miRNAs are generated in a similar fashion from endogenously expressed RNAs containing short hairpin structures, using a related Dicer-like protein. They are capable of similarly silencing gene expression but can also direct post-transcriptional silencing by blocking translation of a targeted host mRNA. This later effect typically depends on binding to a partially complementary target sequence near the 3' end of the mRNA.

RNAi is highly specific and remarkably potent (only a few dsRNA molecules per cell are required for effective interference), and the interfering activity can occur in cells and tissues far removed from the site of introduction.

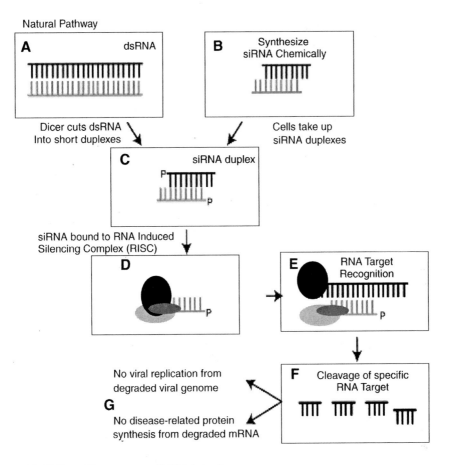

FIGURE 3-3 The process of RNA interference.
SOURCE: Steven Block, presentation to the committee, April 2004.

Current State of the Art

The technology is expected to prove particularly valuable in cases where the targeted RNA encodes genes and protein products inaccessible to conventional drugs (i.e., protein, small-molecule, and monoclonal antibody therapeutics). However, clinical delivery poses a significant challenge, as does the likelihood of undesirable silencing of nontargeted genes.[98] Yet several recent experiments indicate that investigators are well on their way to overcoming these challenges and creating an emerging dual-use risk in the form of bioengineered RNAi-based pathogens. In 2003, a German research team announced the successful lentivirus vector delivery of in vivo gene silencing with RNAi.[99] Also in 2003, researchers announced the successful use of high-pressured, high-volume intravenous injection of synthetic siRNA.[100] Other studies have demonstrated the potential to deliver RNAi to specific organs, such as the eyes,[101] lungs,[102] and central nervous system.[103] Although human trials of RNAi have begun for the treatment of age-related macular degeneration,[104] a systemic mode of delivery would arguably have greater clinical utility. Substantial progress is being made toward this aim, however, using liposome and lipid nanoparticle formulations of chemically modified, and hence stabilized, siRNAs. Scientists at Sirna, a small biotech company working for well over a decade on nucleic-acid-based therapies, have recently described a 1,000-fold reduction in the amount of hepatitis B virus present in the blood of mice replicating this virus in the liver, following a series of three separate intravenous inoculations of a lipid nanoparticle formulated, chemically modified, siRNA.

In November 2004, researchers from Alnylam Pharmaceuticals used chemically modified siRNAs to silence genes encoding Apolipoprotein B (ApoB) in mice, resulting in decreased plasma levels of ApoB protein and reduced total cholesterol.[105] The study thus demonstrated systemic activity following a conventional clinical mode of delivery. Importantly, the delivery did not inadvertently impact nontargeted genes. Still, there are questions about the specificity of the siRNA, given that the investigators did not evaluate all proteins and given that they collected measurements over a relatively short period of time.[106] A longer, more comprehensive study would be necessary to evaluate more fully the specificity of the technique. However, while "off- target" effects of siRNAs are certainly of concern to regulators and industry proponents as well, it is likely they can be managed in much the same way that "off target" effects (i.e., unexpected toxic effects) of small-molecule therapeutics have been in the past.

Potential Applications

Observations that RNAi works in vivo in mammals has not only created opportunities for the development of new therapeutic tools but also spawned a new generation of genetic research in mammals.[107] For example, the vast majority of mammalian RNAi systems are driven by a polymerase III promoter, which can be manipulated such that the experimenter has the ability to turn the expression of a gene on and off at will, allowing for novel experimental designs. One could temporarily switch off a tumor suppressor gene suspected of providing genome protection (e.g., a checkpoint gene) and then turn it on again, allowing the experimenter to determine whether the gene is necessary for the initiation or maintenance of tumorigenesis and whether it might be a good target for late-stage cancer treatments.

It is reasonable to expect significant additional advances in the formulation of siRNAs for use as pharmacological agents, particularly with contributions from the field of nanotechnology. As with so many of the technologies outlined in this chapter, just as RNAi promises new therapeutic options for cancer and other diseases, it could also be used to manipulate gene expression with the intent to do harm.

High-Affinity Binding Reagents (Aptamers and Tadpoles)

Description

Aptamers are short, single-stranded nucleic acid or peptidic ligands that fold into well-defined three-dimensional shapes, allowing them to inhibit or modulate their protein targets with high affinity and specificity. Since their discovery in the early 1990s,[108] aptamers have been used in target validation, detection reagents, and functional proteomic tools.[109] Over the past decade, several studies have explored the potential of aptamers for therapeutic intervention, including the inhibition of targets associated with inflammatory processes, cancer, and other disorders.[110] Aptamers have been compared to monoclonal antibodies but with the added advantage that they are neither toxic nor immunogenic.

Current State of the Art

One of the first aptamers tested in an animal model was an antithrombin agent that blocks the proteolytic activity of thrombin, a protein involved in thrombosis (blood clot formation in a blood vessel).[111] In June 2004, Archemix Corp. (Cambridge, MA) and Nuvelo, Inc. (San Carlos, CA) announced that an Investigational New Drug application had been submitted to the FDA to begin a Phase I clinical trial with an antithrombin aptamer, ARC183, for potential use in coronary artery bypass graft surgery.[112] In another clinical trial, Eyetech Pharmaceuticals, Inc. (New York, NY) is testing Macugen, an aptamer that targets VEGF (vascular endothelial growth factor) as a treatment for age-related macular degeneration and diabetic macular edema.[113]

In January 2005, scientists reported that they had created a new type of high-affinity binding reagent—"tadpoles"—that bind to specific targets, such as *Bacillus anthracis* protective antigen and the enzyme cofactor biotin, as examples.[114] Tadpoles are protein-DNA chimeras that contain a protein head coupled to an oligonucleotide tail. The head has an affinity for a specific target molecule; the tail, which contains a region for PCR amplification, mediates detection. Tadpoles represent another type of high-affinity binding reagent with the power to not only detect but, with its DNA tail, "count" small numbers of proteins and other molecules in a precise fashion.

Future Applications

Their sensitivity, dynamic range, and, in the case of tadpoles, precise quantification make these high-affinity binding molecules potentially very useful tools for disease diagnosis and environmental detection, including pathogen and other biological agent detection in the event of a naturally occurring or deliberate biological attack.

Despite their promise as therapeutic agents, aptamers are very expensive to synthesize and are still a largely unknown entity (with respect to administration, formulation, adverse effects, etc.). So although several compounds have entered clinical trial, their future as biopharmaceuticals is unclear.[115] More certain is their role as valuable lead structures in small-molecule drug discovery (because they can be so readily modified and adapted to almost any kind of high-throughput readout format) and as molecular detection reagents (because of their high specificity).

Computational Biology and Bioinformatics[116]

Description

Life scientists have exploited computing for many years in some form or another. But what is different today—and will be increasingly so in the future—is that the knowledge of computing and mathematical theory needed to address many of the most challenging biological problems can no longer be easily acquired but requires instead a fusion of the disciplines of biology, computation, and informatics. A National Research Council (NRC) report entitled *Catalyzing Inquiry at the Interface of Computing and Biology* (December 2005) has pointed out that the kinds and levels of expertise needed to address the most challenging problems of contemporary biology stretch the current state of knowledge of the field. The report identifies four distinct but interrelated roles of computing for biology:

• *Computational tools* are artifacts—usually implemented as software but sometimes hardware—that enable biologists to solve very specific and precisely defined problems. Such biologically oriented tools acquire, store, manage, query, and analyze biological data in a myriad of forms and in enormous volume for its complexity. These tools allow biologists to move from the study of individual phenomena to the study of phenomena in biological context, to move across vast scales of time, space, and organizational complexity and to utilize properties such as evolutionary conservation to ascertain functional details.

- *Computational models* are abstractions of biological phenomena implemented as artifacts that can be used to test insight, to make quantitative predictions, and to help interpret experimental data. These models enable biological scientists to understand many types of biological data in context, and even at very large volumes, and to make model-based predictions that can then be tested empirically. Such models allow biological scientists to tackle harder problems that could not readily be posed without visualization, rich databases, and new methods for making quantitative predictions. Biological modeling itself has become possible because data are available in unprecedented richness and because computing itself has matured enough to support the analysis of such complexity.

- A *computational perspective or metaphor* on biology applies the intellectual constructs of computer science and information technology as ways of coming to grips with the complexity of biological phenomena that can be regarded as performing information processing in different ways. This perspective is a source for information and computing abstractions that can be used to interpret and understand biological mechanisms and function. Because both computing and biology are concerned with function, information and computing abstractions can provide well-understood constructs that can be used to characterize the biological function of interest. Further, they may well provide an alternative and more appropriate language and set of abstractions for representing biological interactions, describing biological phenomena, or conceptualizing some characteristics of biological systems.

- *Cyberinfrastructure and data acquisition* are enabling support technologies for 21st century biology. Cyberinfrastructure—high-end general-purpose computing centers that provide supercomputing capabilities to the community at large; well-curated data repositories that store and make available to all researchers large volumes and many types of biological data; digital libraries that contain the intellectual legacy of biological researchers and provide mechanisms for sharing, annotating, reviewing, and disseminating knowledge in a collaborative context; and high-speed networks that connect geographically distributed computing resources—will become an enabling mechanism for large-scale, data-intensive biological research that is distributed over multiple laboratories and investigators around the world. New data acquisition technologies such as genomic sequencers will enable researchers to obtain larger amounts of data of different types and at different scales, and advances in information technology and computing will play key roles in the development of these technologies.

Current State of the Art

A new level of sophistication in computing and informatics is required for interpretation of much of the data generated today in the life sciences. These data are highly heterogenous in content and format, multimodal in collection, multidimensional, multidisciplinary in creation and analysis, multiscale in organization, and international in collaborations, sharing, and relevance.[117] Such data may consist of sequences, graphs, geometric information, scalar and vector fields, patterns of organization, constraints, images, scientific prose, and even biological hypotheses and evidence. These data may well be of very high dimension, since data points that might be associated with the behavior of an individual unit must be collected for thousands or tens of thousands of comparable units. The size and complexity of the data sets being generated require novel methods of analysis, which are being provided by computational biologists. For example, scientists at the U.S. Department of Energy's Pacific Northwest National Laboratory have developed a new computational tool—called ScalaBLAST—that is a sophisticated "sequence alignment tool" and can divide the work of analyzing biological data into manageable fragments, so that large data sets can run on many processors simultaneously. The application of this technology means that large-scale problems—such as the analysis of an organism—can be solved in minutes rather than weeks.[118]

The NRC report notes that these data are windows into structures of immense complexity. Biological entities (and systems consisting of multiple entities) are sufficiently complex that it may well be impossible for any human being to keep all of the essential elements in his or her head at once. Thus, advances in computational biology will be driven by the need to understand how complex biological systems operate and are controlled and will contribute fundamentally to the development of a systems view in biology.

Future Applications

The NRC report emphasizes that the life sciences of the future will be an information science and will "use computing and information technology as a language and a medium in which to manage the discrete, nonsymmetric, largely nonreducible, unique nature of biological systems and observations. In some ways, computing and information will have a relationship to the language of 21st century biology that is similar to the relationship of calculus to the language of the physical sciences. Computing itself can provide biologists with an alternative and possibly more appropriate language and sets of intellectual abstractions for creating models and data representations of higher-order interactions, describing biological phenomena, and conceptualizing some characteristics of biological systems." This potential is nowhere more evident than in the nascent field of systems biology.

Systems Biology

Description

Systems biology—also known as integrative biology—uses high-throughput, genome-wide tools (e.g., microarrays) for the simultaneous study of complex interactions involving networks of molecules, including DNA, RNA, and proteins. It is, in a sense, classical physiology taken to a new level of complexity and detail. The term "systems" comes from systems theory or dynamic systems theory: Systems biology involves the application of systems- and signal-oriented approaches to understanding inter- and intracellular dynamic processes.[119] Systems-level problem solving in biology is based on the premise that cellular behavior is a complex system of dynamically interacting biomolecular entities. A systems biologist seeks to quantify all of the molecular elements that make up a biological system and then integrate that information into graphical network models that can serve as predictive hypotheses.

A growing number of researchers in the life sciences community are recognizing the usefulness of systems biology tools for analyzing complex regulatory networks (both inside the cell, and the regulatory networks that integrate and control the functions of distinctly different cell types in multicellular organisms like humans) and for making sense of the vast genomic and proteomic data sets that are so rapidly accumulating.[120] These efforts draw heavily on computational methods to model the biological systems, as described earlier. Systems biology is being seen as a valuable addition to the drug discovery toolbox.[121] In medicine, where disease is being viewed as a perturbation of the normal network structure of a system (i.e., disease-perturbed proteins and gene regulatory networks differ from their healthy counterparts, because of genetic or environmental influences), a systems biology approach can provide insights into how disease-related processes interact and are controlled, guide new diagnostic and therapeutic approaches, and enable a more predictive, preventive, personalized medicine.[122]

Current State of the Art

This field is rapidly evolving, with the computational tools still in an immature state and inadequate for handling the reams of data derived from microarray assays and their functional correlates. Unconventional means of recording experimental results and conveying them rapidly to others in the field using an Internet-based approach are being pursued in an effort to manage the scale of data collection and analysis required for this effort.[123] Whereas scientists previously may have examined only a single facet of a signal transduction pathway involved, for example, in control of a cellular response to infection, they are now looking more broadly at the effect of a particular stimulus on multiple different pathways, including what happens at common nodes and the counter-regulatory pathways that are activated in response to a particular signal. They are coming to realize that many novel molecular mechanisms are involved in controlling these signaling pathways, not only phosphorylation and kinase activation as classically recognized in signal transduction but also specific protein conformational changes, the translocation of proteins to different cellular compartments, proteolytic cleavage of signaling partners and latent transcription factors, and the binding and release of modulatory proteins from key signaling intermediates. A similar multiplicity of mechanisms exists within the extracellular regulatory networks, that must ultimately take their cues from intracellular events. In all of these signaling networks, tremendous specificity of responses stems from the timing, duration, amplitude, and type of signal generated and the pathways from which it emanates. At present, perhaps it could be said that while the magnitude and nature of the challenge posed by systems biology are increasingly well recognized, it remains unclear exactly how these challenges will be met, or how successful such attempts to do so will be.

Future Applications

The rise of systems biology is expected to have profound implications for research, clinical practice, education, intellectual property, and industrial competitiveness. As computational technologies advance, simulation of complex biological systems will have more predictive accuracy, aspects of laboratory experimentation will replaced by more cost-effective computational approaches, and physicians will have new decision support tools to help them identify the best preventative and therapeutic approaches for individual genotypes and phenotypes.

Just as systems biology will profoundly alter the way scientists and physicians conduct their analyses, the same global problem-solving approach could serve as a tool for the identification of ways to deliberately manipulate biological systems with the intent to do harm.

Genomic Medicine

Description

Genomic, or personalized, medicine refers to potential patient-tailored therapies made possible by improved molecular characterization of disease, technologies that allow for rapid genomic and proteomic analyses of individual patients, and advances in information technology that allow practitioners to access this information in meaningful ways. Scientists have known for a long time that human genetic variation is associated with many diseases and questions. With recent advances in technology that allow for quick, affordable genotypic assessments (i.e., from PCR to high-throughput sequencing), researchers have begun to understand the implications of human genetic variation for the treatment of disease.[124] Patient-tailored therapies hold forth great promise as a new way of treating, or preventing, disease and are an active area of research and investment.

Current State of the Art

Recent accomplishments in the field include the use of an epidermal growth factor receptor (EGFR) tyrosine kinase inhibitor, gefitinib (i.e., Iressa), for use in the treatment of non-small cell lung cancer. Scientists have found that certain EGFR mutations may predict a patient's sensitivity to the drug, meaning that some patients are more likely to benefit than others. Moreover, in one study the mutations (and benefits of treatment) were more prevalent among Japanese patients than U.S. patients, raising general questions about the ethnic or geographic specificity of this and other cancer drugs.[125]

Herceptin provides another more publicized example of the potential for genomic medicine. In 1997, this drug became the first gene-based therapeutic licensed and marketed for use against breast cancer. Women with metastatic breast cancer whose cells produce the proteins HER2 and HER2/neu were given new hope in the form of this monoclonal antibody drug developed and manufactured by San Francisco-based Genentech.[126] Herceptin, an erbB2 monoclonal antibody, is now licensed for use in the 20 to 30 percent of breast cancer patients who overexpress this tyrosine kinase receptor.[127] Although mechanism-based cardiotoxicity has been observed, response rates of up to 60 to 70 percent have been reported for Herceptin in combination with paclitaxel or doxorubicin.[128] Similar "proof of principle" is now emerging for the clinical activity of small-molecule inhibitors of oncogenic tyrosine kinases such as Glivec (imatinib) against chronic myeloid leukemia[129] and preclinical activity in tumor models driven by the tyrosine kinase activity of the platelet-derived growth factor and *c*-kit receptors.[130]

Future Applications

Understanding and harnessing genomic variation are expected to contribute significantly to improving the health of people worldwide, including the developing world.[131] In recognition of this, Mexico is in the process of delivering one of the first genomic medicine platforms in Latin America, one that is expected to serve as a regional model for other countries in their efforts to ease health and financial burdens. The Mexican government and medical and biomedical research communities view the present time as a window of opportunity for investing in this emerging technological trend, so as to minimize the likelihood of needing to depend on foreign aid and sources in the future.[132] Likewise, genomic medicine activities in Singapore represent another national effort to gain leverage in this field. Already, high-tech manufacturing and financial services serve as the fulcrum of the Singaporean economy. Strengthening biotechnological capacity, including genomic medicine capacity, is viewed as the next high-tech step forward to accelerated economic growth.[133]

Integrating personalized, or genomic, medicine into regular health care (in any country) will require overcoming two major challenges. First, it will be necessary to make the "$1,000 genome" a reality. The $1,000 genome refers to the cost of sequencing an individual's entire genomic sequence and, although a somewhat arbitrary threshold, has come to represent the point at which the technology is finally affordable enough for widespread use. It is not clear how the $1,000 genome hurdle will be jumped, although biotech companies are trying. Some experts believe it will require a new technology. The second and arguably more significant challenge will be making the philosophical jump from the highly interventional, British-style school of medicine to a preventative, predictive health care paradigm. Genomic medicine is expected to revolutionize human medicine by altering the nature of diagnosis, treatment, and prevention. In traditional medicine, diagnosis is based on clinical criteria, treatment is population-based, and prevention is based on late-stage identification of disease. In genomic medicine, diagnosis is based on molecular criteria (e.g., the use of microarrays in cancer diagnosis), treatment is highly individualized (i.e., genomic based), and prevention is based on early-stage identification of who is at risk.

The same genomic sequences that will one day allow health care providers to identify and provide genotype-specific (and phenotype-specific) treatment may some day be exploited as targets for novel biological agents. Knowledge generated from genomic medicine could potentially be used to target specific ethnic, racial, or other population characteristics. Such weapons need not be hugely effective or even completely selective; proportional selectivity would be sufficient since, in addition to the direct effect of the weapon itself, the social tension, erosion, and (potential) fragmentation resulting from headlines of the "Mystery Virus Strikes Blacks, Spares Whites" variety would be likely to trigger effects far in excess of those from the disease itself.

While knowledge spreading from the various genome projects has fueled speculation in this area, two points should be kept in mind when considering this topic. First, the hugely large number of point mutations and other polymorphisms within the genome are not likely to lead to any selective targeting in the near future. Although techniques such as RNAi, as discussed previously, certainly have the capability to inhibit the expression of key genes with relevant single nucleotide polymorphisms (SNPs) within them, the proportion of such mutations lying in functionally important areas of the genome is small and the technical difficulties associated with exploiting them are real. Second, the idea is not new; South Africa's Project Coast reportedly conducted experiments on vaccines designed to target fertility.[134]

The technology to construct such weapons exists. For almost two decades, researchers have been using adenoviruses to target tumor cells in individuals and steadily refining their techniques for directing viral entry into cells. For example, it is now possible to modify through genetic approaches the fibers used by the virus for cellular attachment so that the virus attaches to particular cell types.[135, 136] Studies have also shown that preferential attachment and infection of target cells can be markedly elevated.[137, 138]

Interestingly, while the availability of the complete human genome sequence has revealed numerous SNPs and other polymorphic elements—and has consequently raised greater concern about the possibility of using biological weapons to target specific racial or ethnic populations—the ability to identify[139] and exploit genetic differences among such populations does not require this new information. Adenoviruses could be used to deliver antibodies that target distinct ethnic groups with characteristic cell surface molecules, without needing to identify population-specific SNPs.[140] For example, human leukocyte antigens have distinctive distributions that vary with geographic origin (e.g., the common haplotype B8-DR3 is distributed almost uniquely among Northern European Caucasians; in the Mediterranean basin, this is replaced by B18-DR3).

Modulators of Homeostatic Systems

Description

The stability and integrity (homeostasis) of the molecular circuits, pathways and networks responsible for diverse body functions are altered by disease and by exposure to noxious environmental pollutants and toxins (xenobiotics). The quest to identify the molecular circuits and control systems in each specialized cell type in the body, and to understand the perturbations that give rise to disease, is a dominant research theme in contemporary biology. Mapping the "molecular signatures" of the body's biocircuits in health and disease is the primary technological catalyst in the development of new molecular diagnostic tests for detecting disease and the emergence of new disease classification schemes based on causal molecular pathologies rather than clinical symptoms. Analysis of the disease-induced perturbations in biocircuits also provides the intellectual foundation of modern drug discovery, which is based increasingly on rational design therapeutic agents directed against the specific molecular lesions responsible in disease etiology.

Burgeoning knowledge about the composition and regulation of homeostatic molecular circuits in the body's cells, tissues, and organs, and their dysregulation in disease, epitomizes the dual-use dilemma created by rapid advances in systems biology. The life sciences are undergoing a profound transformation from their historical reliance on descriptive and phenomenological observations to now focus on the detailed underlying mechanisms of disease and identification of the "rule sets" that govern the assembly and function of biological systems in both health and disease. These insights hold great promise for future advances in medicine, agriculture, ecology, and the environmental sciences. But the very same knowledge about the homeostatic control of body biocircuits can be usurped for less beneficent intentions.

The rapid pace of research progress in revealing the detailed molecular circuit diagrams and control processes for every body function, dictates that the risk of evolution of new threats will escalate in parallel. In this context, the concept of a "biothreat agent" will expand beyond the current limited perspective of biothreats as being only "bugs" (i.e., pathogenic organisms) to include an entirely new category of threats—the biological circuit disruptors.[141]

Current State of the Art

The commercial availability of large libraries of bioactive chemical compounds, together with automated high-throughput screening methods, allows the biological activities of thousands of chemical compounds to be assayed rapidly. Combinatorial chemistry and the directed evolution methods described earlier in this chapter are now used to routinely generate chemical libraries containing 10^4 to 10^7 compounds at relatively low cost (tens of thousands of dollars). In addition to "random" screening to identify compounds with the desired biological properties, robust knowledge about how the structure of a chemical compound correlates with causation of specific biological perturbations will permit increasingly accurate predictions about how the tertiary structure of bioactive molecules correlates with their affinity for, and reactivity with, specific molecular targets in the various cell lineages of the body.

The emerging field of toxicogenomics involves profiling the changes in gene and protein expression induced by chemicals found in the industrial workplace to assess potential risk from exposure to occupational and environmental hazards. The pharmaceutical industry and drug regulatory agencies such as the FDA (and their international counterparts) also have recognized the value of toxicogenomic profiling as a new tool to detect how investigational drugs might adversely affect genes important in drug metabolism or affect homeostatic genes that may lead to acute or chronic side effects. The current heightened public and legislative concern over drug safety will likely intensify pressures for the adoption of toxicogenomics as a routine part of the drug approval process. The benefits of toxicogenomics are self-evident. Once again, however, research that reveals structure-activity relationship (SAR) correlations between chemical structure(s) and specific toxicity events provide useful grist for the design of biological circuit disruptors in malevolent hands.

More robust correlations between chemical structure and therapeutic activity and absorption, distribution, metabolism, excretion, and toxicology (ADMET) properties will also come from research in the new field of chemical genomics (also referred to as chemogenomics or chemical biology). This emerging area of research seeks to establish the SAR rule of how chemical structure defines the selective interaction of different structural classes of molecules with various families of cellular proteins.

The Chemical Genomics Center, established in June 2004, by the Molecular Libraries and Imaging Implementation Group, as part of the NIH New Pathways to Discovery theme is but one example of an initiative that may eventually lead to potent new dual-use information. The center will be part of a consortium of chemical genomics screening centers to be located across the country whose purpose will be to identify small-molecule inhibitors of every important human cellular protein or signaling pathway. Part of the rationale for the chemical genomics initiative(s) is that, in contrast to researchers in the pharmaceutical industry, many academic and government scientists do not have easy access to large libraries of small molecules (i.e., organic chemical compounds that are smaller than proteins and that can be used as tools to modulate gene function).

The database will give academic and government researchers an opportunity to identify useful biological targets and thereby contribute more vigorously to the early stages of drug development. With plans to screen more than 100,000 small-molecule compounds in its first year of operation, one of the goals of the Chemical Genomics Center network is to explore the areas of the human genome for which small-molecule chemical probes have yet to be identified. Data generated by the network will be deposited in a comprehensive database of chemical structures (and their biological activities). The database, known as PubChem, will be freely available to the entire scientific community. In addition to screening and probe data, it will list compound information from the scientific literature.

Should this come to pass, it will offer enormous opportunities for industry and academic scientists alike to pursue novel "drugable" targets in a search for small-molecule inhibitors of certain pathways that could offer substantial clinical benefit. However, the availability of information and reagents that enable one to disrupt critical human physiological systems has profound implications for the nature of the future biological and chemical threat spectrum. The difference between the NIH and industrial efforts resides in the fate of the information produced from these large-scale screening programs. Companies view their screening data and the accompanying SARs to be proprietary assets. Their data are viewed as a source of corporate competitive advantage and are not typically placed in the public domain. In contrast, the NIH data will be placed in the public domain, with the unavoidable accompanying complication of creating a rich source of SAR information that could potentially be exploited for malevolent use.

Future Applications

In the past, the dual-use risk of bioregulators was considered minimal because of their lack of suitability for aerosolization unless microencapsulated, their limited shelf life after atmospheric release, the fact that proteins denature at very high temperatures and lose activity at low temperatures, and high purchase costs. However, new knowledge and advancing technologies, particularly encapsulation technologies (as discussed elsewhere in this chapter), have raised concerns about the dual-use risk of bioregulators.[142]

A greater understanding of how small molecules and naturally occurring bioregulatory peptides function in higher organisms will open up novel opportunities to design agents—for good or bad—that target particular physiological systems and processes, such as the brain and the immune system, in very precise ways. Scientists' understanding of neuropeptides and their role in diverse physiological processes has advanced considerably over the last several decades.[143] This new knowledge, combined with the almost limitless size of the consumer market for pharmaceutical compounds that alleviate pain, depression, sleep disorders, and a wide range of other mental disorders, suggests that many new potentially dual-use psychoactive compounds will be discovered in the near future, including novel compounds that affect perception, sensation, cognition, emotion, mood, volition, bodily control, and alertness.[144] Several so-called "smart drugs"—brain-boosting medications that enhance memory or cognition—are already being sold or are in development.[145]

4. PRODUCTION, DELIVERY, AND "PACKAGING"

The ability to manipulate "biological systems" in a defined, deliberate manner—for either beneficial or malevolent purposes—depends on the ability to produce and deliver such interventions. Technologies that allow for such production and delivery are evolving very quickly, driven by the goals and needs of the pharmaceutical, agricultural, and healthcare sectors. Some of these technologies, which clearly have immense potential future impact on biology, have not been traditionally viewed as biotechnologies or as having relevance to future biological threats. A prime example is the potential now offered by developments in nanoparticle science for the creation of novel and highly efficient delivery systems for previously difficult-to-deliver biologically-active compounds.

These technologies can be subdivided into those concerned with production, packaging, and delivery. Examples of production technologies with relevance to biology include microreactor technology (as used in the chemical engineering industrial sector), microfluidics and microfabrication technologies (e.g., currently being employed for next-generation detection tools), and transgenic plants. Examples of packaging technologies with relevance to biology include microencapsulation and nanotechnology. Examples of delivery technologies with relevance to biology include aerosol technology and gene therapy and gene vector technology.

Plants as Production Platforms—"Biopharming"

Description

"Biopharming," also called "molecular pharming," is the harvest of bioactive molecules from mass-cultured organisms and crops for use as ingredients in industrial products and pharmaceuticals. Transgenic crop plants, into which genes for bioactive compounds from other species have been inserted, serve as the basis of biopharming. (Biopharming differs from bioprospecting in that the latter is sourced in wild populations.) A novel advantage of biopharming is the crop-based production of vaccines and antibodies otherwise not possible or too expensive to produce using conventional methods.[146]

Current State of the Art

As described in Chapter 1, many different genetically engineered crop varieties with genes for therapeutic products have been developed: transgenic rice (beta carotene, human milk proteins, higher iron content, higher zinc content, low phytic acid, high phytase); transgenic potato (gene from grain amaranth for high protein content, antigens of cholera and diarrheal pathogens, and hepatitis B vaccine); transgenic maize (AIDS antigens, higher content of lysine and tryptophan, nutritive value equivalent to that of milk); transgenic fruits and vegetables (bananas, melons, brinjals [*Solanum melongena*], and tomatoes with subunit vaccines against rabies; AIDS antigens in tomatoes; and human glycoprotein in tomatoes to inhibit *Helicobacter pylori* against ulcers and stomach cancer); transgenic tobacco (human hemoglobin, human antibody against hepatitis B virus, and 50 percent lower nicotine), and genetically engineered coffee (decaffeinated by gene splicing). However, despite the existence of functional prototypes and evidence that the technology works, there are some technical, delivery, and regulatory challenges that are slowing progress in the field.[147]

Future Applications

Plant manufacturing platforms may provide a cost-effective means to produce vaccines, offering the ability to address some of the problems associated with global vaccine manufacture and delivery.[148] They are also being used to experiment with plant-derived microbicides, with the goal of finding a cost-effective way to block HIV transmission, and they are being explored as a possible cost-effective way to produce antibodies for use against potential biowarfare agents.[149]

However, transgenic plants could also be engineered to produce large quantities of bioregulatory or otherwise toxic proteins, which could either be purified from plant cells or used directly as biological agents. As with legitimate production, using transgenic plants as bioreactors would eliminate the need for mechanical equipment normally associated with the process. The technology would be limited to producing protein-based agents (transgenic plants would be largely indistinguishable from nontransgenic crops), but it could potentially provide a covert means for producing large amounts of product.[150]

Microfluidics and Microfabrication

Description

Microfluidics and microfabrication are rapidly growing technologies in which a wide variety of processes and manipulations are carried out at miniaturized scales (e.g., nanoliter volumes) and in automated fashion. Microfluidic, or "lab-on-a-chip," technology underlies many recent advances in point-of-care diagnostics, including DNA analysis, immunoassays, cell analysis, and enzyme-activity measurements.[151] Microfabrication involves building functional devices at the molecular size.

Fabricated on glass or plastic chips ranging in size from a microscope slide to a compact disk, microfluidic arrays require only very small (on the order of picoliters, 10^{-12} liters) sample and reagent volumes. The most sophisticated systems are completely integrated, with sample introduction, preprocessing (e.g., cell lysis, dilution), reagent addition, and detection all conducted on the same chip. But most systems are bulkier and rely on external detector and other devices. Limitations of the current technology include reagent stability (or instability) and the need for liquid reagent reservoirs.

Microelectromechanical systems (MEMS) are a similar miniaturized technology. Unlike microfluidic systems, MEMS devices are self-contained and do not require reagents. Swallowed-capsule technology is a popular example of a MEMS: patients swallow a capsule containing all of the miniaturized equipment necessary for taking images in the gastrointestinal tract.

Current State of the Art

Nanotechnological advances are decreasing the size of microfluidic and other miniature diagnostic systems even further. For example, Biotrove, Inc. (Waltham, MA) has developed a nanoliter sample size real-time PCR machine that, when commercially available, will allow users to analyze thousands of samples simultaneously and for a much lower per-sample cost than with currently available high-throughput microarray systems. Other sampling problems come into play at smaller volumes (e.g., the small volume may not be representative of the whole sample or population).[152]

Future Applications

As stated in a recent *Science* review on miniaturized diagnostic systems: "Farther down the road may be personalized health care with diagnosis and disease-monitoring occurring in the home with easy-to-use miniature devices." Although this possibility may be farther into the future than the scope of this report covers, for regulatory as much as technical reasons, steps are being taken in this direction. For example, there have been several recent advances in convenient sampling methods, including breath and saliva sampling, that would be necessary before personalized diagnostic devices become a widely accepted component of personal health care.

Nanotechnology

Description

Nanotechnology, which was defined in Chapter 2, started off as little more than a clever means of making incredibly small things. In 1990, IBM scientists made headlines by painstakingly arranging 35 xenon atoms to spell out the company's three-letter name, creating the world's smallest corporate logo. Other scientists followed with an invisibly small "nanoguitar." Its strings, each just a few atoms across, could be plucked by laser beams to play notes 17 octaves higher than those produced by a conventional guitar—well above the human hearing range. Novelties though they were, these feats proved that,with new tools in hand, scientists could arrange atoms as methodically as masons arrange bricks—and in doing so build materials never made in nature.

Last year alone, hundreds of tons of nanomaterials were made in U.S. labs and factories. Microscopically thin sheets of tightly woven carbon atoms are being wrapped around the cores of tennis balls to keep air from escaping; new fabrics have been endowed with nanofibers that keep stains from setting; some sunscreens have ultraviolet-absorbing nanoparticles so small they cannot reflect light, making them invisible; and tennis rackets and airplane bodies are being made with nanomaterials whose atoms have been carefully arranged to make them especially strong.

Current State of the Art[153]

An intriguing feature of the nanoscale is that it is the scale on which biological systems build their structural components, like microtubules, microfilaments, and chromatin.[154] In other words, biochemistry is a nanoscale phenomenon. Even more intriguingly, a key property of these biological structural components—including, of course, the DNA double helix—is self-assembly. In their quest to emulate these biological phenomena, scientists have created the field of DNA nanotechnology[155] and the closely related field of DNA-based computation by algorithm self-assembly.[156]

Some of the most interesting nanotech research being conducted today falls within the realm of so-called DNA nanotechnology. DNA nanotechnology is the design and development of objects, lattices, and devices made of synthetic DNA. Since the DNA helix is naturally linear (i.e., unbranched), the assembly of structures or devices built with synthetic DNA requires constructing branched molecules that can then be connected to form structural networks, or motifs. The DNA motifs are combined by sticky-end cohesion, a high-specificity DNA reaction.

The DNA can be used as either "brick" and "mortar" in the construction of various kinds of nano-objects (so-called "high structural resolution DNA nanotech") or as just mortar to join non-DNA particles ("compositional DNA nanotech"). The latter, which laboratories worldwide are involved with, can be used in many ways to organize large complexes. There are only about a dozen labs worldwide involved in high structural resolution DNA nanotech, the potential applications—which are many and varied—include architectural control and scaffolding (e.g., DNA-based computation), nanomechanical devices (e.g., nanorobotics and nanofabrication), and self-replicating nano-systems.

Self-assembling systems are completely autonomous devices that do not require the input of a person (or a robot) in order to function (i.e., as nanomechanical devices do). Last year an investigator at Purdue University made one of the first self-assembling nano-devices, in this case a DNAzyme, which can bind and cleave RNA molecules one by one.[157] Unimaginable just a couple of years ago, the creation of this device epitomizes the progress that the field of DNA nanotech has achieved in just a few years.

Future Applications

The future trajectory of the field, particularly the convergence of nanotechnology and molecular biology, is unclear, although it will almost certainly have multiple medical applications, including therapeutic delivery by nanoparticles.[158] In October 2004, scientists from the Institute of Bioengineering and Nanotechnology (Singapore) reported having invented a contact lens capable of releasing precise amounts of medication to treat glaucoma and other eye diseases.[159] Nanobiotechnology also promises multiple new approaches to molecular detection and diagnostics.[160]

Just as nanotubes and other nanodevices promise novel advantageous means of drug delivery, there is considerable concern that the very same devices and particles could have inadvertent dangerous (i.e., toxic) consequences. Several recent studies have examined the possible toxicity of nanotechnology-derived products.[161] Likewise, the field opens up an entirely new means of potential deliberate misuse.

Aerosol Technology

Description

Very broadly, aerosol science is an interdisciplinary field focused on the study of the presence and movement of biological particles in the earth's atmosphere, including the impact of such particles of human populations, agriculture, and animals (including insect control).[162] The widespread aerial spraying of the *Bacillus thuringiensis* var. *kurstaki* (Btk), to protect forests from damage and defoliation caused by the spruce budworm, is a good example of how aerosol technology is being used and optimized.[163] Other examples of recent research in this field include a study on the use of animal models for understanding the threat to human health caused by inhalation of toxic airborne particulate matter;[164] a study on wind as a potential aerosolization mechanism for dispersing microorganisms at flooded wastewater irrigation sites (reuse of partially treated domestic wastewater is increasingly being done worldwide for agricultural irrigation purposes);[165] studies of plume characteristics of bioaerosols generated during the application of liquid biosolids to farmland, and the microbial risk to human health associated with this practice;[166] and studies on the aerial spraying of insecticides.[167]

In biomedical research, aerosol science revolves around the study of the use of inhaled particulate matter as a means to treat human disease. Although its current widespread use is for local treatment of asthma and chronic obstructive pulmonary disease, direct administration of drugs to the respiratory tract has been effectively used or is being tested to treat bacterial lung infections, cystic fibrosis, and lung carcinoma. The effectiveness of aerosol delivery for systemic action is also being explored, as a novel, injection-free way to control pain and deliver various therapeutics for the treatment of diabetes, human growth hormone deficiency (in children), prostate cancer, and endometriosis.[168] Compared to oral delivery, advantages of aerosolized delivery[169] include its rapid speed of onset and even biodistribution.[170]

Current State of the Art

In the drug delivery industry the three most common types of aerosol delivery devices currently in medical use are propellant metered-dose inhalers (pMDIs), dry powder inhalers (DPIs), and nebulizers.[171] Propellant MDIs are the most popular, since they are small, convenient, and self-powered (i.e., by the high-pressured contents of the "metering chamber"). The aerosol is drawn into a metering chamber, followed by propulsion of the solution as droplets into the lung. In the past, most pMDIs utilized suspensions or solutions of drugs in chlorofluorocarbons (CFCs). But CFC propellants are being phased out in favor of non-ozone-depleting hydrofluoroalkane inhalers. But the latter require device components and formulations that are different than those of CFC pMDIs, which has necessarily led to the creation of novel delivery means.[172]

Many new DPI devices and technologies have been developed and patented since the first one was introduced in the 1960s.[173] DPIs deliver powdered dry particles into the lungs, relying on the energy produced by the forces of the inhaled airflow. Most powder products are mixtures of drug particles and large lactose carrier particles. The smaller particles are delivered to the lungs while the larger particles, which help with dispersion, are deposited on the mouthpiece. A variety of different technologies have been used in the development of DPIs, and performance varies widely among different types of inhalers. Attempts to improve the delivery of respirable dry products to the lower airways and lungs remains an active area of research.[174]

Although air-jet nebulizers are inconvenient devices, due to their utilization of compressed gas (and thus requiring an air compressor) and their comparatively long aersolization time, their capability to deliver a high dose over an extended time period is widely considered an advantage over pMDIs and DPIs.[175] Nebulizers work by passing an air-jet stream (which is created by using compressed gas or piezoelectric ceramics) through a capillary tube that runs through a reservoir containing the drug; the drug solution is drawn out of the reservoir and deposited into the lungs in droplet form. In addition to their inconvenience, other limitations of the technology include the partial loss of drug dose during exhalation (since nebulizers generate aerosol continuously) and the large size of some of the devices. A variety of newer, "next-generation" nebulizers, which overcome some of these limitations, are being developed and produced.

In addition to the quality and features of the delivery device, critical to the delivery of the drugs to the lungs is the preparation of particles of correct size and shape for incorporation into aerosol products. Advances in powder technology and particle engineering play a significant role in improving powder production and aerosol drug formulation (e.g., by improving particle dispersibility, control of particle morphology, physical and chemical stability). For example, supercritical fluid (SCF) processing has recently emerged as an alternative technology for designing particles to use in metered-dose and dry powder inhalers.[176] SCFs are substances that exist as a single phase but have the properties of both liquids and gases (at certain temperatures and pressures), and they can extract compounds from complex substrates much more quickly than liquid organic solvents can.

In addition to its pharmaceutical applications, SCF technology is being used in the food industry (for decaffeination of coffee and tea and extraction of edible oils); the flavor and fragrance industry (for extraction of aromas and flavors); the nutraceutical industry (for the extraction of active ingredients for nutraceuticals and purification of antioxidants for nutraceuticals), the paint/coating industry (for the production of small particles for paint coating applications); and for a variety of other industrial purposes (e.g., purification of natural and synthetic materials and polymers and production of small particles for explosives).

Future Applications

Biomedical advances in aerosol delivery technology are expected to improve drug delivery and patient adherence. Several companies are pursuing aerosolized insulin delivery as a non-invasive alternative to injectable insulin. It is widely believed that, once proven safe for prolonged use, aerosolized insulin delivery will stimulate further activity in this already very active field. Aerosol delivery is also being explored as a means of gene therapy.

Advances in drug delivery technology, including aerosol delivery, have raised concerns about the use of bioregulators for nefarious purposes. In the past, bioregulators have not generally been viewed as potential dual-use agents, largely because of the lack of effective delivery technology.[177]

The dual-use risk of bioregulators was considered to be minimal due to their lack of suitability for aerosolization unless microencapsulated, their limited shelf life after atmospheric release, the fact that proteins denature at very high temperatures and lose activity at low temperatures, and high purchase costs. However, new knowledge and advancing technologies, particularly delivery technologies, have raised concerns about the dual-use risk of bioregulators. Potential delivery platforms include the use of bacterial plasmids or viral vectors for cloning the genes that encode bioregulators; use of transgenic insects (i.e., to secrete and inoculate the bioregulators); nanoscale delivery systems (e.g., engineered proteins either within or bound to nanotubes); and microencapsulated delivery systems (i.e., incorporating vectors or the proteins themselves into biodegradable microspheres or liposomes for controlled release).[178] Given that anything less than 3 microns in diameter is respirable across what amounts to a 75-square-meter absorptive surface, miniaturization of respiratory delivery systems comes with considerable dual-use risk. Moreover, transgenic plants could be put to dual use as bioregulator-production factories.

Microencapsulation Technology

Description

Microencapsulation is the envelopment of small solid particles, liquid droplets, or gas bubbles with a protective coating derived from any of a number of compounds (organic polymer, hydrocolloid, sugar, wax, fat, metal, or inorganic oxide). The capsules, which are basically miniature containers that protect their contents from evaporation, oxidation, and contamination and can be engineered with any of a variety of unique release mechanisms (e.g., from controlled, delayed, targeted release to biodegradable or salt-induced release), have countless applications.

Microencapsulation is not a new technology. Between the late 1940s and early 1960s, the concept of chemical microencapsulation generated interest in the pharmaceutical industry as an alternative mode of drug delivery that could offer sustained controlled release. Researchers and entrepreneurs continue to utilize and investigate advances in microencapsulation technology in efforts to make dosages more palatable, make active ingredients more stable and/or soluble, and otherwise improve drug delivery.[179] In the decades since the technology first emerged, many other life sciences industrial sectors have benefited tremendously from non-pharmaceutical applications of microencapsulation. In fact, it was partly in response to potential agrochemical applications of encapsulation technology that the Controlled Release Society, an international organization with 3,000 members from more than 50 countries, was formed in the mid-1970s (microencapsulation is the most common but not the only form of controlled release). As defined on its Web site, controlled release is "the field of scientific activity concerned with the control in time and space of the biological effects of therapeutic agents in human and animal health, and of other active agents in environmental, consumer and industrial applications."[180]

According to data provided by the Southwest Research Institute, the number of U.S. patents for encapsulation processes increased from about 1,250 during 1976-1980 to about 8,500 during 1996-2001.[181] U.S. patents for nanoencapsulation grew from near zero to about 1,000 over the same time period.

There are two general categories of microencapsulation processes: physical (e.g., spray drying, fluid bed coating, co-extrusion, rotary disk atomization) and chemical (e.g., polymerization, phase separation, solvent evaporation, coacervation). Between 1996 and 2002, polymerization was the most commonly used process, based on U.S. patent data, followed by spray drying and rotary disk atomization.

Current State of the Art

Today, microencapsulation technology is being used in water treatment (to remove emulsified oils, heavy metals, phosphates, and suspended solids from wastewater), food and agriculture (to improve taste and mask odor; stabilize thermal, oxidative, and shelf-life properties of ingredients; and allow for more effective absorption of nutrients and vitamins), and in the cosmetics industry (to create "eye appeal" or a specific or special feel in a wide range of personal care products). It has also been used as a way to manage mercury-contaminated and other hazardous wastes.[182]

Examples of recent use and exploration of this technology include an investigation by University of Saskatchewan researchers into the use of microencapsulated engineered cells as an alternative approach to cancer treatment.[183] The cells had been engineered to release a compound that kills tumor cells (i.e., functional necrosis factor-alpha). Implantation of encapsulated cells (into a mouse model system) led to tumor regression and slower tumor growth. In another study, researchers from the Netherlands tested the release, upon chewing, of flavored microencapsulates in Gouda cheese (the microencapsulates contained sunflower oil, lemon, and orange oil flavors).[184] Japanese researchers recently demonstrated the use of a novel nanoencapsulation drug delivery method for the external treatment of photo-damaged skin.[185] Advanced BioNutrition Corporation (Columbia, Maryland) was recently awarded a National Science Foundation grant to further develop its proprietary microencapsulation technology for the incorporation of functional ingredients—such as enzymes, fatty acids, probiotics, even vaccines—into its animal and human food products.[186] The company will use the money to scale up its microencapsulation technology production process.

Future Applications

An exciting future application is the transplantation of encapsulated live cells for therapeutic purposes.[187] Other future applications range from teddy bears that release a scent to help children sleep to novel military applications. In January 2005, for example, Northrop Grumman (San Diego, CA) announced the development of new encapsulation technology that allows non-marinized[188] weapons and vehicles to be released by submarines.

Gene Therapy Technologies

Description

Gene therapy is an experimental technique that uses "healthy" genes to treat or prevent disease. In most gene therapy studies, a "normal" gene is inserted into the genome to replace an "abnormal," disease-causing gene. A carrier molecule—called a vector—must be used to deliver the "healthy" gene to a recipient's target cells. Currently, the most commonly used vectors are viruses (including retroviruses, adenoviruses, adeno-associated viruses, and herpes simplex viruses) that have been genetically altered to carry normal human DNA. Nonviral options for gene delivery include the direct introduction of therapeutic DNA into target cells, although direct administration can only be used with certain tissues and requires large amounts of DNA (see Figure 3-4).

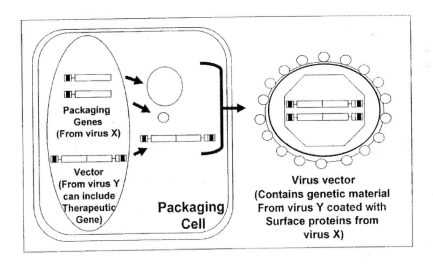

FIGURE 3-4 Viral vectors.
SOURCE: James Benjamin Petro, presentation to the committee, February 2004.

State of the Art

Gene therapy is still experimental, and most of the research performed to date has been conducted in animal trials (from rodents to primates). For example, in a study that appeared in *Nature Medicine* in March 2005, using a guinea pig model system, researchers from the University of Michigan and Kansai Medical University, Japan, reported that they had used gene therapy to restore hearing in mature deaf animals.[189] The evidence suggests that gene therapy can be used to regenerate functional hair cells, which are necessary to restore hearing, by using (in this case) an adenovector to deliver the "healthy" gene into nonsensory cells that reside in the deaf cochlea. The introduced gene, Atoh1 (also known as Math1), encodes a basic helix-loop-helix transcription factor and key regulator of hair cell development. Upon delivery, hearing is substantially improved.

The few human clinical trials that have been conducted have not been as successful as originally hoped.[190] Although substantial progress has been made, and some clinical successes seem to be on the horizon, further vector refinement and/or development is required before gene therapy will become standard care for any individual disorder.

Future Applications

When gene therapy does become a clinical reality, it will be used to correct faulty or defective disease-causing genes. But just as it will be used to delivery "healthy" genes into cells and tissues, gene therapy could potentially be used to deliver harmful genes.

Targeting Biologically-Active Materials to Specific Locations in the Body

The efficacy and safety of medical drugs, imaging agents, and vaccines depend on the ability to deliver these agents to the right location in the body and, ideally, with precision targeting only to the cells of interest. Selectivity in drug delivery reduces the exposure of nontarget tissues to the drug, thereby reducing the risk of unwanted drug actions and adverse events. However, this obvious therapeutic need is far from easy to achieve in practice. Selective targeting of bioactive molecules remains a largely unfulfilled objective in clinical therapeutics. The pharmaceutical and biotechnology industries, including companies that specialize only in the design of ways to optimize drug delivery, are investing substantial sums in research and development to achieve this attractive, yet elusive, goal.

Considerable ingenuity has been exhibited in designing "targeting" vehicles and "homing" systems for precision delivery of drugs and imaging agents. These range from efforts to deliver materials to specific zones in the body (e.g., aerosol delivery to the lungs, selective drug delivery to different regions of the gastrointestinal tract) to the more challenging objectives of targeted delivery to specific cell types (e.g., cancer cells versus their normal counterparts) or delivery of a drug or other bioactive agents to a specific compartment inside the cell (e.g., nuclear uptake of genes into chromosomal DNA for gene therapy or targeted therapeutic ablation of deleterious genes).

A broad repertoire of targeting vehicles have been examined in this research effort. These include carrier particles containing encapsulated drugs (e.g., liposomes, nanoparticles, dendrimers); exploitation of the "homing" ability of microorganisms to bind selectively to specific cells (e.g., viruses or bacteria as vectors for targeted delivery of genes and proteins); and the coupling of drugs to cognate carrier molecules designed to recognize only the desired cell type and then release their therapeutic payload. A unifying theme linking these different approaches lies in engineering suitable "molecular recognition" systems whereby cognate molecules in/on the carrier system recognize and attach to molecules expressed exclusively on the desired target cell, tissue, or organ. Additional cognate molecular interaction systems can be designed to enhance the efficiency of drug uptake by cells once selective targeting has occurred and for directing the delivered drug or gene to the correct location inside the cell.

Two different technical approaches underpin technical strategies for targeted drug delivery. The first incorporates the targeting (homing) property into the drug itself so that it will interact only with target cells that bear a "receptor" molecule that recognizes a structural region (domain) on the drug molecule. In the second approach the cognate properties required for recognition and binding to target cells are engineered into a drug carrier rather than the drug itself. Drugs are associated with the carrier either via passive encapsulation (e.g., particulate carriers) or by chemical coupling to the carrier. Both approaches exploit cognate molecular interactions as a common design principle. Targeted delivery is achieved as a consequence of molecular recognition events that limit the interaction of the drug and/or the drug carrier to only those cells that express a specific molecular determinant that interacts with the drug or drug-carrier complex.

As emphasized in the earlier section on how knowledge of the body's biocircuits can be used for both constructive and abusive purposes, the technical platforms for precision drug targeting pose a similar dual-use problem. Knowledge of how to target bioactive materials to specific cells can be usurped to disrupt or destroy vital functions in humans, animals, or plants. However, the technical ease with which such assaults could be mounted will be influenced by the location of the target in the targeted host and the anatomic barriers that a targeting system must breach in order to reach its molecular locus of action.

One of the more concerning assaults, yet attainable even with today's delivery technology, could arise from the use of targeted delivery systems to insert genes into chromosomal DNA. For example, viral delivery vectors developed for human gene therapy exploit the ability of viruses to bind selectively to specific cell types as a way to deliver genes encapsulated inside the viral particle into the target cells. The question of whether these viral delivery systems are applied to beneficent or malevolent goals is defined solely by the nature of the genetic payload incorporated into the vector. Although therapeutic gene therapy has yet to attain routine clinical utility, the extensive research literature on gene transfection technologies using viruses and various particulate carriers has demonstrated the feasibility of inserting exogenous genes in multiple cell types in the body. Future improvements in the efficiency of these delivery technologies can be confidently expected, with accompanying expansion in the horizons of both therapeutic and nefarious utility.

The delivery and expression of genes that code for the uncontrolled production of highly potent hormones and other natural bioactive mediators involved in homeostasis offer the simplest example of how this knowledge could be abused and used to expand the emerging threat spectrum. Alternatively, rather than using a transfected gene directly to produce a bioactive product to perturb body function, the transfected gene could act as a trigger for the abnormal expression or destruction of other genes vital to body homeostasis. In either of these settings the introduced gene is designed to integrate into the chromosomal DNA of the host. The disruptive effects could be manifest immediately as an acute event or the gene could lie silent in the genome for activation at a later time by a second external trigger.

An aphorism frequently cited in the design of drug delivery systems is that "the opportunities are limited only by the imagination of the inventor." Theoretically, the ability to design drugs and carrier vehicles endowed with cognate molecular properties that enable them to home selectively to the desired target in the body is limited only by the availability of suitable molecular recognition molecules that can be incorporated into the delivery system to confer recognition and binding by molecules unique to the desired target cell. The availability of relevant molecular cognate pairs for the delivery system and for the target is an obligate prerequisite for targeting. However, this is but one component in the engineering of targeted delivery systems. For therapeutic applications, the tar-

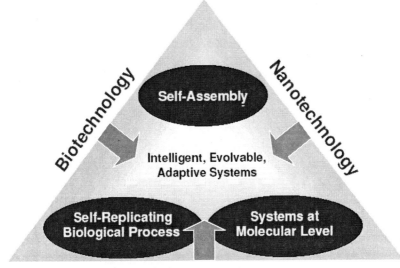

Information Technology

FIGURE 3-5 Converging technologies. Biotechnology, nanotechnology, and information technology are converging in ways that will enable humans to do things never dreamt of until now.
SOURCE: Michael Morgan, presentation during the Cuernavaca workshop, September 2004.

geting system must also exhibit suitable absorption, distribution, metabolism, excretion, and toxicology (ADMET) properties.

Complementarity and Synergy of Technologies

Some futurists consider the convergence of bio-, nano-, and information technologies, along with the neuro- and cognitive sciences, a transformation that will prove as powerful as the Industrial Revolution (Figure 3-5). However, the details and impact of possible convergent events are unclear at this time.

Enabling technologies are those that interact with each other to create novel products that would otherwise be impossible to achieve. Nanotechnology enables other technologies by providing a common hardware for molecular engineering and allowing for the realization of desirable architectures. Nanotechnology enables biotechnology by developing new imaging techniques, probes, and sensors; it also contributes to the miniaturization demands of information technologies. Biotechnology enables other technologies by identifying chemical and physical processes and algorithmic structures in living systems that have a genetically based material organization. It enables nanotechnology by providing a paradigm that nanotechnologists use in developing systems; much of the work in nanotechnology involves mimicking biotechnological processes while simultaneously redesigning them to fit particular purposes. Biotechnology enables information technology by providing new systems of computing, some of which may be based on DNA. Information technology enables other technologies through its ability to represent physical states as information and model processes. It provides the computing power that is essential to all research; it enables nanotechnology through precision control of patterning and intervention; and it enables biotechnology by providing the means to model complex processes and thereby solve difficult research problems.

In addition to convergence, which leads to the emergence of entirely new disciplines such as DNA nanotechnology and bioinformatics, technologies combine and converge on a smaller, less dramatic scale all the time. In terms of future potential threats, one should note the importance of combinations or interactions involving technologies in any of the first three categories—the acquisition of biological or molecular diversity, directed design, and understanding and manipulation of biological systems—and technologies in the fourth category: production, delivery, and packaging. In other words, the impact, both beneficial and detrimental, of a small-molecule agent, synthetic agent, or an agent bred through "DNA shuffling" is enhanced by appropriate packaging and delivery. Indeed, growing concerns about the dual-use risk of bioregulators are partly in response to advances in microencapsulated delivery systems, which make the use of bioregulators for either beneficial or nefarious purposes more feasible.

Based on extensive deliberations on a wide range of advancing technologies with relevance to the life sciences, including many technologies and fields of knowledge not traditionally viewed within the rubric of *bio*technology, the committee was particularly struck by the extent to which various tools and technologies are interacting and converging[191] — both additively and synergistically—and creating unanticipated opportunities for these technologies to be used for either beneficial or malicious intent (or with beneficial intent but unintended consequences). As already mentioned, the convergence of nanotechnology and molecular biology serves as a prime example of how an entirely new discipline, DNA nanotechnology, can emerge unexpectedly and with profound consequences. Nanotechnology is also merging with encapsulation and microfluidic technologies, providing the means for further miniaturization of already very low-volume biological sampling, detection, delivery, and other processes.

As one example, synthetic biologists are using their new tools in conjunction with nanotechniques to program cells with decision-making therapeutic power. For example, researchers have designed a prototype "DNA computer" with the capacity to logically analyze mRNA disease indicators in vitro (i.e., in this case, early signs of prostate and lung cancer) and control the administration of biologically active ssDNA molecules, including drugs.[192] The procedure is relatively innocuous, requiring the injection of a very small amount of fluid containing billions of nanoparticles, each of which operates as a tiny computer by effectively interrogating the cell and detecting the presence of diagnostic DNA markers (e.g., mutated mRNA sequences or underexpressed or overexpressed mRNA). If the markers are present, the nanoparticle sends out a therapeutic short nucleic acid that can affect the levels of gene expression.

The field of bioinformatics represents another key example of converging technologies—in this case biology, computer science, and information technologies—all of which have merged to form what is now a single discipline. Over the past 10 years, major advances in the field of molecular biology, coupled with advances in genomic technologies, have led to an explosive growth in biological information generated by the life sciences community. This deluge of genomic information, in turn, has led to an absolute requirement for computerized databases to store, organize, and index the data and for specialized tools to view and analyze the data. These databases and tools comprise the field of bioinformatics. Increasingly, biological studies begin with a scientist surveying databases to formulate specific hypotheses or design large-scale experiments, representing a dramatic shift in biology from a purely lab-based to an information-based science. Moreover, the growing availability of vast amounts of biological and other relevant information (e.g., small-molecule libraries) will also allow nonspecialists to tinker with or design constructs that, in the past, would have required years of education or training.

CONCLUSION

"During the century just begun, as our ability to modify fundamental life processes continues its rapid advance, we will be able not only to devise additional ways to destroy life but will also be able to manipulate it—including the processes of cognition, development, reproduction, and inheritance."—Matthew Meselson[193]

It is difficult to predict what the global technology landscape will look like in 20, 10, or even 5 years into the future. But it is not difficult to anticipate that as advances are made, so too will opportunities for misuse. This chapter summarizes information on emerging technologies that are expected to have significant economic, societal, and dual-use risk impacts in the near future. As highlighted during the committee's international workshop in Cuernavaca, prominent among these are advances in knowledge and delivery technology that have increased the dual-use potential and risk of nonlethal bioregulators and the convergence of nano- and biotechnology in the form of DNA nanotechnology.

A major theme that emerged from the committee's deliberations in Mexico was the notion that pathogens are not the only problematic agents of biological origin. Some argue that bioregulators,[194] which are nonpathogenic organic compounds, may pose a more serious dual-use risk than had previously been appreciated, particularly as improved targeted delivery technologies have made the potential dissemination of these compounds much more feasible than in the past. This shift in focus highlights the reality that the materials, equipment, and technology necessary for disseminating and delivering the agents to their intended recipient(s) are equally, if not more, important than the agents themselves in terms of their dual-use risk.

The immune and neuroendocrine systems[195] are particularly vulnerable to bioregulator modification. In fact, the capacity to develop bioweapons that can be aimed at the interaction of the immune and neuroendocrine systems again points to a shift in focus from the agents to, in this case, how a range of agents can be exploited (or created) to affect the human body in targeted, covert, and insidious ways.

A controversial issue that arose from these discussions is how all research on immune system evasion could be considered potentially dangerous, thus highlighting the very important need to uphold the norms of the Biological and Toxin Weapons Convention. Another important theme that emerged from discussions of the material presented here is the notion of time and how the advancing technology landscape has an uncertain future and unpredictable dual-use risk implications. This unpredictability poses a significant challenge for developing and implementing a strategy to manage these risks. These challenges—and potential solution sets—are discussed in the following chapter.

ENDNOTES

[1] Stemmer, W.P. 1994. Rapid evolution of a protein in vitro by DNA shuffling. *Nature* 370(6488):389-391.

[2] With approximately 10^{33} microorganisms on earth today, even with a 10-

minute fission time, only about 10^{45} have existed over the history of the earth, which is tiny compared to the number of possible 10^8 base pair DNA sequences.

[3] See discussion of virulence and evolution of pathogens in Chapter 1.

[4] Agarwal, K.L., et al. 1974. Total synthesis of the gene for an alanine transfer ribonucleic acid from yeast. *Nature* 227(5253):27-34.

[5] Cello, J., et al. 2002. Chemical synthesis of poliovirus cDNA: generation of infectious virus in the absence of natural template. *Science* 297(5583):1016-1018.

[6] Smith, H.O., C.A. Hutchison, III, C. Pfannkoch, and J.C. Venter. 2003. Generating a synthetic genome by whole genome assembly: phiX174 bacteriophage from synthetic oligonucleotides. *Proceedings of the National Academy of Sciences* 100(26): 15440–15445.

[7] Wade, N. 2005. "A DNA success raises bioterror concern. *New York Times* (January 12). Many experts in the field consider this view alarmist, since not only is the smallpox virus longer, but it cannot self-generate from its nucleotide sequence alone.

[8] Carlson, R. 2003. The pace and proliferation of biological technologies. *Biosecurity and Bioterrorism: Biodefense Strategy, Practice, and Science* 1(3):203-214.

[9] Carr, P.A., et al. 2004. Protein-mediated error correction for de novo DNA synthesis. *Nucleic Acids Research* 32(20); Richmond, K.E., et al. 2004. Amplification and assembly of chip-eluted DNA (AACED): a method for high-throughput gene synthesis. *Nucleic Acids Research* 32(17):5011–5018; Tian, J. et al. 2004. Accurate multiplex gene synthesis from programmable DNA microchips. *Nature* 432(7020):1050-1054.

[10] Tian, J. et al. 2004. Accurate multiplex gene synthesis from programmable DNA microchips. *Nature* 432(7020):1050-1054.

[11] Carr, P.A. et al. 2004. Protein-mediated error correction for *de novo* DNA synthesis. *Nucleic Acids Research* 32(20):e162.

[12] Elowitz, M.B. and S. Leibler. 2000. A synthetic oscillatory network of transcriptional regulators. *Nature* 403(6767):335-338.

[13] Martin, V.J., et al. 2003. Engineering a mevalonate pathway in *Escherichia coli* for production of terpenoids. *Nature Biotechnology* 21(7):796-802.

[14] Hutchinson, C.A., et al. 1999. Global transposon mutagenesis and a minimal *Mycoplasma* genome. *Science* 286(5447):2165-2169.

[15] Martin, V.J. et al. 2003. Engineering a mevalonate pathway in Escherichia coli for production of terpenoids. *Nature Biotechnology* 21(7):796-802.

[16] Church, G. 2004. A synthetic biohazard non-proliferation proposal. Updated May 21, 2005. Available online at arep.med.harvard.edu/SBP/Church_Biohazard04c.htm [accessed January 5, 2006]. The Alfred P. Sloan Foundation recently funded an joint activity by the Massachusetts Institute of Technology, the Venter Institute, and the Center for Strategic and International Studies to examine the benefits and risks of synthetic genomics and develop and analyze policy options for governance of the relevant technologies. A press release issued by the three institutions describing this study may be found online at www.csis.org/press/pr05_23.pdf.

[17] Mann, C.C. 1999. Crop scientists seek a new revolution. *Science* 283(5400): 310-314.

[18] Crameri, A. et al. 1998. DNA shuffling of a family of genes from diverse species accelerates directed evolution. *Nature* 391(6664):288-291.

[19] Ness, J.E. 1999. DNA shuffling of subgenomic sequences of subtilisin. *Nature Biotechnology* 17(9):893-896.

[20] Zhang, Y.Z. et al. 2002. Genome shuffling leads to rapid phenotypic improvement in bacteria. *Nature* 415(6872):644-646.

[21] Pekrun, K. et al. 2002. Evolution of a human immunodeficiency virus type 1 variant with enhanced replication in pig-tailed macaque cells by DNA shuffling. *Journal of Virology* 76(6):2924-2935.

[22] Leong, S.R. et al. 2003. Optimized expression and specific activity of IL-12 by directed molecular evolution. *Proceedings of the National Academy of Sciences* 100(3):1163-1168.

[23] Soong, N.W., et al. 2000. Molecular breeding of viruses. *Nature Genetics* 25 (4):436-439; Powell, S.K. et al. 2000. Breeding of retroviruses by DNA shuffling for improved stability and processing yields. *Nature Biotechnology* 18(12):1279-1282.

[24] Much of the information in this section is adapted from Strobel, G. and B. Daisy. 2003. Bioprospecting for microbial endophytes and their natural products. *Microbiology and Molecular Biology Reviews* 67(4):491-502. Available online at www.pubmedcentral.nih.govarticlerender.fcgi?tool=pubmed&pubmedid=14665674, [accessed March 24, 2005].

[25] Grabley, S. and R. Thiericke, eds. 1999. *Drug discovery from nature*. Berlin: Springer-Verlag; 3-33.

[26] Concepcion, G.P. et al. 2001. Screening for bioactive novel compounds. In Pointing, S.B. and K.D. Hyde, eds. 2001. *Bio-exploitation of filamentous fungi*. Hong Kong: Fungal Diversity Press; 93-130

[27] Wani, M.C. et al. 1971. Plant antitumor agents, VI. The isolation and structure of taxol, a novel antileukemic and antitumor agent from Taxus brevifolia. *Journal of the American Chemical Society* 93:2325-2327.

[28] Pace, N.R. 1997. A molecular view of microbial diversity and the biosphere. *Science* 276(5313):734-740; Venter, J.C. et al. 2004. Environmental genome shotgun sequencing of the Sargasso Sea. *Science* 304(5667):66-74.

[29] Demain, A.L. 2000. Microbial natural products: a past with a future. In Wrigley, S.K., M.A. Hayes, R. Thomas, E.J.T. Chrystal, and N. Nicholson, eds. *Biodiversity: new leads for pharmaceutical and agrochemical industries*. The Royal Society of Chemistry, Cambridge, United Kingdom; 3-16.

[30] Findlay, J.A. et al. 1997. Insect toxins from an endophyte fungus from wintergreen. *Journal of Natural Products* 60:1214-1215.

[31] Strobel, G. and B. Daisy. 2003. Bioprospecting for microbial endophytes and their natural products. *Microbiology and Molecular Biology Reviews* 67(4):491-502.

[32] Lorenz, P. and J. Eck. 2005. Metagenomics and industrial applications. *Nature Reviews. Microbiology* 3(6):510-516.

[33] Tyson, G.W. et al. 2004. Community structure and metabolism through reconstruction of microbial genomes from the environment. *Nature* 428(6978):37-43; Venter, J.C. et al. 2004. Environmental genome shotgun sequencing of the Sargasso Sea. *Science* 304(5667):66-74; Tringe, S.G. et al. 2005. Comparative Metagenomics of Microbial Communities. *Science* 308(5721):554-557.

[34] This search for novel microbial genomes to identify useful products is

achieved through the use of laboratory methods and queries of bioinformatics "libraries."

[35] Marshall, W.F. 3rd et al. 1994. Detection of *Borrelia burgdorferi* DNA in museum specimines of *Peromyscus leucopus*. *Journal of Infectious Diseases* 170:1027-1032; Mills, J.N. et al. 1999. Long-term studies of hantavirus reservoir populations in the southwestern United States: A synthesis. *Emerging Infectious Diseases* 5(1):135-142; Monroe, M.C. et al. 1999. Genetic diversity and distribution of *Peromyscus*-borne hantaviruses in North America. *Emerging Infectious Diseases* 5(1):75-86.

[36] Relman, D.A. 2002. Mining the natural world for new pathogens. *American Journal of Tropical Medicine and Hygiene* 67(2):133-134.

[37] Parola, P. et al. 2002. First molecular evidence of new *Bartonella* s in fleas and a tick from Peru. *American Journal of Tropical Medicine and Hygiene* 67(2):135-136.

[38] Breitschwerdt, E.B. and Kordick, D.L. 2000. Bartonella infection in animals: carriership, reservoir potential, pathogenicity, and zoonotic potential for human infection. *Clinical Microbiology Reviews* 13(3):428-438.

[39] In higher eukaryotes, biological processes such as cellular growth and organogenesis are mediated by differential gene expression. To understand molecular regulation of these processes, differentially expressed genes of interest must be identified, cloned, and studied in detail. Subtractive cDNA hybridization has been a powerful tool in the identification and analysis of differentially expressed cDNAs. See www.evrogen.com/t6.shtml.

[40] Merrifield, R.B. 1963. Solid phase peptide synthesis: the synthesis of a tetrapeptide. *Journal of the American Chemical Society* 85: 2149-2154.

[41] A more detailed understanding of how the technology works requires understanding the basic chemistry of polypeptide formation: the general chemical formula for amino acids is $H_2NCH(R)CO_2H$. Amino acids can be linked together to form peptides by reacting the $-NH_2$ group of one amino acid with the $-CO_2H$ group of another, thus forming an amide bond. Solid-phase synthesis involves reacting the $-CO_2H$ group with a CH_2Cl group on the resin, thereby leaving the $-NH_2$ group free to form an amide bond with the second amino acid. The second amino acid is structurally modified, prior to mixing with first amino acid, in order to render its $-NH_2$ group incapable of participating in an amide-forming reaction. The now protected second amino acid is added to the reaction mixture and a dipeptide, attached to the solid support, is created. The protecting group of the new dipeptide is removed, and a third protected amino acid is added to the mixture, resulting in a tripeptide. The process is continued until the desired product is created.

[42] Geyson, M.H. et al. 1984. Use of peptide synthesis to probe viral antigens for epitopes to a resolution of a single amino acid. *Proceedings of the National Academy of Sciences* 81(13):3998-4002.

[43] Houghton, R.A. 1985. General method for the rapid solid-phase synthesis of large numbers of peptides: specificity of antigen-antibody interaction at the level of individual amino acids. *Proceedings of the National Academy of Sciences* 82(15): 5131-5135.

[44] The general approach devised by Geyson and Houghten was modified further in the early 1990s, when Kit Lam developed a rapid method for producing and evaluating random libraries of millions of peptides. Initially applied to pep-

tides, solid phase synthesis was gradually extended to produce libraries of druglike small molecules, which were of greater interest to the drug discovery industry. In the early 1990s, Jonathan A. Ellman, University of California, Berkeley, used Geyson's multi-pin approach to create a library of 192 structurally diverse benzodiazepines. Concurrently, Sheila H. DeWitt, then at Parke-Davis Pharmaceutical Research, Michigan, reported a technique and apparatus for the multiple, simultaneous synthesis of so-called "diversomers" (collections of organic compounds, including dipeptides, hydantoins, and benzodiazepines). These studies represented some of the earliest techniques for generating small molecule libraries.

[45] Sanchez-Martin, R.M. et al. 2004. The impact of combinatorial methodologies on medicinal chemistry. *Current Topics in Medicinal Chemistry* 4(7): 653-669.

[46] Needles, M.C. et al. 1993. Generation and screening of an oligonucleotide-encoded synthetic peptide library. *Proceedings of the National Academy of Sciences* 90(22):10700-10704.

[47] Ohlmeyer, M.H.J. et al. 1993. Complex synthetic chemical libraries indexed with molecular tags. *Proceedings of the National Academy of Sciences* 90(23):10922-10926; Moran, E. J., et al. 1995. Radio frequency tag-encoded combinatorial library method for the discovery of tripeptide-substituted cinnamic acid inhibitors of the protein tyrosinase phosphatase PTP1B. *Journal of the American Chemical Society* 117(43):10787-10788; Nicolau, K.C. et al. 1995. Radiofrequency encoded combinatorial chemistry. *Angewandte Chemie International Edition* 34(20): 2289-2291.

[48] Reader, J. C. 2004. Automation in medicinal chemistry.*Current Topics in Medicinal Chemistry* 4(7):671-686.

[49] Matzger, A.V. et al. 2000. Combinatorial approaches to the synthesis of vapor detector arrays for use in an electronic nose. *Journal of Combinatorial Chemistry* 2(4):301-304.

[50] Wheelis, M. 2002. Biotechnology and biochemical weapons. *The Nonproliferation Review* Spring:48-53.

[51] Austin, C.P., L.S. Brady, T.R. Insel, and F.S. Collins. 2004. NIH Molecular Libraries Initiative. *Science* 306(5699):1138-1139.

[52] See also, discussion of this issue in Chapter 1 "The NIH Roadmap."

[53] Major Histocompatibility Complex (protein complexes that present antigens to lymphocytes).

[54] Berman, H.M., J. Westbrook, Z. Feng, G. Gilliland, T.N. Bhat, H. Weissig, I.N. Shindyalov, and P.E. Bourne. 2004. The Protein Data Bank. *Nucleic Acids Research* 28(1):235-242. See pdbbeta.rcsb.org/pdb/static.do?p=general_information/pdb_statistics/content_growth_graph.html [accessed January 5, 2006].

[55] The Smallpox Research Grid project distributed a screensaver to thousands of home computer owners to perform these calculations to identify drugs that might interfere with the enzyme that unwinds variola DNA to permit replication. The project is described at www.chem.ox.ac.uk/smallpox/news.html . (Altogether over 39,000 years of computer time were devoted to the project in less than six months, screening 35 million molecules against eight models of the target protein.)

[56] Yokobayashi, Y. et al. 2002. Directed evolution of a genetic circuit. *Proceedings of the National Academy of Sciences* 99(26):16587-16591.

[57] Registry of Standard Biological Parts. The Endy Lab, Massachusetts Institute of Technology. See parts.mit.edu/ [accessed January 5, 2006].

[58] Elowitz, M.B. and S. Liebler. 2000. A synthetic oscillatory network of transcriptional regulators. *Nature* 403(6767):335-338.

[59] Atkinson, M.R. et al. 2003. Development of genetic circuitry exhibiting toggle switch or oscillatory behavior in *Escherichia coli. Cell* 113(5):597-607.

[60] Looger, L.L. et al. 2003. Computational design of receptor and sensor proteins with novel functions. *Nature* 423 (6936):185-90; DeGrado, W.F. 2003. Biosensor Design. *Nature* 423(6936):132-133.

[61] Benenson, Y. et al. 2004. An autonomous molecular computer for logical control of gene expression. *Nature* 429(6990):423-429.

[62] Ferber, D. 2004. Microbes made to order. *Science* 303(5655):158-161.

[63] The Center for Strategic & International Studies (CSIS), the J. Craig Venter Institute (Venter Institute), and the Massachusetts Institute of Technology (MIT) have initiated a project, funded by the Alfred P. Sloan Foundation, to examine the societal implications of synthetic genomics, exploring risks and benefits as well as possible safeguards to prevent abuse, including bioterrorism. See further description online at www.csis.org/press/pr05_23.pdf.

[64] Racaniello, V.R. and Baltimore, D. 1981. Cloned poliovirus complementary DNA is infectious in mammalian cells. *Science* 214(4523):916-919.

[65] Ahlquist, P. et al. 1984. Multicomponent RNA plant virus infection derived from cloned viral cDNA. *Proceedings of the National Academy of Sciences* 81(22):7066–7070.

[66] Rice, C.M., et al. 1989. Transcription of infectious yellow fever RNA from full-length cDNA templates produced by in vitro ligation. *New Biologist* 1(3):285-296.

[67] Rice, C.M. et al. 1987. Production of infectious RNA transcripts from Sindbis virus cDNA clones: mapping of lethal mutations, rescue of a temperature-sensitive marker, and in vitro mutagenesis to generate defined mutants. *Journal of Virology* 61(12): 3809–3819.

[68] Satyanarayana, T. et al. 1999. An engineered closterovirus RNA replicon and analysis of heterologous terminal sequences for replication. *Proceedings of the National Academy of Sciences* 96(13):7433-7438.

[69] Van Dinten, L.C. et al. 1997. An infectious arterivirus cDNA clone: Identification of a replicase point mutation that abolishes discontinuous mRNA transcription. *Proceedings of the National Academy of Sciences* 94(3):991–999.

[70] Masters, P.S. 1999. Reverse genetics of the largest RNA viruses. *Advances in Virus Research* 53:245-64.

[71] Almazán, F., et al. 2000. Engineering the largest RNA virus genome as an infectious bacterial artificial chromosome. *Proceedings of the National Academy of Sciences* 97(10):5516–5521.

[72] Yount, B., et al. 2003. Reverse genetics with a full-length infectious cDNA of severe acute respiratory syndrome coronavirus. *Proceedings of the National Academy of Sciences* 100(22):12995–13000.

[73] Negative-stranded RNA viruses have a genome consisting of one or more molecules of single-stranded RNA that is of opposite polarity (i.e., complementary) to the positive-sense mRNA that encodes their proteins.

[74] Enami, M. et al. 1990. Introduction of site-specific mutations into the genome of influenza virus. *Proceedings of the National Academy of Sciences* 87(10):3802–3805; Luytjes, M. et al. 1989. Amplification, expression, and packaging of foreign gene by influenza virus. *Cell* 59(6):1107-1113.

[75] Fodor, E. et al. 1999. Rescue of influenza A virus from recombinant DNA. *Journal of Virology* 73(11):9679–9682; Neumann, G. et al. 1999. Generation of influenza A viruses entirely from cloned cDNAs. *Proceedings of the National Academy of Sciences* 96(16): 9345-9350.

[76] Hatta, M. et al. 2001. Molecular basis for high virulence of Hong Kong H5N1 influenza A viruses. *Science* 293(5536):1840-1842.

[77] Fodor, E., et al. 1999. Rescue of influenza A virus from recombinant DNA. *Journal of Virology* 73(11):9679-9682.

[78] Schnell, M.J. et al. 1994. Infectious rabies viruses from cloned cDNA. *EMBO Journal* 13(18):4195-4203.

[79] Lawson, N.D. et al. 1995. Recombinant vesicular stomatitis virus from DNA. *Proceedings of the National Academy of Sciences* 92(10):4477–4481;Whelan, S.P. et al. 1995. Efficient recovery of infectious vesicular stomatitis virus entirely from cDNA clones. *Proceedings of the National Academy of Sciences* 92(18):8388-8392.

[80] Collins, P.L. et al. 1995. Production of infectious human respiratory syncytial virus from cloned cDNA confirms an essential role for the transcription elongation factor from the 5¢ proximal open reading frame of the M2 mRNA in gene expression and provides a capability for vaccine development. *Proceedings of the National Academy of Sciences* 92(25):11563-11567; Jin H. et al. 1998. Recombinant human respiratory syncytial virus (RSV) from cDNA and construction of subgroup A and B chimeric RSV. *Virology* 251(1):206-214.

[81] Radecke, F. et al. 1995. Rescue of measles virus from cloned DNA. *EMBO Journal* 14(23):5773-5784.

[82] Garcin, D. et al. 1995. A highly recombinogenic system for the recovery of infectious Sendai paramyxovirus from cDNA: generation of a novel copy-back non-defective interfering virus. *EMBO Journal* 14(24):6087; Kato, A. et al. 1996. Initiation of Sendai virus multiplication from transfected cDNA or RNA with negative or positive sense. *Genes Cells* 1(6):569-579.

[83] Durbin, A.P. et al. 1997. Recovery of infectious human parainfluenza virus type 3 from cDNA. *Virology* 235(2):323–332; Hoffman, M.A. and A.K. Banrjee. 1997. An infectious clone of human parainfluenza virus type 3. *Journal of Virology* 71(6):4272-4277.

[84]Baron, M.D. and T. Barrett. 1997. Rescue of rinderpest virus from cloned cDNA. *Journal of Virology* 71(2):1265-1271.

[85] He, B. et al. 1997. Recovery of infectious SV5 from cloned DNA and expression of a foreign gene. *Virology* 237(2):249-260.

[86] Buchholz, U.J. et al. 1999. Generation of bovine respiratory syncytial virus (BRSV) from cDNA: BRSV NS2 is not essential for virus replication in tissue culture, and the human RSV leader region acts as a functional BRSV genome promoter. *Journal of Virology* 73(1):251-259.

[87] Peeters, B.P.H. et al. 1999. Rescue of Newcastle disease virus from cloned cDNA: evidence that cleavability of the fusion protein is a major determinant for virulence. *Journal of Virology* 73(6):5001-5009.

[88] Bridgen, A. and R. Elliott. 1996. Rescue of a segmented negative-strand RNA virus entirely from cloned complementary DNAs. *Proceedings of the National Academy of Sciences* 93(26):15400-15404.

[89] Kobasa, D. et al. 2004. Enhanced virulence of influenza A viruses with the haemagglutinin of the 1918 pandemic virus. *Nature* 431(7009):703-707.

[90] Tumpey, T.M., C.F. Basler, P.V. Aguilar, H. Zeng, A. Solórzano, D.E. Swayne, N.J. Cox, J.M. Katz, J.K. Taubenberger, P. Palese, and A. García-Sastre. 2005. Characterization of the Reconstructed 1918 Spanish Influenza Pandemic Virus. *Science* 310(5745):77-80; Taubenberger, J.K., A.H. Reid, R.M. Lourens, R. Wang, G. Jin and T.G. Fanning. 2005. Characterization of the 1918 influenza virus polymerase genes. *Nature* 437(7060):889-893.

[91] Kaiser, J. 2005. Resurrected influenza virus yields secrets of deadly 1918 pandemic. *Science* 310(5745):28-29.

[92] Ibid.

[93] Sharp, P.A. 2005. 1918 flu and responsible science. *Science* 310(5745):17.

[94] Snijder, E.J. et al. 2003. Unique and conserved features of genome and proteome of SARS-coronavirus, an early split-off from the coronavirus group 2 lineage. *Journal of Molecular Biology* 331(5):991-1004; Yount, B. et al. 2003. Reverse genetics with a full-length infectious cDNA of severe acute respiratory syndrome coronavirus. *Proceedings of the National Academy of Sciences* 100(22):12995–13000.

[95] Krug, R.M. 2003. The potential use of influenza virus as an agent for bioterrorism. *Antiviral Research* 57(1-2):147-150.

[96] "Systems biology" is not a technology in the classic sense. It is an attempt to draw many disparate technologies together in the service of a new field, or perhaps in a new way of doing biology.

[97] Napoli, C. et al. 1990. Introduction of a chimeric chalcone synthase gene into petunia results in reversible co-suppression of homologous genes in trans. *Plant Cell* 2(4):279-289.

[98] Jackson, A.L. et al. 2003. Expression profiling reveals off-target gene regulation by RNAi. *Nature Biotechnology* 21(6):635-37; Scacheri, P.C. et al. 2004. Short interfering RNAs can induce unexpected and divergent changes in the levels of untargeted proteins in mammalian cells. *Proceedings of the National Academy of Sciences* 101(7):1892-1897.

[99] Scherr, M. et al. 2003. Inhibition of GM-CSF receptor function by stable RNA interference in a NOD/SCID mouse hematopoietic stem cell transplantation model. *Oligonucleotides* 13(5):353-363.

[100] Song, E. et al. 2003. RNA interference targeting Fas protects mice from fulminant hepatitis. *Nature Medicine* 9(3):347-351.

[101] Reich, S.J. et al. 2003. Small interfering RNA (siRNA) targeting VEGF effectively inhibits ocular neovascularization in a mouse model. *Molecular Vision* 9: 210-216.

[102] Zhang, X. et al. 2004. Small interfering RNA targeting heme oxygenase-1 enhances ischemia-reperfusion-induced lung apoptosis. *Journal of Biological Chemistry* 279(11)10677-10684.

[103] Dorn, G. et al. 2004. siRNA relieves chronic neuropathic pain. *Nucleic Acids Res.* 32(5):e49.

[104] See phx.corporate-ir.net/phoenix.zhtml?c=141787&p=irol-newsArticle&ID=610478&highlight=

[105] Soutschek, J. et al. 2004. Therapeutic silencing of an endogenous gene by systemic administration of modified siRNAs. *Nature* 432(7014):173-178.

[106] Check, E. 2004. Hopes rise for RNA therapy as mouse study hits target. *Nature* 432(7014):136.

[107] Voorhoeve, P.M. and R. Agami. 2003. Knockdown stands up. *Trends in Biotechnology* 21(1):2-4.

[108] Ellington, A.D. and J. W. Szostak. 1990. *In vitro* selection of RNA molecules that bind specific ligands. *Nature* 346(6287):818-822; Tuerk, C. and L. Gold. 1990. Systematic evolution of ligands by exponential enrichment: RNA ligands to bacteriophage T4 DNA polymerase. *Science* 249(4968):505-510.

[109] Block, C. et al. 2004. Photoaptamer arrays applied to multiplexed proteomic analysis. *Proteomics* 4(3):609-618; Jayasena, S.D. 1999. Aptamers: an emerging class of molecules that rival antibodies in diagnostics. *Clinical Chemistry* 45(9):1628-1650; Mayer, G. and A. Jenne. 2004. Aptamers in research and drug development. *Biodrugs* 18(6): 351-359.

[110] Mayer, G. and A. Jenne. 2004. Aptamers in research and drug development. *Biodrugs* 18(6):351-359.

[111] Wang, K.Y. et al. 1993. A DNA aptamer which binds to and inhibits thrombin exhibits a new structural motif for DNA. *Biochemistry* 32(8):1899-1904; Li , W.X., et al. 1994. A novel nucleotide-based thrombin inhibitor inhibits clot-bound thrombin and reduces arterial platelet thrombus formation. *Blood* 83(3):677-82.

[112] See www.archemix.com/press/pr_jun04.html [accessed March 27, 2005].

[113] See www.eyetk.com/clinical/clinical_index.asp [accessed March 27. 2005].

[114] Burbulis, I. et al. 2005. Using protein-DNA chimeras to detect and count small numbers of molecules. *Nature Methods* 2(1):31-37.

[115] Mayer, G. and A. Jenne. 2004. Aptamers in research and drug development. *Biodrugs* 18(6):351-359.

[116] The following section is taken from the Executive Summary of an NRC report entitled: *Catalyzing Inquiry at the Interface of Computing and Biology* (December 2005).

[117] National Research Council. 2005. *Catalyzing Inquiry at the Interface of Computing and Biology*. Washington, DC: The National Academies Press.

[118] Pacific Northwest National Laboratory. 2005. Genomic sequences processed in minutes, rather than weeks. *The Daily Nonproliferator*, June 21. Available online at www.pnl.gov/news/2005/05-45.stm.

[119] Wolkenhauer, O. et al. 2005. The dynamic systems approach to control and regulation of intracellular networks. *FEBS Letters* 579(8):1846-1853.

[120] Goldbeter, A. 2004. Computational biology: a propagating wave of interest. *Current Biology* 14(15):601-602; Uetz, P. and R.L. Finley, Jr. 2005. From protein networks to biological systems. *FEBS Letters* 579(8):1821-1827; Aloy, P. and R.B. Russell. 2005. Structure-based systems biology: a zoom lens for the cell. *FEBS Letters* 579(8):1854-58; Rousseau, F. and J. Schymkowitz. 2005. A systems biology perspective on protein structural dynamics and signal transduction. *Current Opinion in Structural Biology* 15(1):23-30.

[121] Apic, G. et al. 2005. Illuminating drug discovery with biological pathways.

FEBS Letters 579(8):1872-1877; Young, J.A. and E.A. Winzeler. 2005. Using expression information to discover new drug and vaccine targets in the malaria parasite Plasmodium falciparum. *Pharmacogenomics* 6(1):17-26.

[122] Hood, L. et al. 2004. Systems biology and new technologies enable predictive and preventative medicine. *Science* 306(5696):640-643.

[123] AfCS Nature; The Signalling Gateway; See www.signaling-gateway.org/update/updates/200201/nrn714.html.

[124] Balakrishnan, V.S. et al. 2005. Genomic medicine, gene polymorphisms, and human biological diversity. *Seminars in Dialysis* 18(1):37-40; Carr, K.M. et al. 2004. Genomic and proteomic approaches for studying human cancer: prospects for true patient-tailored therapy. *Human Genomics* 1(2):134-140.

[125] Guillermo Paez, J. et al. 2004. EGFR mutations in lung cancer: correlation with clinical response to gefitinib therapy. *Science* 304(5676):1497-1500.

[126] Marietti, C. 1999. Body Language; Health Care Informatics. *Healthcare Informatics* [Online]. Available online at www.healthcare-informatics.com/issues/1999/01_99/body.htm [accessed January 5, 2006].

[127] Workman, P. 2001. New drug targets for genomic cancer therapy: Successes, limitations, opportunities and future challenges. *Current Cancer Drug Targets* 1(1):33-47.

[128] Ibid.

[129] Schinder, T., W. Bornmann, P. Pellicena, W.T. Miller, B. Clarkson, J. Kuriyan. 2000. Structural mechanism for STI-571 inhibition of Abelson tyrosine kinase. *Science* 289(5486):1938-1941.

[130] Workman, P. 2001. New drug targets for genomic cancer therapy: Successes, limitations, opportunities and future challenges. *Current Cancer Drug Targets* 1(1):33-47.

[131] Daar, A.S. and P.A. Singer. 2005. Pharmacogenetics and geographical ancestry: implications for drug development and global health. *Nature Reviews. Genetics* 6(3):241-246.

[132] National Research Council/Institute of Medicine. 2005. *An International Perspective on Advancing Technologies and Strategies for Managing Dual-Use Risks.* Washington, DC: The National Academies Press.

[133] Ibid.

[134] While the vaccine was not one that would specifically target black as opposed to white people, it was clearly intended to be used to limit fertility in black women.

[135] Glasgow, J.N. et al. 2004. An adenovirus vector with a chimeric fiber derived from canine adenovirus type 2 displays novel tropism. *Virology* 324(1): 103-16.

[136] Nettelbeck, D.M. et al. 2004. Retargeting of adenoviral infection to melanoma: combining genetic ablation of native tropism with a recombinant bispecific single-chain diabody (scDb) adapter that binds to fiber knob and HMWMAA. *International Journal of Cancer* 108(1):136-45.

[137] Suzuki, T. et al. 2000. Adenovirus-mediated ribozyme targeting of HER-2/neu inhibits in vivo growth of breast cancer cells. *Gene Therapy* 7(3):241-248.

[138] Rein, D.T. et al. 2004. Gene transfer to cervical cancer with fiber-modified adenoviruses. *International Journal of Cancer* 111(5):698-704.

139 A company called DNAprint Genomics has identified a number of genetic markers that correlate highly with racial or ethnic designations, many of them having to do with metabolizing toxins found in foods that are indigenous to certain areas. The markers identified by this firm are used to provide quantitative measures of an individual's ancestry, according to four different "anthropological groups"—Native American; East Asian; Sub-Saharan Africa; and European. "European" can be broken down into Northern European; Southeastern European, Middle Eastern, and South Asian. For additional information on this company's "products" see www.dnaprint.com/welcome/, and in particular a related site, www.ancestrybydna.com/welcome/home/.

140 The Sunshine Project. 2003. Emerging Technologies: Genetic Engineering and Biological Weapons. Background Paper #12. Available online at www. sunshine-project.org/publications/bk/bk12.html#sec6 [accessed January 5, 2006].

141 Kagan, E. 2001 Bioregulators as instruments of terror. *Clinics in Laboratory Medicine* 21(3): 607-618. See also, Wheelis, M. 2004. Will the new biology lead to new weapons? *Arms Control Today* 34(July/August):6-13; and, Dando, M. 1999. *Biotechnology, Weapons, and Humanity.* British Medical Association. Amsterdam: Harwood Academic Publishers, especially Chapter 4 on "Genetic weapons." See also Dando, M. 1996. *A New Form of Warfare: The Rise of Non-Lethal Weapons.* Dulles, VA: Potomac Books, Inc., especially Chapters 5 and 8.

142 Wang, D. et al. 1999. Encapsulation of plasmid DNA in biodegradable poly(D, L-lactic-co-glycolic acid) microspheres as a novel approach for immuno-gene therapy. *Journal of Controlled Release* 57(1): 9-18; National Research Council/ Institute of Medicine. 2005. *An International Perspective on Advancing Technologies and Strategies for Managing Dual-Use Risks.* Washington, DC: The National Academies Press.

143 Neuropeptides, a type of bioregulator found in nervous system tissue, have a powerful modulatory effect on the nervous and immune systems.

144 Wheelis, M. 2002. Biotechnology and biochemical weapons. *The Nonproliferation Review* 9(Spring):48-53. Available online at cns.miis.edu/pubs/npr/vol09/ 91/91whee.pdf [accessed January 5, 2006].

145 Healy, M. 2004. Sharper minds. *Los Angeles Times* (December 20): F1; Tully, T. et al. 2003. Targeting the CREB pathway for memory enhancers. *Nature Reviews. Drug Discovery* 2(4):267-77.

146 National Research Council/Institute of Medicine. 2005. *An International Perspective on Advancing Technologies and Strategies for Managing Dual-Use Risks.* Washington, DC: The National Academies Press.

147 Ibid.

148 Arntzen, C. et al. 2005. Plant-derived vaccines and antibodies: potential and limitations. *Vaccine.* 23(15):1753-1756; Huang, Z. et al. 2005. Virus-like particle expression and assembly in plants: hepatitis B and Norwalk viruses. *Vaccine* 23(15): 1851-1858; Thanavala, Y. et al. 2005. Immunogenicity in humans of an edible vaccine for hepatitis B. *Proceedings of the National Academy of Sciences* 102(9):3378-3382.

149 National Research Council/Institute of Medicine. 2005. *An International Perspective on Advancing Technologies and Strategies for Managing Dual-Use Risks.* Washington, DC: The National Academies Press.

150 Petro, J.B. et al. 2003. Biotechnology: Impact on biological warfare and

biodefense. *Biosecurity and Bioterrorism: Biodefense Strategy, Practice, and Science.* 1(3):161-168.

[151] Walt, D.R. 2005. Miniature analytical methods for medical diagnostics. *Science* 308(5719):217-219.

[152] Ibid.

[153] This section is based on the workshop presentation of N. Seeman, in National Research Council/Institute of Medicine. 2005. *An International Perspective on Advancing Technologies and Strategies for Managing Dual-Use Risks.* Washington, DC: The National Academies Press.

[154] Seeman, N.C. and A.M. Belcher. 2002. Emulating biology: building nanostructures from the bottom up. *Proceedings of the National Academy of Sciences* 99(Suppl 2):6451-6455.

[155] Seeman, N.C. 1982. Nucleic acid junctions and lattices. *Journal of Theoretical Biology* 99(2): 237-247; Seeman, N.C. 1999. DNA engineering and its application to nanotech-nology. *Trends in Biotechnology* 17(11):437-443.

[156] Winfree, E. 1995. On the computational power of DNA annealing and ligation. In Lipton, R. and E. Baum, eds. 1995. *DNA Based Computers Volume 27, Proceedings of a DIMACS Workshop.* Am. Math. Society:199-215; and Adleman, L. 1994. Molecular computation of solutions to combinatorial problems. *Science* 266(5187): 1021-1024.

[157] Chen, Y. and C. Mao. 2004. Putting a brake on autonomous DNA nanomotor. *Journal of the American Chemical Society* 126(28):8626-8627; Emerich, D.F. 2005. Nanomedicine—prospective therapeutic and diagnostic applications. *Expert Opinion in Biological Therapy* 5(1):1-5.

[158] Kohli, P. and Martin, C.R. 2005. Smart nanotubes for biotechnology. *Current Pharmaceutical Biotechnology* 6(1):35-47; Kubik, T. et al. 2005. Nanotechnology on duty in medical applications. *Current Pharmaceutical Biotechnology* 6(1):17-33.

[159] www.ibn.a-star.edu.sg/news_interface_article.php?articleid=54.

[160] Fortina, P. et al. 2005. Nanobiotechnology: the promise and reality of new approaches to molecular recognition. *Trends in Biotechnology* 23(4):168-173; Patolsky, F. et al. 2004. Electrical detection of single viruses. *Proceedings of the National Academy of Sciences* 101(39):14017-14022.

[161] Warheit, D. et al. 2004. Comparative pulmonary toxicity assessment of single-wall carbon nanotubes in rats. *Toxicological Sciences* 77:117-125; Oberdorster, E. 2004. Manufactured nanomaterials (fullerenes, C60) induce oxidative stress in the brain of juvenile largemouth bass. *Environmental Health Perspectives* 112(10):1058-62.

[162] Main, C.E. 2003. Aerobiological, ecological, and health linkages. *Environment International* 29(2-3):347-349.

[163] Bauce, E. et al. 2004. *Bacillus thuringiensis* subsp. *kurstaki* aerial spray prescriptions for balsam fir stand protection against spruce budworm (Lepidoptera: Tortricidae). *Journal of Economic Entomology* 97(5):1624-1634.

[164] Martonen, T.B. and J.D. Schroeter. 2003. Risk assessment dosimetry model for inhaled particulate matter: I. human subjects. *Toxicology Letters* 138(1-2): 119-132.

[165] Paez-Rubio, T. et al. 2005. Source bioaerosol concentration and rRNA gene-

based identification of microorganisms aerosolized at a flood irrigation wastewater reuse site. *Applied and Environmental Microbiology* 71(2):804-810.

[166] Tanner, B.D. et al. 2005. Bioaerosol emission rate and plume characteristics during land application of liquid class B biosolids. *Environmental Science and Technology* 39(6):1584-90; Brooks, J.B. et al. 2005. Estimation of bioaerosol risk of infection to residents adjacent to a land applied biosolids site using an empirically derived transport model. *Journal of Applied Microbiology* 98(2):397-405.

[167] Brown, J.R. et al. 2005. Aerial optimization and canopy penetration study of Dibrom 14 Concentrate. *Journal of the American Mosquito Control Association* 21(1): 106-113.

[168] Chan, H.K. 2003. Inhalation drug delivery devices and emerging technologies. *Expert Opinion on Therapeutic Patents* 13(9):1333-1343.

[169] Edwards, D. 2002. Delivery of biological agents by aerosols. *American Institute of Chemical Engineers Journal* 48(1):2-6.

[170] LiCalsi, C., M.l Maniaci, T. Christensen, E. Phillips, G.H. Ward, and C. Witham. 2001. A powder formulation of measles vaccine for aerosol delivery. *Vaccine* 19(17-19):2629-2636. The authors describe a method to deliver live, attenuated, measles vaccine via the lungs. "In this study, live attenuated measles vaccine is micronized by jet milling to generate particle sizes appropriate for pulmonary delivery (1-5 μm). Milling does not induce detectable physical change and significant viral potency is maintained. . . . The measles vaccine formulation is dispersible . . ."

[171] Clark, A.R. 1995. Medical aerosol inhalers: past, present, and future. *Journal of Aerosol Science and Technology* 22:374-381.

[172] Chan, H.K. 2003. Inhalation drug delivery devices and emerging technologies. *Expert Opinion on Therapeutic Patents* 13(9):1333-1343.

[173] Crowder, T.M. 2002. Fundamental effects of particle morphology on lung delivery: predictions of Stokes' Law and the particular relevance to dry powder inhaler formulation and development. *Pharmaceutical Research* 19(3):239-245.

[174] Garcia-Contreras, L. and H.D.C. Smyth. 2005. Liquid-spray or dry-powder systems for inhaled delivery of peptide and proteins? *American Journal of Drug Delivery* 3(1):29-45.

[175] Chan, H.K. 2003. Inhalation drug delivery devices and emerging technologies. *Expert Opinion on Therapeutic Patents* 13(9):1333-1343.

[176] Tan, H.S. and S. Borsadia. 2001. Particle formation using supercritical fluids: pharmaceutical applications. *Expert Opinion on Therapeutic Patents* 11(5):861-872.

[177] Based on Elliott Kagan's presentation at the Cuernavaca workshop. See National Research Council/Institute of Medicine. 2005. *An International Perspective on Advancing Technologies and Strategies for Managing Dual-Use Risks.* Washington, DC: The National Academies Press.

[178] Wang, D. et al. 1999. Encapsulation of plasmid DNA in biodegradable poly(D,L-lactic-co-glycolic acid) microspheres as a novel approach for immunogene delivery. *Journal of Controlled Release* 57(1):19-18.

[179] Dai, C. et al. 2005. Microencapsulation peptide and protein drugs delivery system. *Colloids and Surfaces B: Biointerfaces* 41(2-3):117-20.

[180] See www.controlledrelease.org/about/index.cgi [accessed May 12, 2005].

[181] See www.swri.org/4org/d01/microenc/microen/default.htm [accessed May 12, 2005].

[182] Randall, P. and S. Chattopadhyay. 2004. Advances in encapsulation technologies for the management of mercury-contaminated hazardous wastes. *Journal of Hazardous Materials* 114(1-3):211-223.

[183] Hao, S. et al. 2005. A novel approach to tumor suppression using microencapsulated engineered J558/TNF-alpha cells. *Experimental Oncology* 27(1):56-60.

[184] Weinbreck, F. et al. 2004. Microencapsulation of oils using whey protein/gum Arabic coacervates. *Journal of Microencapsulation* 21(6):667-679.

[185] Yamaguchi, Y. et al. 2005. Successful treatment of photo-damaged skin of nano-scale atRA particles using a novel transdermal delivery. *Journal of Controlled Release* 104(1):29-40.

[186] See www.advancedbionutrition.com/html/news_press.html#2005_5 [accessed May 12, 2005].

[187] Chang, T.M. 2005. Therapeutic applications of polymeric artificial cells. *Nature Reviews. Drug Discovery* 4(3):221-235; and Orive, G. et al. 2004. History, challenges and perspectives of cell microencapsulation. *Trends in Biotechnology* 22(2):87-92.

[188] A new technology, which will allow weapons and vehicles to be released from submarines even if they were not originally designed for undersea use.

[189] Izumikawa, M. et al. 2005. Auditory hair cell replacement and hearing improvement by Atoh1 gene therapy in deaf mammals. *Nature Medicine* 11(3):271-276.

[190] Parsons, D. 2005. Airway gene therapy and cystic fibrosis. *Journal of Paediatrics and Child Health* 41(3):94-96.

[191] Although "convergent technology" is a common term often used to refer to the convergence of specific types of technologies, we use it here loosely to refer to the convergence of *any* technologies.

[192] Benenson, Y. et al. 2004. An autonomous molecular computer for logical control of gene expression. *Nature* 429(6990):423-429.

[193] Kagan, E. 2001 Bioregulators as instruments of terror. *Clinics in Laboratory Medicine* 21(3): 07-618. See also, Wheelis, M. 2004. Will the new biology lead to new weapons? *Arms Control Today* 34(July/August):6-13; and Dando, M. 1999. *Biotechnology, Weapons, and Humanity* British Medical Association. Amsterdam: Harwood Academic Publishers, especially Chapter 4 on "Genetic weapons." See also Dando, M. 1996. *A New Form of Warfare: The Rise of Non-Lethal Weapons.* Dullas, VA: Potomac Books, Inc., especially Chapter 8: "An assault on the brain?" and Chapter 5: "Lethal and non-lethal chemical agents."

[194] Ibid.

[195] Nixdorff, K. and W. Bender. 2002. Ethics of university research, biotechnology and potential military spin-off. *Minerva* 40:15-35. See also Nixdorff, K., N. Davison, P. Millett, and S. Whitby. 2004. Technology and biological weapons: Future threats. *Science and Technology Report*, Number 2, University of Bradford, Department of Peace Studies. Available online at www.brad.ac.uk/acad/sbtwc/ST_Reports/ST_Report_No_2.pdf [accessed January 5, 2006].

4

Conclusions and Recommendations

Future applications of advances in the life sciences and related technologies are likely to have a profound impact on human health and well-being, as well as promote the efficiency of crop production and animal husbandry. Continuing advances in biotechnology hold promise for improved nutrition, a cleaner environment, a longer, healthier life span, and cures for many once-formidable diseases. Even older technologies, such as classic methods for vaccine manufacture, have enabled the eradication or reduction of many once-dreaded diseases such as smallpox, poliomyelitis, diphtheria, tetanus, and whooping cough. Newer reverse genetic technologies for RNA viruses may facilitate the rapid, rational development of vaccines for such agents. In the developing world, broader application of biotechnology may make it economically feasible for resource-limited countries to produce vaccines locally that are capable of protecting their populations against endemic infectious diseases but for which there is little or no economic incentive for large multinational vaccine producers. In addition to improved health, world agriculture stands to benefit greatly from new discoveries in the life sciences and growing technological capabilities.

To a considerable extent, new advances in the life sciences and related technologies are being generated not just domestically, but also internationally. The preeminent position that the United States has enjoyed in the life sciences has been dependent on the flow of foreign scientific talent to its shores and is now threatened by the increasing globalization of science and the international dispersion of a wide variety of related technologies. The increasing pace of scientific discovery abroad and the fact that the United States may no longer hold a monopoly on these leading technologies means that this country is, as never before, dependent on international collaboration, a theme explored in depth in Chapter 2.

Although this report is concerned with the evolution of science and technology capabilities over the next 5 to 10 years with implications for next-generation threats, it is clear that today's capabilities in the life sciences and related technologies have already changed the nature of the biothreat "space." The accelerating pace of discovery in the life sciences has fundamentally altered the threat spectrum. Some experts contend that bioregulators, which are small, biologically active compounds, pose an increasingly apparent dual-use risk. This risk is magnified by improvements in targeted-delivery technologies that have made the potential dissemination of these compounds much more feasible than in the past. The immune, nervous, and endocrine systems are particularly vulnerable to bioregulator modification.

The growing concern regarding novel types of threat agents does not diminish the importance of naturally occurring threat agents—for example, the "classic" category A select agents—or "conventionally" genetically engineered pathogenic organisms. However, it does mandate the need to adopt a broader perspective in assessing the threat, focusing not on a narrow list of pathogens, but on a much wider spectrum that includes biologically active chemical agents. The potential threat spectrum is thus exceptionally broad and continuously evolving—in some ways predictably, in other ways unexpectedly. The viruses, microbes, and toxins listed as "select agents" and on which our biodefense research and development activities are so strongly focused today are just one aspect of this changing landscape of threats. Although some of them may be the most accessible or apparent threat agents to a potential attacker, particularly one lacking a high degree of technical expertise, this situation is likely to change as a result of the increasing globalization and international dispersion of the most cutting-edge aspects of life sciences research.

The committee has proposed a conceptual framework in Chapter 3 for how to think about the future threat landscape. The task here will be never ending, and as the world becomes more competent and sophisticated in the biological sciences, it is vitally important that the national security, public health, and biomedical science communities have the knowledge and tools to address both beneficial and harmful applications of advances in the life sciences.

.   In interpreting its charge, the committee sought to examine current trends and future objectives of research in the life sciences, focusing particularly on applications that might be relevant to the development of "next-generation" agents of biological origin 5 to 10 years into the future.

While the committee understood that readers of this report might hope to find a well-defined list or set of lists of future threats, perpetrators, and timelines for the acquisition and exploitation of certain technologies for malevolent purposes, the committee also realized the futility of this approach. The global technology landscape is shifting so dramatically and rapidly that it was simply not possible for this committee—or any committee—to devise a formal risk assessment of the future threat horizon, based on the possible exploitation of dual-use technologies by state actors, nonstate actors, or individuals. Given that within just the past few years the global scientific community has witnessed the unexpected development and proliferation of important new technologies, such as RNA interference, nanobiotechnology, and synthetic biology, biological threats of the next 5 to 10 years could extend well beyond those that can be predicted today. The useful life span of any such list of future threats developed in 2006 would likely be measured in months, not years. Instead, the committee sought to define more broadly how continuing advances in technologies with applications to the life sciences' enterprise can contribute to the development of novel biological weapons and to develop a logical framework for analysts to consider as they evaluate the evolving technology threat spectrum.

While evaluating the rapidly evolving global landscape of knowledge and capability in the life sciences and associated technologies, the committee agreed on five key findings and recommendations that it believes are strongly supported by the information presented in this report, as summarized in Box 4-1, that build on and reinforce the findings and recommendations put forward in earlier National Research Council reports, including, but not limited to, *Biotechnology Research in an Age of Terrorism*.[1] Because it believes that continuing advances in science and technology are essential to countering terrorism, the committee's recommendations affirm policies and practices that promote the free and open exchange of information in the life sciences (Recommendation 1). The committee also recognized the need to adopt a broader perspective on the nature of the "threat spectrum" (Recommendation 2) and to strengthen the scientific and technical expertise available to the security communities so that they are better equipped to anticipate and manage a diverse array of novel threats (Recommendation 3). The recommendations call for the global community of life scientists to adopt a common culture of awareness and a shared sense of responsibility and include specific actions that would promote such a culture (Recommendation 4). Finally, the committee recognized that no set of measures can ever provide complete protection against the malevolent use of life sciences technologies, and its recommendations reaffirm previous calls to strengthen the public health infrastructure and the nation's existing response and recovery capabilities (Rec-

BOX 4-1
Recommendations

1. **The committee endorses and affirms policies and practices that, to the maximum extent possible, promote the free and open exchange of information in the life sciences.**

1a. Ensure that, to the maximum extent possible, the results of fundamental research remain unrestricted except in cases where national security requires classification, as stated in National Security Decision Directive 189 (NSDD-189) and endorsed more recently by a number of groups and organizations.

1b. Ensure that any biosecurity policies or regulations implemented are scientifically sound and are likely to reduce risks without unduly hindering progress in the biological sciences and associated technologies.

1c. Promote international scientific exchange(s) and the training of foreign scientists in the United States.

2. **The committee recommends adopting a broader perspective on the "threat spectrum."**

2a. Recognize the limitations inherent in any agent-specific threat list and consider instead the intrinsic properties of pathogens and toxins that render them a threat and how such properties have been or could be manipulated by evolving technologies.

2b. Adopt a broadened awareness of threats beyond the classical "select agents" and other pathogenic organisms and toxins, so as to include, for example, approaches for disrupting host homeostatic and defense systems and for creating synthetic organisms.

3. **The committee recommends strengthening and enhancing the scientific and technical expertise within and across the security communities.**

3a. Create by statute an independent science and technology advisory group for the intelligence community.

3b. The best available scientific expertise and knowledge should inform the concepts, plans, activities, and decisions of the intelligence, law enforcement, homeland security, and public policy communities and the national political leadership about advancing technologies and their potential impact on the development and use of future biological weapons.

3c. Build and support a robust and sustained cutting-edge analytical capability for the life sciences and related technologies within the national security community.

3d. Encourage the sharing and coordination, to the maximum extent possible, of future biological threat analysis between the domestic national security community and its international counterparts.

4. The committee recommends the adoption and promotion of a common culture of awareness and a shared sense of responsibility within the global community of life scientists.

4a. Recognize the value of formal international treaties and conventions, including the 1972 Biological and Toxin Weapons Convention (BWC) and the 1993 Chemical Weapons Convention (CWC).

4b. Develop explicit national and international codes of ethics and conduct for life scientists.

4c. Support programs promoting beneficial uses of technology in developing countries.

4d. Establish globally distributed, decentralized, and adaptive mechanisms with the capacity for surveillance and intervention in the event of malevolent applications of tools and technologies derived from the life sciences.

5. The committee recommends strengthening the public health infrastructure and existing response and recovery capabilities.

5a. Strengthen response capabilities and achieve greater coordination of local, state, and federal public health agencies.

5b. Strengthen efforts related to the early detection of biological agents in the environment and early population-based recognition of disease outbreaks, but deploy sensors and other technologies for environmental detection only when solid scientific evidence suggests they are effective.

5c. Improve the capabilities for early detection of host exposure to biological agents, and early diagnosis of the diseases they cause.

5d. Provide suitable incentives for the development and production of novel classes of preventative and therapeutic agents with activity against a broad range of biological threats, as well as flexible, agile, and generic technology platforms for the rapid generation of vaccines and therapeutics against unanticipated threats.

ommendation 5). All of the insight and capabilities generated by advances in the life sciences and related technologies must be brought to bear on the problem of building a more robust public health defense.

The committee could not envision any sort of "silver bullet" capable of providing absolute protection against the malevolent application of new technologies. Rather, the actions and strategies recommended here are intended to be complementary and synergistic. An effective system for managing the threats that face society will require a broad array of mutually reinforcing actions in a manner that successfully engages the variety of different communities that share stakes in the outcome. As in fire prevention, where the best protection against the occurrence of damage from catastrophic fires comprises a multitude of interacting preventive and mitigating actions (e.g., fire codes, smoke detectors, sprinkler systems, fire trucks, fire hydrants, fire insurance), rather than any single "best" but impractical or improbable measure (e.g., stationing a fire truck on every block), the same is true here. The committee envisions a broad-based, intertwined network of steps—a *web of protection*—for reducing the likelihood that the technologies discussed in this report will be used successfully for malevolent purposes.

While recognizing that all of its recommended measures, taken together, provide no guarantee that continuing advances in the life sciences and the new technologies they spawn will not be used with the intent to cause harm, the committee members agreed that implementation of these recommendations in aggregate will likely decrease the risk of inappropriate application or unintended misuse of increasingly widely available knowledge and technologies, favor the early detection of malevolent applications, and mitigate the loss of life or other damage sustained by society in both the short and the long term, should the worst-case scenario occur.

CONCLUSION 1: THE COMMITTEE CONCLUDES THAT THERE IS A NEED TO MAINTAIN FREE AND OPEN EXCHANGE OF SCIENTIFIC AND TECHNOLOGICAL INFORMATION.

In general, restrictive regulations and the imposition of constraints on the flow of information are not likely to reduce the risks that advances in the life sciences will be utilized with malevolent intent in the future. In fact, they will make it more difficult for civil society to protect itself against such threats and ultimately are likely to weaken national *and* human security.[2] Such regulations and constraints would also limit the tremendous potential for continuing advances in the life sciences and its related technologies to improve health, provide secure sources of food and energy, contribute to economic development in both resource-rich and resource-poor parts of the world, and enhance the overall quality of human life. In the past, society has gained from advances in the life sciences because of the open exchange of data and concepts.

One of the main challenges to the committee was to formulate measures that would continue to benefit human development[3] while taking into account legitimate national security needs. The goal of the committee was to ensure that scientific progress and industrial development advance expeditiously while not unduly aiding state or nonstate actors that may wish to exploit these tools and technologies for malevolent purposes. The recommendations put forth in this section consider policies and actions that are balanced with respect to national security needs and the multiple and varied beneficial applications of science and technology.

Recommendation 1

The committee endorses and affirms policies and practices that, to the maximum extent possible, promote the free and open exchange of information in the life sciences.

The many ways that biological knowledge and its associated technologies have improved and can continue to improve biosecurity, health, agriculture, and other life sciences industries are highlighted in Chapter 2. Reducing or restricting the open exchange of information would over time reduce the ability of the United States to remain competitive in the global marketplace and to build robust defenses against future potential biological threats. Equally important, it would deny many individuals, both within and outside the United States, the tremendous health, agricultural, and other benefits likely to be derived from advanced technologies. The committee's recommendation has three components. The first focuses on the openness of information generated from fundamental scientific research, the second concerns policies and regulations, and the third relates to international exchanges between scientists who are working in the life sciences.

Recommendation 1a. Ensure that the results of fundamental research remain unrestricted except in cases where national security requires classification, as stated in National Security Decision Directive (NSDD)-189 and endorsed more recently by a number of groups and organizations.

Like all sciences, the life sciences have relied on a culture of openness in research, where the free exchange of information and ideas allows researchers to build on the results of others, while simultaneously opening scientific results to critical scrutiny so that mistakes can be recognized and corrected sooner rather than later. Recent and proposed changes in the existing classification system threaten this culture in ways that are potentially harmful to national and human security. For example, the recent extension of classification authority to agencies not previously involved in these matters (e.g., the U.S. Department of Agriculture,[4] the Environmental Protection Agency,[5] and the U.S. Department of Health and Human Services[6]) raises questions about the criteria for classification that might be applied to federally funded research. Under the current system, in most agencies the task of applying classification standards is so large that information classification authority has been delegated to literally thousands of government officials. While detailed guides purport to offer classification standards, the usefulness of these standards as clear and objective tools fails and subjectivity intervenes when the subject matter and associated risks—and the direct and indirect costs of overclassification—are less well defined or understood.[7]

In 2002, a draft U.S. Department of Defense regulation, if enacted, would have required researchers "to obtain DoD approval to discuss or publish findings of all military-sponsored unclassified research."[8] Such a process would have made it possible for the department to prevent any of its funded life sciences research that it considered "sensitive" (because it could theoretically aid terrorists or be used in the production of biological weapons) from entering the public domain, thus in effect allowing the department to treat it as secret. The draft was withdrawn in the face of considerable criticism from the scientific research community.[9] This proposal reflected the current opinion of some that government control should go further than the regulations imposed by "The Uniting and Strengthening America by Providing Appropriate Tools Required to Intercept and Obstruct Terrorism of October 2001" Act (i.e., the PATRIOT Act[10]), and "The Public Health Security and Bioterrorism Preparedness and Response Act" (i.e., the Bioterrorism Response Act[11]), and that it should include broad controls on the dissemination of the results of scientific research. In fact, current controls on dissemination of research now in place within the pharmaceutical industry or at many federal institutions, including, for example, Los Alamos National Laboratory, do extend beyond current regulations for publicly funded research. There are many reasons why such restrictions may be imposed, such as a desire to confirm scientific validity of reports through an independent review process prior to public release or a desire to protect intellectual property important to commercialization. Restrictions based on a desire to prevent the potential misuse of information are more problematic. Proposals to institute such controls on basic, fundamental research reflect a long-standing tension between those who believe that limiting the dissemination of such information may provide a margin of safety and others who believe that the free and open exchange of fundamental research results is critical for maintaining the technological and scientific prowess and agility required for a robust national security enterprise.[12]

As discussed in Chapter 1, it would be beyond the scope of this committee's charge to evaluate and articulate recommendations regarding the U.S. system of data and information classification and other means of data and information control (e.g., categorizing information as "sensitive but unclassified"). However, the committee did recognize the limits of any such system (as it currently exists or otherwise) with respect to its ability to control the practically immeasurable amount of data and information already extant in the public domain (e.g., freely available on the Internet) and/or generated by non-U.S.-funded sources. The U.S. classification system primarily applies to work done in government laboratories or that is funded by the government. It does not extend to the vast, growing, and increasingly accessible global knowledge base being built by private interests or in foreign countries. For example, as detailed in Chapter 2, China, France, Germany, Japan, the United Kingdom, and other countries are now making proportionately greater contributions to the scientific literature and knowledge base than they did in the past (i.e., the entire scientific literature and knowledge base, including but not limited to the life sciences).

Although many consider that a restrictive approach has been largely successful in slowing the proliferation of nuclear weapons technology, many of the same conditions do not apply regarding matters involving the potential "dual use" of life sciences research and living organisms. Various arguments suggest that overly restrictive regulations on the conduct and funding of research, the dissemination of research results, and the industrial development of biotechnology will not prevent state and non-state actors from gaining access to and using advances in research to develop novel agents of biological origin. Some of these arguments are presented below.

First, and perhaps most important, efforts to restrict the flow of information in the life sciences are likely to impede the ability of the scientific establishment to keep ahead of potential threats. Not only is an open exchange necessary for the recognition of potential threats in advance of their realization, but it is also essential for the creation of effective countermeasures. Science does not advance in a linear fashion, and typically it is a development in an unrelated field that suggests a novel approach to a particular problem. Great advances often come from the seemingly random blending of technical approaches and theoretical insights from different fields, as, for example, the application of semiconductor chip manufacturing technologies to the development of ultradense DNA oligonucleotide microarrays synthesized in situ. Another example is the application of knowledge concerning transcriptional silencing in plants to the development of novel therapeutics for humans. Such associations, in effect a convergence of technologies and unrelated hypotheses, cannot be predicted effectively in advance, nor managed in any directed fashion. They require the dissemination of the requisite information bits into the scientific ferment, and the maintenance of an open environment for research—including the open publication of research results. Such an environment will hasten the development of effective countermeasures against biological threats. An open approach to the dissemination of information may also aid intelligence and law enforcement agencies in their efforts to predict, assess, and deter the malevolent applications of new tools and technologies as they arise.

Second, unlike research relevant to the design of nuclear weapons, the fields of research in the life sciences with potential dual-use applications cover a very broad range of disciplines (see Chapter 3 for some examples) and a large number of individuals and institutions. Potential dual-use applications may only become apparent long after an initial discovery. Undoubtedly, the majority of life sciences research would probably be of little interest to state-level offensive weapons programs or nonstate actors. Nonetheless, life sciences research is being pursued for a variety of purposes: improved prevention, diagnosis, and treatment of human and animal diseases; enhanced production of food and energy; environmental remediation; and even microfabrication of electronic circuits. It is likely that some work in each of these diverse areas offers significant dual-use possibilities. Thus, the range and number of scientists and institutions that would be affected by any attempt to impose new information controls would be vast and difficult to list, let alone monitor. The magnitude of the task becomes even more daunting given the lack of any international body that is in a position to assume responsibility for this on a global scale, even if all parties involved were agreeable with such controls, which is highly unlikely to be the case.

Third, the financial costs associated with any regime aimed at restricting the flow of information would be very high. An estimate of the costs of the U.S. nuclear weapons program between 1940 and 1996 suggests a rough figure of $1 trillion for protecting the secrecy of classified information.[13] The costs that would be involved in any attempts to control information related to biological research would include those involved in the screening of personnel; acquisition of secure storage facilities, guards, materials management, and maintenance of routine inventories of controlled material.[14] These financial costs would need to be borne by academic, commercial, and governmental institutions complying with government regulations. While U.S. institutions engaged in research on category A select agents already bear the costs of security screening of personnel, maintenance of secure storage facilities, guards, and routine maintenance of inventories of controlled material to meet current U.S. regulations, any program aimed at regulating the dissemination of results of potential dual-use research in the biological sciences would of necessity need to be much, much broader in scope, and thus would be enormously more expensive. Importantly, as discussed below in Recommendation 2, any regime designed to control information related to research on the currently listed select agents only would miss most potential dual-use developments that are likely to emerge in the life sciences over the next decade or more. It is unlikely that many foreign nations would consider such costs justifiable, and any regime adopted only within the United States in isolation from the remainder of the global life sciences would be futile and likely counterproductive.

Fourth, history has demonstrated that efforts to impose restrictions on the flow of information are generally unrealistic and may lead to a black market that is much more difficult to monitor and oversee than an open market.[15, 16] In particular, this is very likely to be the case in the life sciences where large international networks of scientists in many specific fields of research have been accustomed to the free and rapid exchange of information. A recent request by the U.S. Department of Health and Human Services (HHS) to suppress or alter the publication of a recent paper accepted by the *Proceedings of the National Academy of Sciences USA* concerning the risks of botulinum toxin being introduced into the U.S. milk supply rapidly led to a much larger awareness of the manuscript on the part of the biomedical research community and the greater public.[17, 18] Indeed, some of the information contained in the manuscript in question was published six weeks earlier in the *New York Times*,[19] demonstrating the determination of some authors to share the results of their research with the public and the willingness of the lay press to disseminate such putatively "sensitive" information if it is considered sufficiently newsworthy (as many new dual-use developments are likely to be). While some aspects of this particular incident may be unusual, it demonstrates the difficulties inherent in attempting to control information generated by the academic sector in the United States, from which many if not most novel developments in the life sciences currently emerge.

The nature of the biological research enterprise is very different from that of research related to nuclear weapons and even more different than fundamental nuclear physics research: Most of the world already has access to and cannot possibly be denied further access to the knowledge, materials, and equipment necessary for developing or disseminating potentially "dual use" knowledge in the life sciences. Unlike the relative U.S. monopoly on nuclear knowledge and technology during the early years of the Cold War, today's advancing technologies with biological dual-use potential are for the most part beyond the reach of U.S. regulations and influence. In many ways, the "genie is out the bottle" and it is difficult to envision how it can be put back in.

Fifth, unlike nuclear physics, there is no accepted culture of secrecy or control on the flow of research information in the life sciences. The very real dual-use potential of atomic physics was made forcefully evident 60 years ago with the explosion of the first nuclear device at the Trinity Test site in Alamagordo, New Mexico. In contrast, despite a long global history of limited attempts to use biological agents as weapons, realization of the magnitude of the dual-use implications of advances in biotechnology and the life sciences has come only much more recently. No doubt, this reflects the pace and timing of advances in these two fields over the past century. High-end research in the life sciences is also generally much less capital intensive than research in nuclear physics. The result is that the research culture in the life sciences is generally one that has been historically open, international in scope, and widely distributed.

Few molecular biologists or biomedical research institutions have any experience with classified, or secret, research. There are nationally operated biological research laboratories with missions focusing on defense against potential biological agents [such as the United States Army Medical Research Institute for Infectious Diseases (USAMRIID[20]) at Fort Detrick in the United States or Porton Down in the United Kingdom], but no laboratories exist that have openly declared missions to develop, test, or stockpile biological weapons—all of which of would be in flagrant violation of the biological weapons convention. This is in stark contrast to research activities in nuclear physics, which while also encompassing a great deal of open, fundamental research historically have included the use of facilities that are openly acknowledged to have missions related to the development and testing of nuclear weapons. Such laboratories, involved in the production and processing of special nuclear materials, are operated with special security clearances that restrict access to a small number of scientists and technicians who, in effect, constitute a closed society.[21] To be clear, the work at these laboratories is a small subset of all research in nuclear physics, which as a field has strongly embraced openness in the conduct and dissemination of its science.

Current U.S. laws and policies have the potential of creating de facto "closed" facilities focusing on research with category A select agents. Although these select agent laboratories, both in the federal government and the academic sector, have been and would be focused on the development of purely defensive countermeasures against perceived biological threat agents, the restrictions surrounding access to the laboratories has raised concerns in some quarters that some research may be more offensive in nature.[22] Such fears and misperceptions can only be alleviated by policies that promote transparency and encourage widespread dissemination of the research results generated in such facilities.

Although the principles that underlie the design of nuclear warheads are well understood by scientists around the world, details of nuclear weapons design remain largely classified. Except during the period following World War II, when the United States managed a mostly classified offensive biological weapons program, biology has enjoyed a long history of openness and free exchange of materials, personnel, and ideas. As the evidence presented in Chapter 2 strongly demonstrates, this open exchange of information is necessarily and increasingly global. Both basic research and the development of commercial products utilizing life sciences technology increasingly involves international collaboration and partnerships, many of which are outside of formal bilateral controls. Changing the open nature of the life sciences culture, or attempting to change it, could have unintended consequences by discouraging graduate students and postdoctoral researchers—in many cases the best minds engaged in rapidly developing fields—from becoming involved in restricted fields or even communicating with people who are involved in those fields, as major universities that accept classified research usually create separate facilities where access is limited and controlled.[23]

While not classified, research on category A select agents now requires special security safeguards that are both unusual and unsettling for the academic research centers that are being asked to pursue this research by the National Institutes of Health (NIH) and other federal agencies in the search for better countermeasures against possible bioterrorism attacks.[24] These include background security checks on all personnel involved, tightly restricted access to the laboratories involved, and in some institutions the presence of an armed security force "24/7." Such measures, irrespective of their degree of merit or utility, are likely to segregate a group of research scientists from their peers and perhaps make the recruitment of the best and brightest to an important enterprise more difficult.

Despite all of the above, the committee recognized that in relatively rare instances there may still exist a need for the U.S. government and the larger scientific community to impose some restrictions on the conduct of research and/or on the publication of results, a point made by the Fink committee in its 2004 report, *Biotechnology Research in an Age of Terrorism.*[25] The committee labored without much success to define such circumstances. Explicit, specific, detailed "recipes" concerning how to make and deliver a weapon might certainly be worthy of attempts to suppress dissemination. However, defining what specifically constitutes such a "recipe" is difficult. Research designed to create or exploit a critical host vulnerability for which no countermeasures are available would trigger review under recommendations 2, 3, and 4 of the previously cited Fink committee report. The potential value of focusing on such "functional" criteria for defining problematic research should be further explored (e.g., research that deliberately seeks to exploit critical public health vulnerabilities). Of course, in some cases, proprietary interests may dictate that information be kept confidential.

The recent *PNAS* publication alluded to above has been considered by some to be a roadmap for the introduction of botulinum toxin into the U.S. milk supply. Although opinions are split, there were cogent reasons to support its publication. While only time will tell whether the work by Wein et al. was beneficial or detrimental to the security of the milk supply, its formal publication was likely moot. The analysis described in the manuscript had been previously presented, and the manuscript itself was widely disseminated in advance of its publication, highlighting once again the difficulties inherent in attempts to suppress information in the "Internet era."

The scientific and intelligence communities will need to define narrowly those "special circumstances" when classification is warranted and allow public scrutiny of the process used to arrive at those definitions. The scientific and intelligence communities will also need to devise effective methods to keep a close hold on information that truly needs to be kept secret. At the same time, these communities need to maintain an environment that promotes the advancement of science and technology both domestically and globally. As the committee completes its report, it notes that HHS Secretary Leavitt has formally established the National Science Advisory Board for Biosecurity (NSABB), following a recommendation in the Fink report, and that the board has begun its work. The NSABB has among its charges the development of specific guidelines to meet these challenges.

The committee, therefore, strongly reaffirms the principles embodied in NSDD-189 (Box 4-2), which defines the national policy for controlling the flow of science, technology, and engineering information produced in federally funded fundamental research at academic institutions, governmental and nongovernmental facilities, and private laboratories receiving federal funds. Issued by President Reagan on September 21, 1985, NSDD-189 has not been superseded and continues to be the official U.S. government policy. Indeed, then Assistant to the President for National Security Affairs, Condoleeza Rice, reaffirmed NSDD-189 on November 1, 2001, in a letter to Harold Brown of the Center for Strategic and International Studies. As she stated, "This Administration will review and update as appropriate the export control policies that affect basic research in the United States. In the interim, the policy on the transfer of scientific, technical, and engineering information set forth in NSDD-189 shall remain in effect." The director of the Office of Science and Technology Policy, John

---

BOX 4-2
NSDD-189

NSDD-189 states that, "to the maximum extent possible, the products of fundamental research remain unrestricted. It is also the policy of this Administration that, where the national security requires control, the mechanism for control of information generated during federally funded fundamental research in science, technology and engineering at colleges, universities and laboratories is classification. Each federal government agency is responsible for: a) determining whether classification is appropriate prior to the award of a research grant, contract, or cooperative agreement and, if so, controlling the research results through standard classification procedures; b) periodically reviewing all research grants, contracts or cooperative agreements for potential classification. No restriction may be placed upon the conduct or reporting of federally funded fundamental research that has not received national security classification, except as provided in applicable U.S. Statutes."

NSDD-189 defines fundamental research as "basic and applied research in science and engineering, the results of which ordinarily are published and shared broadly within the scientific community, as distinguished from proprietary research and from industrial development, design, production, and product utilization, the results of which ordinarily are restricted for proprietary or national security reasons."

Marburger, reaffirmed this position in a talk at the National Academy of Sciences on January 9, 2003.[26] A number of recent publications and statements by other organizations also endorse the principles set forth in NSDD-189.[27]

Recommendation 1b. Ensure that any biosecurity policies or regulations implemented are scientifically sound and are likely to reduce risks without unduly hindering progress in the life sciences and associated technologies.

Although the regulatory environment for life sciences research has evolved over the course of several decades, the United States is witnessing a rapid transition from a scientific environment based on voluntary compliance with recommended practices to one based on the imposition and enforcement of statutes and regulations, particularly with respect to the control of biological materials and personnel, leading in some cases to the imposition of criminal penalties and sanctions. The high-profile case brought against an infectious disease research scientist, Dr. Thomas

Butler, by the federal government following his self-disclosure of missing plague bacillus inventory provides a stark example of the changes wrought since the terrorist attacks of September 11, 2001.[28] Now serving a prison sentence following his conviction on several counts unrelated to his handling of *Yersinia pestis*, his actions, however inappropriate, are unlikely to have prompted such a response in prior years. Although the transition started before the terrorist attacks of 9/11 (e.g., the 1996 Antiterrorism and Effective Death Penalty Act enacted new regulatory controls regarding transfers of dangerous pathogens), two major pieces of relevant legislation were passed into law less than a year after the attacks on the World Trade Center and the Pentagon, and subsequent anthrax mailings: "The Uniting and Strengthening America by Providing Appropriate Tools Required to Intercept and Obstruct Terrorism of October 2001" Act (i.e., the PATRIOT Act)[29] and "The Public Health Security and Bioterrorism Preparedness and Response Act" (i.e., the Bioterrorism Response Act).[30] These new security provisions have radically transformed the research environment for those who work with category A select agents in the United States from one that was traditionally open to one that is highly restricted and regulated in a number of ways. Of note, the PATRIOT and Bioterrorism Response Acts represent only 2 of 17 bioterrorism bills introduced by the 107th Congress (2001-2002) with potential ramifications for the research scientists who are funded by NIH to work on these agents and on whom the nation is dependent, in part, for the development of effective vaccines, therapies, and related diagnostics.[31]

Here, the committee emphasizes that these and any additional related proposed policies or regulations must be carefully and scientifically evaluated to ensure that they do more good than harm. Examples of regulations and policies that may potentially do more harm than good include the extension of the "deemed export" regulations under the Export Administration Act to information exchanges in the life sciences and unnecessarily onerous VISA requirements for foreign scientists to study and work in the United States.

An additional example is the extension of security provisions for select agents in the PATRIOT Act to foreign laboratories funded by NIH, often under subcontract to an American academic institution. While this is consistent with the treatment of other federal policies and regulations in such contracts, such provisions may be impossible for many foreign laboratories to meet or unpalatable to local authorities in countries where the restricted select agents are endemic and readily available in the environment or in other research or clinical settings. The net result is likely to be a reduction in the number of foreign collaborators with U.S. scientists, with the result that the nation's ability to understand the epidemiology and evolution of these biological agents in their native settings is degraded. It is questionable whether such a policy effectively promotes global awareness of the "culture of responsibility" sought by many in the study of such agents. The potential adverse impacts of policies such as these[32] need to be studied in an evidence-based manner, and decisions concerning continued or future implementation should be made on the basis of the balance between the harm done to the scientific establishment charged with protecting society against such threats and any additional direct security such policies may provide.

In addition to the many beneficial applications of life sciences knowledge and technologies that were highlighted in Chapter 2, the promises offered by the 13-year Human Genome Project provide an exemplary case of a recent advance in life sciences made possible by the unrestricted exchange of information and technology. The International Human Genome Sequence Consortium involved hundreds of scientists from 20 sequencing centers in China, France, Germany, Japan, the United Kingdom, and the United States.[33] The ability of the scientific community to respond rapidly and rationally to the SARS epidemic was based in large part on recognition of the offending etiologic agent, SARS coronavirus (SARS-CoV). Within six weeks, the virus that causes SARS, SARS-CoV, had been isolated and its complete 29,727-nucleotide sequence determined and posted on the Internet.[34] The rapidity with which this happened was dependent in part on technology developed to advance the Human Genome Project and on the sharing of data as they were generated in multiple laboratories on different continents. In the months that followed, dozens more SARS-CoV isolates were sequenced and published. Not only did availability of the sequence data put to rest fears that SARS was the result of a laboratory-fabricated agent, such data allowed researchers in open laboratories worldwide to begin immediately to analyze the virus' structure, function, and molecular pathogenic mechanisms, as well as develop rapid nucleic acid-based diagnostic tests and identify potentially useful antiviral lead compounds targeting the viral protease that were already on the shelf.[35] The use of these sequences by scientists addressing the SARS crisis globally is a prime example of the crucial role that the free exchange of international information and technology can play with respect to, in this case, a rapid response to a public health crisis. It would likely be much the same should a manmade infectious disease threat be unleashed.

Restrictive policies and regulations that unduly hinder scientific and technological progress would keep scientists and society from achieving important goals, like sequencing the human genome or developing a rapid response to a new disease outbreak, like SARS, not to mention the development of effective countermeasures for bioterrorism.

Recommendation 1c. Promote international scientific exchange(s) and the training of foreign scientists in the United States.

Foreign scientific exchange is an integral and essential component of the culture of science. As technological growth becomes increasingly dependent on international exchange, it is also an increasingly vital component of U.S. technological capacity, including biodefense technological capacity. Weakening this link by prohibiting or discouraging foreign scientific exchange—including the engagement of foreign students and scientists in U.S. laboratories, meetings, and business enterprises *and* vice versa—could impede scientific and technological growth and have counterproductive, unintended consequences for the biodefense enterprise. As described in Chapter 2, international scientific exchanges and the training of foreign scientists in the United States have played integral roles in the scientific and technological development of this country over the past few decades. Such exchanges will continue to play important roles in maintaining the international linkages that are so vital (and are only becoming more so) for both basic and applied research and development in the life sciences. Moreover, from the perspective of enhancing biosecurity, these exchanges will be essential for the development of a shared global culture of awareness and responsibility with respect to the dual use potential of many future advances in the life sciences.

The implementation of the regulatory regime imposed by the PATRIOT and Bioterrorism Response Acts on the life sciences community have raised concerns that qualified individuals may be discouraged from conducting biomedical and agricultural research of value to the United States because of the apparent infringement of these rules and regulations on individual liberties. Included among these measures are policies directed at individuals based on their country of birth, rather than current citizenship. As emphasized in Chapter 2, foreign interest in U.S. graduate education in science and technology is waning, as the increased competitiveness of graduate schools elsewhere in the world attracts gifted students who, in the past, may have emigrated to the United States to study and because of perceived and actual difficulties with obtaining entry to the United States. For example, according to a February 24, 2004 General Accounting Office report, between April and June 2003, it took an average of 67 days to complete the security checks associated with visa applications, due to the wait time for required interviews (as long as 12 weeks in India and 6 weeks in China) and Visas Mantis clearance.[36] (The committee notes, however, that by November 2004, review time had reportedly dropped to 15 days.[37])

Moreover, there have been recent indications that other steps are being taken, or pressures exerted, which may curtail foreign national participation in U.S. scientific activity. For example, in March 2004 the Inspector General of the U.S. Department of Commerce issued recommendations for regulatory changes that would affect existing requirements and policies for "deemed export" licenses. A deemed export occurs when a foreign national working in the United States gains access to or uses export-controlled technology or information, including many types of standard laboratory equipment. The recommendations include regulatory or other administrative action that would clarify the definition of "use," base the requirement for a deemed export license on the foreign national's country of birth; and modify regulatory guidance on the licensing of technology to foreign nationals involved with government-sponsored or university research.[38] In March 2005, the Bureau of Industry and Security (BIS) solicited comments on the proposed requirements (through May 27, 2005). In a letter sent to Peter Lichtenbaum, assistant secretary of commerce for the Export Administration, the presidents of the National Academy of Sciences, Institute of Medicine, and National Academy of Engineering, provided formal comments on the effect that this Advanced Notice of Proposed Rulemaking would have on the scientific enterprise (Annex 4-1).

On May 6, 2005, the National Academies hosted a workshop on the proposed changes and their implications.[39] On May 18, 2005, the presidents of the National Academies, along with the presidents and executive directors/CEOs of leading domestic and international scientific and educational associations including, but not limited to, the Association of American Universities, the American Association for the Advancement of Science, the National Association of State Universities and Land Grant Colleges, the American Council on Education, the Council on Competitiveness, the American Physical Society, NAFSA: Association of International Educators, the Council of Graduate Schools, and the Institute of International Education jointly issued six recommendations for enhancing the U.S. visa system to advance America's national security interests while promoting its economic and scientific competitiveness. The text of the announcement can be found in Annex 4-2.

CONCLUSION 2: THE COMMITTEE CONCLUDES THAT A BROADER PERSPECTIVE MUST BE ADOPTED WHEN CONSIDERING THE SPECTRUM OF PRESENT AND FUTURE THREATS.

U.S. national biodefense programs currently focus on a relatively small number of specific agents or toxins, chosen as priorities in part because of their history of development as candidate biological weapons agents by some countries during the 20th century. The committee believes that a much broader perspective on the "threat spectrum" is needed. While current biodefense programs[40] do consider the future potential for specific pathogenic agents to be manipulated in ways that make them, for example, more virulent or more resistant to available antimicrobial drugs, even this approach is too narrowly focused. Recent advances in understanding the mechanisms of action of bioregulatory compounds, signaling processes, and the regulation of human gene expression—combined with advances in chemistry, synthetic biology, nanotechnology, and other technologies—have opened up new and exceedingly challenging frontiers of concern. Future advances that cannot now be described will continue to extend these frontiers.

Recommendation 2

The committee recommends adopting a broader perspective on the "threat spectrum."

Recommendation 2a. Recognize the limitations inherent in any agent-specific threat list and consider instead the intrinsic properties of pathogens and toxins that render them a threat and how such properties have been or could be manipulated by evolving technologies.

Lists are inherently problematic. As explained in detail in Chapter 1, the spectrum of threats is much broader than the U.S. select agents list might suggest. As one example, the current select agents list does not include the uncounted numbers of biologically active molecules identified annually through industrial or federal government-sponsored (NIH Roadmap) drug discovery processes, many of which could be construed as potential threats. Nor does it include synthetic molecules or life forms, such as those that could be created using a variety of emerging techniques as described in Chapter 3, for example, through the application of reverse genetic engineering of RNA viruses, the use of purely synthetic biology, or DNA nanotechnology. Moreover, as described in Chapters 2 and 3 of this report, propelled by a variety of powerful economic and scientific drivers, biotechnology is developing, diversifying, and proliferating rapidly and globally, in largely unpredictable ways, with all its attendant, potential, dual-use applications. New capabilities, for either good or bad purposes, including the manipulation of gene expression in mammals through the use of RNA interference, have achieved prominence even within the life span of the present committee. Committee members were repeatedly reminded to "expect the unexpected." Against this central reality, it is doubtful that any authority could enumerate a "select agents list" that is sufficiently comprehensive, robust, or of enduring relevance, although most currently listed agents, such as smallpox, are likely to remain a potential menace even as new threats emerge. The select agents list had its origins in the Antiterrorism and Effective Death Penalty Act of 1996, which required the secretary of HHS to establish and enforce safety procedures for the transfer of biological agents considered to be the greatest threats to human health—that is, "select agents"—including measures to ensure proper training and appropriate skills to handle such agents and proper laboratory facilities to contain and dispose of such agents. The PATRIOT and Bioterrorism Response Acts imposed additional physical security requirements and regulatory obligations for laboratories working with these select agents. The Bioterrorism Response Act made it a criminal offense for any person to possess knowingly any biological agent, toxin, or delivery system of a type or in a quantity that, under the circumstances, is not reasonably justified by prophylactic, protective, bona fide research, or other peaceful purpose.[41] In addition, the new laws prohibited transfer or possession of a listed biological agent or toxin by a "restricted person."[42] Among other requirements, the Bioterrorism Response Act added new criteria for consideration by the secretary in listing agents[43] among other things, requiring that the secretary ensure the appropriate availability of biological agents and toxins for research, education, and other legitimate purposes.

On February 7, 2003, the CDC's interim final rule (42 CFR 73) for the possession, use, and transfer of select agents went into effect, changing the way that select agents and toxins are managed in the United States. Originally, the CDC was authorized to require laboratories transferring select agents to register, as a way to ensure the safe transfer and shipment of lethal pathogens and not with the intent to collect any specific information. In accordance with the PATRIOT and Bioterrorism Response Acts, the new regulations established additional requirements for those who may possess select agents as well as those who might send and receive those agents (e.g., the new regulations involved the U.S. Department of Justice in performing background checks on individuals who may have access to or conduct research on select agents). An expanded list of pathogens and toxins, including agricultural plant and animal pathogens, went into effect on February 11, 2003.[44]

The interim final rules were initially met with many protests by scientists and universities who argued that some of the rules were ambiguous, would be expensive to implement, did not offer significant protection to the public (because of the availability of some agents in nature), and could delay or impede research.[45] With respect to the list of category A select agents covered under the interim final rule, the life sciences research community raised many concerns about the extent to which decisions to list particular pathogens, toxins, and nucleic acids had been reached on the basis of the best scientific advice, as opposed to perceived or hypothetical risks. For example, several rickettsial agents have been employed by state-sponsored biological warfare programs in the past—they can be readily disseminated, are highly pathogenic, and may not be easily diagnosed by physicians in the United States today. While treatable, the number of infected persons could easily overwhelm antibiotic supplies. Moreover, these agents can be readily engineered today for antibiotic resistance. Thus, many rickettsial agents share these features and are not on the list. More importantly, the select agents list does not include many classes of future potential dual-use agents. In addition, there is considerable uncertainty about the risks that many of the currently listed items actually pose to public health and safety and whether those risks are great enough to warrant such restrictions. Smallpox and anthrax are obvious concerns, but are the filoviruses worthy of their position on the list given the dangers and difficulties inherent in working with them?

On March 18, 2005, the CDC issued the Final Rule for (42 CFR Part 73), Possession, Use, and Transfer of Select Agents and Toxins, which implements the provisions of the Bioterrorism Response Act and updates the interim final rule. Although some changes were made in response to submitted comments, the select agents list was not modified (although some language was clarified), as many scientists and scientific organizations had requested (e.g., the American Society for Microbiology[46]). Concerns remain also about the status of cDNA clones of select RNA viruses—the generation of infectious virus from such clones, when they represent the complete genome sequence, is becoming increasingly facile. Yet their status as select agents in the absence of overt infectivity remains poorly defined and represents a potential illogical loophole in the control regime.

The main concern with the select agents list remains the extent to which decisions to list particular pathogens, toxins, and nucleotide sequences are based on the best scientific evidence, with respect to their risk of being used for malevolent purposes or the danger they pose to public (or plant or animal) health should they fall into the wrong hands. A list may also provide an unwarranted sense of security because of what is not on it. Moreover, while any approach to meeting the diversity of biosecurity threats that society faces today will require prioritization in the application of resources, and this requires the development of a "list," the intelligence and scientific communities must be careful not to let any established, agreed-upon list act to retard continued, intense surveillance of the technological horizon for newly emerging threats. In fact, multiple lists may be necessary for the disparate purposes of research prioritization, public health surveillance and response to outbreaks, development of practical countermeasures, and intelligence activities.

Recommendation 2b. Adopt a broadened awareness of threats beyond the classical "select agents" and other pathogenic organisms, to include, for example, approaches for disrupting host homeostatic systems and/or the creation of synthetic organisms.

The limitations of the current select agents list, and indeed any list, point to the need for a broadened awareness of the threat spectrum. Mechanisms must be put in place that ensure regular and deliberate reassessments of advances in science and technology and identification of those advances with the greatest potential for changing the nature of the threat spectrum. The process of identifying potential threats needs to be improved. This process needs to incorporate newer scientific methodologies that permit more rigorous assessment of net overall risks. Rather than adopting a static perspective, it will be important to identify and reassess continually the degree to which scientific advances or current or future biological "platforms" hold the potential for being put to use by potential adversaries. This will require engagement of the scientific community in new ways and an expansion of the science and technology expertise available to the intelligence community (as outlined in Recommendation 3).

In addition to the importance of relying on the best-available science and technology expertise for assessing the nature of the future threat spectrum and for integrating such expertise within and across the national security communities, there is an equally important need for providing the same kind of expertise to the public policy community and senior decision makers in the U.S. government. The structure and charge of the entity that might fill this role are beyond the purview of this committee; however, the committee recommends that further serious discussions be held to consider how the following goals might be accomplished:

• Regular, independent peer review of policies, rules, and regulations that address future threats, including an independent review of the PATRIOT Act and other related statutes and regulations to ensure their relevance and effectiveness in enhancing biosecurity with respect to intelligence, law enforcement, and homeland security (see Recommendation 1b).
• Establish measures of effectiveness for science and technology-based programs in the intelligence, homeland security, and law enforcement communities that address emerging and future biothreats and technologies.
• Create and evolve a cross-agency strategy and implementation plan for scientific countermeasures and operational capabilities related to emerging and future biological threats and technologies—in essence, an integrated future-oriented national biodefense plan. This plan

should guide policy makers in their long-term investment planning for biodefense.

CONCLUSION 3: THE COMMITTEE RECOGNIZES THE IMPORTANCE OF A PROACTIVE, ANTICIPATORY PERSPECTIVE AND ACTION PLAN THAT RELIES ON AN EVALUATION OF STATE-OF-THE-ART SCIENCE, SO THAT FUTURE BIOLOGICAL THREATS CAN BE BETTER UNDERSTOOD, ADDRESSED, AND MINIMIZED.

A sound defense against the misuse of the life sciences and related technologies is one that anticipates future threats that result from misuse, one that seeks to understand the origins of these threats, and one that strives to prevent the misuse of science and technology before it happens. It would be tragic if society failed to consider, on a continuing basis, the nature of future biological threats, using the best-available scientific expertise, and did not make a serious effort to identify possible methods for averting such threats. Interdiction and prevention of malevolent acts are far more appealing than treatment and remediation. The committee, therefore, urges the adoption of a broader perspective in considering the threat spectrum (Recommendation 2). And the committee urges a proactive, anticipatory perspective and action plan for the national and international security communities.

These perspectives and plans must be based on a current working familiarity with the life sciences and related technologies, especially those that pose a clear and significant opportunity for misuse (Chapter 3), as well as an appreciation for the future trajectories of these sciences and technologies across the globe (Chapter 2). To meet these challenges effectively, the committee recognizes an urgent need to establish new processes, resources, and organizational structures that will enhance the breadth and level of sophistication of the scientific expertise residing in agencies concerned with national security.

Recommendation 3

The committee recommends strengthening and enhancing the scientific and technical expertise within and across the security communities.

Recommendation 3a. Create by statute an independent science and technology advisory group for the intelligence community.

The national security community and its assessments of future biological threats must be informed by the best-available scientific expertise.

Expertise can be acquired through outside collaboration as well as internal investments. With respect to the former, there have been several noteworthy efforts to build useful outside advisory groups for the life sciences, including the Defense Intelligence Agency's (DIA) Bio-Chem 2020. However, as discussed in greater detail below, Bio-Chem 2020 and other existing advisory groups do not have the resources, expertise, administrative charge, independence, and statutory standing that are needed. The committee, therefore, recommends the creation of an independent advisory group that would work closely with the national security community for the purpose of anticipating future biological threats based on an analysis of the current and future science and technology landscape and current intelligence. In proposing the creation of this group, the committee supports Recommendation 13.1 of the Commission on the Intelligence Capabilities of the United States Regarding Weapons of Mass Destruction (March 31, 2005), which suggests the creation of an advisory group similar to the one recommended here.[47]

In making this recommendation the committee did consider other options, including, whether this responsibility could be tasked to an existing entity, such as DIA's Bio-Chem 2020, or the recently created the NSABB. The committee concluded that the mandate, structure, and functions of the proposed advisory group are sufficiently distinct from those of existing entities as to warrant the creation of a new science and technology advisory group for the national security and intelligence communities. While either of these two existing advisory bodies could, in theory, be restructured and provided with a new charter that would accomplish the aims envisioned by the committee in this recommendation, in practice they would be so altered from their present structure and purpose as to render them, in essence, new entities. In addition, while the advisory group proposed here might make the functions of DIA's Bio-Chem 2020 redundant and could possibly supplant this group, it cannot and should not replace the NSABB, which has a large and important charge distinct from that envisioned for the advisory group proposed in this section.

Red Team Bio-Chem 2020 was established by the DIA in 1998, as a group of government and nongovernment experts in the life sciences and related technologies whose mission was to lead and focus the defense intelligence community's assessments of emerging technologies that nation-states or terrorists could use for biological or chemical warfare and to mitigate technological surprise from foreign biological warfare programs. It has met three to four times per year since then and serves as an ad hoc partnership between leading life scientists in academia, industry, government, and science and technology analysts from the intelligence community. It produces analyses on emerging technologies and innovative approaches to threats for use by the broader intelligence community. One of the most important successes of this group has been the establishment of close, productive working relationships between outside scientists and science and technology analysts from within the intelligence community. While Bio-Chem 2020 encompasses some of the features that the committee finds most important for an external advisory group, the committee concludes that major restructuring would be necessary for it to take on the functions that are critical for all relevant stakeholders.

Bio-Chem 2020 operates under several limitations. First, its primary responsibility is to the DIA and the DOD. Even though other agencies participate in Bio-Chem 2020 meetings, the group is not formally charged with addressing the needs of the entire intelligence community.[48] Second, it is not permanent; it exists at the behest of the director of the DIA. Third, the group of outside experts is small and therefore lacks expertise in some important areas. Fourth, it has operated at no higher than the secret classified level and has not engaged in analysis of primary sources and methods, or perform real-time, independent assessments of intelligence pertaining to potential threats in the life sciences arena.

The committee also considered whether the NSABB was an appropriate body for implementing this new advisory function. However, upon further analysis, there were at least two fundamental reasons why the committee concluded that the NSABB could not, and should not, attempt to address this critical unmet need.

In making its recommendation for the creation of the NSABB the Fink committee envisioned that this new advisory body would "provide advice to the government and guidance and leadership for the system of review of life sciences research . . . " The Fink committee encouraged the HHS to model the NSABB after the Advisory Committee on Immunization Practices (ACIP)—an independent advisory body to the federal government.[49]

In implementing this recommendation, however, the Director of the National Institutes of Health created the NSABB as a federal advisory committee under the Federal Advisory Committee Act.[50] As such, the NSABB has no independent budget or staffing authority, serves at the direction of the federal department or agency that created it, and can be "retired" at any time.[51] These structural features prevent the NSABB from establishing the kinds of long-term working relationships and providing the kinds of functions to the necessary stakeholders that would address the crucial needs of the national security and intelligence communities, as described in more detail below.

In addition, the NSABB's charter defines a relatively narrow charge that does not include the type of advisory and ongoing analytical and evaluative functions that this committee envisions the advisory group fulfilling for the national security and intelligence communities, as proposed in this section:

> The NSABB will advise the Secretary of the Department of Health and Human Services (HHS), the Director of the National Institutes of Health (NIH), and the heads of all federal departments and agencies that conduct or support life science research. *The NSABB will advise on and recommend specific strategies for the efficient and effective oversight of federally conducted or supported dual-use biological research*, taking into consideration both national security concerns and the needs of the research community.[52]

Addressing these current responsibilities has and will continue to consume all of this board's resources for the foreseeable future. As has been pointed out, "its role resembles that of the Recombinant DNA Advisory Committee that was established [by law] within the NIH in 1974, and that played an important part in setting guidelines and reviewing research protocols."[53] Restructuring the NSABB would make little sense given that the board's current membership has been selected by the secretary with the current charge in mind.

The advisory group proposed by the committee in this section would be tasked with forecasting the applications and implications of technological developments in the life sciences; providing expert analysis of relevant collection information; providing guidance on intelligence targeting and collection requirements; and providing an independent, outside "reality check" on technical assessments in the life sciences. Not only are these needed functions outside the charge and purview of the NSABB, but to provide them an advisory board will need a membership with expertise and background complementary to but distinct from those of the NSABB.

The committee elaborates further below on the nature and functions of such an advisory body for the national security and intelligence communities.

- This advisory group should operate under the auspices of the national security community leadership and provide direct input at the highest levels of this community. The recommendation of the Weapons of Mass Destruction (WMD) Commission that such a group report to the Director of National Intelligence should be given serious consideration, as this would increase the likelihood that the group serves the entire intelligence community. The advisory group should also have an independent source of funding and a dedicated staff. These latter features will help strengthen its independence and stability, insulate it from short-term budgetary pressures, and enhance the dedication of its members to the demands of membership.

- To provide objective technical assessments, the advisory group should be made independent of any specific agency. The functions of this group should be codified in law, and include self-initiated as well as externally requested analyses of science and technology with special relevance to future potential threats, independent technical review of national security intelligence assessments in the life sciences, and real-time assessments of relevant raw intelligence when deemed to be of special current importance. This group might review and enhance intelligence targeting and collection in the life sciences. It would provide an outside "reality check" on technical assessments in the life sciences. The larger set of members might constitute a network of available experts to whom national security officials and policymakers might turn for technical advice on matters of timely and special importance.

- It should be composed of leading experts from academia, industry, and government in a wide spectrum of disciplines relevant to the life sciences and related technologies. The government members should represent the broad national security community, and include those scientists most familiar with the "state of the art" in these disciplines. Membership should take into account possible future threats to livestock and agriculture, as well as threats to physical or information technology infrastructure, when related to the life sciences and associated technologies.

- The size of the group should be sufficient to represent all important areas of the life sciences and related technology in some depth and yet small enough to allow close working relationships and trust to develop among members of the group. Because these two needs may conflict, the advisory group should consist of a small core of elite experts who are broadly versed in cutting-edge developments with applications to the life sciences enterprise and who meet on a regular basis, much like Bio-Chem 2020, as well as a larger set of members that provide greater in-depth expertise for a more complete set of disciplines and who meet less frequently or on an as-needed basis.

- The advisory group should publish both open and classified reports on current, emerging, and future biological threats. The output of the group should be shared widely with the intelligence, national security, and policy communities and to the maximum degree possible the general scientific and public health communities and, in particular, with the NSABB. The output should inform national decision makers in the relevant areas of science and technology developments and policy options.

- It is critical that members of this Advisory group develop relationships of trust and familiarity among themselves. Preexisting differences in culture between the national security community and the outside science community pose barriers that must be overcome. Frequent and regular meetings among a group with reasonably stable membership would help in this regard. The advisory group should be given access to any and all classified intelligence that is directly relevant to their tasks.

Recommendation 3b. The best-available scientific expertise and knowledge should inform the concepts, plans, activities, and decisions of the intelligence, law enforcement, homeland security, and public policy communities and national political leaders about advancing technologies and their potential impact on the development and use of future biological weapons.

Given the broad and constantly changing nature of future potential biological threats, as highlighted in Conclusion 2 and as illustrated throughout this report, the committee believes there is an urgent need to create an agile, anticipatory system to recognize and, rapidly and effectively, respond to emerging threats. If the national security and public policy communities are to fulfill this mission, they must be well-informed about scientific and technological advances in a variety of disciplines relevant to the life sciences. This committee recognizes several as yet unsolved and ongoing challenges for the national security community in this area and takes note of the expert judgments of recent national investigatory bodies[54] as background for its recommendations here. The power of this science and technology is increasingly wielded by individuals. Understanding the intent of a would-be malfeasant, a "holy grail" of the intelligence community, becomes ever more necessary. The committee fully recognizes that the challenges associated with the collection of useful and actionable intelligence on the potential malevolent use of biological agents are substantial. These challenges will only grow, as the life sciences and their associated enabling technologies evolve, expand, and disseminate at a dizzying rate. However, as one senior intelligence analyst working on biological threats has said, "We have no choice, but to try as hard as we can."

There are several existing problems in the national security community and national political leadership related to the task of anticipating future biological threats. First, these groups have not developed the kinds of working relationships with the "outside" (nongovernmental) science and technology communities that are needed (and feasible). Second, "inside" groups (national security community and national political leadership) have been unable to establish and maintain the breadth, depth, and currency of knowledge and subject matter expertise in the biological sciences and related technologies that are needed. The number of analysts in the national security community who have professional training in the life sciences and technologies is small and insufficient; these analysts often lose touch with the cutting-edge of science and technology over time and tend to be moved from position to position, preventing them from developing any particular depth of expertise and experience. Some of the same problems are also true of intelligence collection and the collectors. And to the degree that the right kinds of expertise do exist within the intelligence community, this expertise is unevenly distributed. Moreover, intelligence assessments are not always shared among the different member agencies of the national security community. Finally, historical, political, and cultural barriers have prevented the national security community from working closely with counterparts from other nations and regions of the world. Yet the life sciences and related technologies are distributed around the globe in a seamless fashion, and future threats that arise from this science and technology will be distributed globally as well. This committee addresses each of these three problem areas with the preceeding subrecommendation, and the following two subrecommendations.

Recommendation 3c. Build and support a robust and sustained cutting-edge analytical capability for the life sciences and related technologies within the national security community.

Analytical capability is a function of both the quality and quantity of the relevant resources. The committee views people as the most important resource for the national security community in building an internal expertise in the life sciences and related technologies. Thus, it is suggested that the national security community be provided the means to hire and sustain significant additional personnel with current expertise in the scientific disciplines discussed in Chapter 3. Open-source intelligence and human intelligence are the most useful kinds of data today for identifying and anticipating future threats from the life sciences and related technologies. Collection and analysis of both kinds of data will require intimate familiarity with the scientific and technology workplace. Researchers with state-of-the-art, hands-on experience in relevant areas of science and technology should be recruited at the completion of their doctoral or postdoctoral training. Retaining these individuals and sustaining their capabilities and currency are not easy tasks. They will need to maintain close contact with the outside scientific world, for example, through attendance at scientific meetings, courses, workshops, and perhaps sabbaticals "at the bench." Their employers should refrain from frequent reassignment of these valuable experts to unrelated jobs and responsibilities.

Scientific expertise must inform intelligence collection in a meaningful manner. In much the way that foreign-language expertise can be critical for some areas of intelligence assessment, working familiarity with the language of modern molecular biology (and other scientific dialects) will be essential for both analyst and collector in assessing potential future biological threats. It goes without saying that if relevant information is not recognized as such, it cannot be collected for analysis. Conversely, an inadequate understanding of today's life sciences can lead to the collection of massive quantities of irrelevant information, resulting in degradation of overall analytical capabilities.

> **Recommendation 3d. To the maximum extent possible, encourage the sharing and coordination of future biological threat analysis between the domestic national security community and its international counterparts.**

As described in Chapter 2 of this report, the future of the life sciences and related technologies reaches all corners of the globe, and the implications of future trajectories in these areas pose potential problems and opportunities for all of us. Not only do potential threats cross national boundaries, but so do potential solutions. The power of international collaborations in addressing future biological threats cannot be underestimated. For these reasons the committee recommends that the analysis and assessments of potential biological threats be shared across international boundaries wherever and whenever possible.

While general concerns about the sensitivity of sources and methods will lead to caution and a reluctance to share data, the open nature of the life sciences enterprise and the important role of open-source material in the assessment of potential threats suggest that sharing of intelligence assessments by the national security community with international counterparts may be more feasible than might have been assumed as well as desirable. In addition, the sharing of biological threat assessments becomes increasingly practical when adopting a later time frame further into the future.

CONCLUSION 4: THERE IS A CRITICAL NEED TO ADVANCE A GLOBAL PERSPECTIVE IN ADDRESSING THE INAPPROPRIATE USE OF EMERGING TECHNOLOGIES IN THE LIFE SCIENCES.

The committee appreciates that the threat posed by the potential dual-use applications of advancing technologies is a global problem, one that can be mitigated successfully only by actions taken in a global context. A purely national policy, executed in the absence of engagement with and participation of the global community, is unlikely to have a significant impact on reducing these dangers. This is made abundantly evident by the global dispersion of advanced technologies in the life sciences, as described in Chapter 2. Recent years have witnessed the rapid growth of biotechnology-related research and commercialization efforts in countries of the Asian-Pacific rim, Latin America, and elsewhere. U.S. preeminence in the life sciences is not only being challenged by other nations but may soon be lost. In October 2005 a National Academies panel delivered a dire warning to Congress: Give science an extra $10 billion annually or watch jobs and national status disappear to Asia. Many people may agree with the message, but details of the panel's ambitious prescriptions are already drawing criticism.[55]

The committee therefore sought to develop an international perspective in formulating its recommendations and recognizes an urgent need to engage the global community further in addressing these issues.

Recommendation 4

The committee recommends the adoption and promotion of a common culture of awareness and a shared sense of responsibility within the global community of life scientists.

Even while considering steps that can and must be taken to strengthen biodefense efforts at the national level, a proactive strategy against next-generation threats will require collective and concerted global action. This four-part recommendation outlines actions that could enhance the global capacity to mitigate the biosecurity risks associated with advanced technologies.

Recommendation 4a. Recognize the value of formal international treaties and conventions, including the 1972 Biological and Toxin Weapons Convention and the 1993 Chemical Weapons Convention.

The biological weapons control regime of the 20th century dates back at least to the 1925 Geneva Protocol, which entered into force in 1928.[56] The protocol, which was supported by one of the most outspoken and ferocious public appeals that the International Committee of the Red Cross has ever made, was drafted in response to the horrific consequences of the extensive use of poison gas in World War I. It prohibits the wartime use of "asphyxiating, poisonous, or other gases, and of all analogous liquids, materials, or devices" and of "bacteriological methods of warfare." The most important international step taken to strengthen the biological weapons regime occurred decades later, with the 1972 Biological and Toxin Weapons Convention (BWC), which entered into force in 1975. The BWC prohibits the development, production, stockpiling, or acquisition of biological agents or toxins of any type or quantity that do not have protective, medical, or other peaceful purposes or any weapons or means of delivery for such agents or toxins.[57] According to the treaty, all such materials had to be destroyed within nine months of its entry into force. As of December 2004, there were 169 signatories, including 153 ratifying and acceding countries.[58]

Despite its relatively long history, beginning with the Geneva Protocol, the biological weapons control regime and the BWC in particular have been fraught with challenges, not the least of which is the lack of a treaty compliance verification protocol.[59] Many of these challenges are related to the unique characteristics of biological weapons, as discussed in Chapter 1 (i.e., unique in comparison to nuclear, chemical, and other weapons of mass destruction).

The chemical weapons control regime, also rooted in the 1925 Geneva Protocol, has been strengthened by the 1993 Chemical Weapons Convention (CWC), which entered into force in April 1997.[60] It is the only multilateral treaty that seeks to eliminate an entire category of weapons of mass destruction within an established time frame (by 2012) and to verify destruction through inspections and monitoring by the Organization for the Prohibition of Chemical Weapons (OPCW). Moreover, the CWC verification regime extends to dual-use industrial facilities judged especially vulnerable to abuse for proliferation purposes. Although the CWC has helped reduce chemical weapons risks, member states are experiencing delays in meeting CWC requirements. For example, neither Russia nor the United States is expected to have completed destruction of their stockpiles until after 2012.[61] Also, only a minority of member states have adopted national legislation to criminalize CWC-prohibited activities, and many have not yet put in place, as the CWC requires, the measures necessary to ensure that toxic chemicals and their precursors are used only for nonprohibited purposes. Moreover, although the OPCW, as of September 2005, had conducted 2,195 inspections in 72 member states over the eight-plus years since the CWC went into force, the organization does not have enough resources to conduct all the inspections that many consider necessary.

Despite the difficulties of implementing the BWC and CWC properly, the two conventions serve as the cornerstones of the global biological-chemical control regime, which has expanded to include rules and procedures rooted in measures ancillary to the two treaties (such as the Australia Group[62] and United Nations Security Council Resolution 1540[63]). The biological-chemical regime as it currently exists—including the BWC, CWC, Australia Group, SCR 1540[64], and other measures—must be recognized for its positive contributions and placed within the overall array of measures taken to prevent biological warfare.

In particular, the committee concluded that the BWC and CWC embody and codify international norms of behavior that should govern all policies, actions, and strategies implemented both nationally and internationally. The biological-chemical regime encompasses more than law: It is based on long-standing taboos stemming from public abhorrence to poison and the deliberate spread of disease. The original BWC and CWC negotiators largely defined the scope of their treaties not in terms of lists of agents or devices that could quickly become outmoded by technological change but in terms of a general-purpose criterion whereby all biological or chemical agents became subject to the constraints of the regime unless they were intended for nonprohibited purposes. Where specific lists were deemed useful, however, they are incorporated into these regimes, as evidenced by the three schedules of materials that are subject to verification within the CWC. Such lists do not, of course, limit the scope of the prohibitions set out in the two treaties, which remains set by the general purpose criterion. It is this device that enables the biological-chemical regime in principle to control dual-use technologies and to keep up with scientific advance.

Such international conventions should not be considered the solution to the issues society confronts today with respect to potential harmful use of advances in the life sciences, nor should they be cast aside and ignored. Despite their limitations, the committee appreciates their value in articulating international norms of behavior and conduct and suggests that these conventions serve as a basis for future international discussions and collaborative efforts to address and respond to the proliferation of biological threats. Important opportunities will arise when states parties conduct their next quinquennial reviews of the operation of the BWC (in 2006) and the CWC (in 2008).

The present report has several times noted that technologies are bringing chemistry and biology closer together. That toxins and synthetic biological agents, including bioregulators, immunoregulators, and small interfering RNAs, fall within the scope of both treaties is one such linkable feature. These two review conferences will as always be dominated by political considerations, but in view of the profound developments now under way in the life sciences, the committee draws attention to the possibilities held out by the 2008 conference for building on the parallel or linkable features of the BWC and the CWC.

Recommendation 4b. Develop explicit national and international codes of conduct and ethics for life scientists.

The committee reviewed the potential for codes of conduct or codes of ethics to mitigate the risk that advances in the life sciences might be applied to the development or dissemination of biological weapons. Codes for professional behavior date back at least two millennia to the Hippocratic Oath, which provided guidance for the conduct of physicians in ancient Greece. A code of conduct (also known as an educational or advisory code) provides relatively specific guidelines with respect to what is considered appropriate behavior. [65] A code of conduct developed for the life sciences could thus assist those working in the field to become sensitive to specific actions in the course of their work or that are carried out by their colleagues. In the absence of a code, such actions might otherwise go unnoticed. In contrast, rather than suggesting how to behave specifically, a code of ethics (also known as an aspirational code), lays forth the ideals to which practitioners should aspire, such as standards of objectivity or honesty. In the case of the life sciences, such a code might call for biologists to consider the ethical implications of their work or to discourage generally the use of biology for malevolent purposes. Clearly, many codes, including the Hippocratic Oath, may address elements of both conduct and ethics.

In considering such codes, the committee concluded that their primary effect would be to create an enabling environment that would facilitate the recognition of potentially malevolent behavior (i.e., experiments aimed at purposefully developing potential weapons of biological origin) or potentially inappropriate experiments that might unwittingly promote the creation of a more dangerous infectious agent. The committee also recognized that such codes could generally be expected to achieve their desired effect only when reinforced by a substantial educational effort and appropriate role modeling on the part of scientific leaders.

In addition to "codes of conduct" and "codes of ethics," there are "codes of practice," also known as enforceable codes. Regulations controlling research with the select agents derived from the PATRIOT Act and other national legislation, including that enacted in response to the BWC as discussed above, may be considered examples of an enforceable code. The desired effects of such codes are to a considerable extent dependent on the ability of enforcing agencies to detect proscribed behavior and the nature of the consequences imposed on the offending individual. In making this recommendation, the committee focuses on the potential utility of "codes of conduct" and "codes of ethics" that may arise primarily from within the life sciences professions, rather than enforceable codes that may arise from legislative or regulatory bodies that are largely outside the life sciences in an attempt to regulate them.

Today, a wide variety of professional organizations, research institutions, and scientific societies active in the life sciences have adopted codes to guide the conduct of their members, and many other societies and institutions are considering what such codes should comprise. Of relevance to research aimed at developing offensive biological weapons, the aspirational 2000 American Society of Microbiology (ASM) Code of Ethics states that "ASM members aspire to use their knowledge and skills for the advancement of human welfare."[66] In 2002, the ASM reaffirmed that bioterrorism and "the use of microbes as biological weapons" violated its code of ethics.[67] The code of ethics of the Australian Society of Microbiology is somewhat more direct: "The Society requires each member . . . not to engage knowingly in research for the production, or promotion of biological warfare agents."[68, 69] The fact that the Australian Society affirmatively "requires" something of its members indicates that it is an "enforceable code," and in fact, members who are found to have violated this or any component of the society's code are subject to expulsion from the Society. The BIOTECanada Statement of Ethical Principles states unequivocally that the organization, "oppose[s] the use of biotechnology to develop weapons."[70] Similarly, the EuropaBio Core Ethical Values document states: "We oppose the use of biotechnology to make any weapons and will not develop or produce biological weapons."[71]

Recently, several international forums have made efforts to construct globally applicable sets of principles guiding the development of specific codes of conduct related to potential dual-use research in the life sciences.[72] For example, in November 2002, at the conclusion of their intersessional meeting of the Fifth Review Conference, States Parties to the Biological and Toxin Weapons Convention agreed that the topic for the 2005 intersessional meetings would be "the content, promulgation, and adoption of codes of conduct for scientists."[73] Also in 2002, the United Nations General Assembly and Security Council endorsed a report by the Policy Work Group on the United Nations and Terrorism recommending the establishment of codes of conduct for scientists related to weapons technologies.[74] The International Centre for Genetic Engineering and Biotechnology is in the process of developing a draft code of conduct, and the International Institute for Strategic Studies/Chemical and Biological Arms Control Institute (CBACI) have already drafted a relevant charter.[75] The International Committee of the Red Cross (ICRC) also has been considering the establishment of a "principles of practice" code that could serve as the life sciences equivalent to the Hippocratic Oath.[76]

However, despite the presumption that ethical codes foster ethical conduct, little is known about the effectiveness of these codes in practice.[77] People may not comply with codes or even consult them.[78, 79, 80] Nor will codes of ethics likely deter anyone who is firmly committed to applying biotechnology for malevolent purposes, such as a disgruntled scientist with a deep-seated animosity and intent to "get even" or a dedicated member of a terrorist group. Nonetheless, codes may be useful in raising awareness, fostering norms, and establishing public accountability.[81, 82, 83, 84] A code may sensitize researchers who might be unknowingly or unwittingly used by such individuals to aid and abet their plans by supplying knowledge or materials and may therefore make it less likely that such aiding and abetting will occur. Moreover, codes may create a climate in which voluntary reporting of suspicious activities on the part of colleagues is more likely to occur and hence change the risk calculus of potential offenders.

It seems clear that a widely promulgated code of conduct could raise the awareness of scientists concerning the risks posed by certain types of experiments, much as the list of the seven types of "experiments of concern" contained in the Fink report have heightened awareness and prompted debate among scientists engaged in microbiological research. A widely accepted code of ethics or conduct would appear to be an integral component of any plan to promote the development of a culture of awareness and responsibility. Of note, HHS Secretary Leavitt has recently charged the NSABB to develop such a code for scientists working in the United States.[85] However, as suggested above, a national code will have little effect on the global behavior of life scientists. While there is thus a need to promote the development of such codes globally, it is unlikely that any single code will be uniformly acceptable, especially if it contains the relatively specific features of a "code of conduct". Thus, the efforts of the international bodies referred to above may be particularly useful in creating sets of principles as guides to the development of such codes.

The risk with any code or policy is that it will sit on the shelf gathering dust. To prevent this, it needs to become part of the lived culture of a social group. The first step to establishing this culture will be to develop educational programs for scientists. Indeed, education may ultimately be more valuable than a formal code of conduct, particularly if it encompasses not just ethical but also legal norms with regard to dual-use agents, information, and technologies.[86] Many scientists today are unaware of the BWC, and the laws and regulations that have been enacted in the United States and elsewhere for the control of biological materials and personnel.[87] It would be relatively straightforward to incorporate the concept that a large proportion of current research in the life sciences has dual-use potential into the formal training in research ethics that the NIH mandates for postdoctoral trainees, for example. Efforts to expand awareness concerning the risks of potential dual-use research and technologies could also be integrated into continuing education courses, licensure courses, or other regular sets of activities in which experts engage as a way to update their credentials or resumes.

However, all the education in the world will not be as important as the role modeling provided by respected figures in the scientific community, both locally, nationally, and internationally. This "informal curriculum" probably drives what students learn and emulate more powerfully than the formal curriculum. Identifying, celebrating, and rewarding senior scientists who through word and deed serve as role models in preventing the malicious application of advances in biotechnology is perhaps the most important element in creating an environment that enables ethical and appropriate behavior. To the extent that such role modeling extends to the training of foreign nationals in the United States, it may also help establish a global culture of awareness and responsibility when such trainees return to their countries of origin to continue their professional careers. Foreign trainees will also be exposed to explicit codes of ethics and/or conduct adopted within the United States, further reinforcing Recommendation 1c that encourages foreign scientific exchanges and the training of foreign nationals in the life sciences here in the United States.

Recommendation 4c. Support programs promoting beneficial uses of technology in developing countries.

As highlighted in Chapter 1, advancing technologies possess a "dark side"—their potential to be used with the intent to cause harm. While this is the focus of much of this report, the "bright side" of advancing technologies holds great promise for health and economic development, especially for people in developing countries. Significantly, there is evidence that developing countries themselves—especially the "innovating developing countries" such as India, China, Brazil, and South Africa—are harnessing biotechnology and other emerging technologies to meet their local health needs. Biotechnology, nanotechnology, and other emerging technologies have the potential to improve human security by addressing threats to human security such as disease and hunger.[88] Moreover, continued progress in this sector, with structural reforms in the science, technology, and innovation systems of developing countries, will be crucial to meeting the UN Millennium Development Goals.[89]

However, this biodevelopment agenda is on a potential collision course with the biosecurity agenda. This is exemplified by the restrictions on U.S. visas for foreign scientists described in Chapter 2 and in the requirement that NIH grantees directing research on emerging infections in a developing country comply with American laws and regulations concerning select agents at their overseas study sites. Many developing nations face urgent public health crises, including outbreaks of emerging infectious diseases, on a daily basis. There are legitimate questions about whether and how such countries should respond to the risk of biological terrorism. Few of these countries are likely to perceive themselves as being at risk or that the risk is significant against the backdrop of the natural infectious disease threats they face daily. Some analysts warn that, in some cases, biodefense policies designed to prevent or mitigate the risk of a bioterrorist attack could create hardships and even be counterproductive—for example, by pressuring countries burdened with other problems to satisfy regulatory and other biodefense-related demands.[90] Requirements to establish a regulatory authority and to promulgate intricate safety and protection measures with respect to select agent pathogens could divert already scarce resources from less formal but more immediately effective operational systems in place for treating sick or vulnerable populations.[91]

Biosecurity should not, and need not, come at the expense of lost potential for promoting health and economic development in developing nations through biotechnology. Efforts to promote the development of peaceful uses of biotechnology in poorer countries can enhance biosecurity by strengthening international relationships. These relationships provide opportunities for building a common culture of awareness and responsibility. As we defend against the dark side of the life sciences, the bright side should continue to shine—not only because the lives of millions in the developing world may depend on it, but also because it is likely to promote a common global approach to the dual-use conundrum.

> **Recommendation 4d. Establish globally distributed, decentralized, and adaptive mechanisms with the capacity for surveillance and intervention in the event of malevolent applications of tools and technologies derived from the life sciences.**

Under this recommendation, the committee envisions the establishment of a decentralized, globally distributed network of informed, concerned scientists who have the capacity to recognize when knowledge or technology is being used inappropriately or with the intent to cause harm. This network of scientists and the tools that they use would be adaptive in the sense that the capacity for surveillance and intervention must evolve along with advances in technology. Such intervention could take the form of informal counseling of the offending scientist when the use of these tools appears unwittingly inappropriate or reporting such activity to national authorities when it appears potentially malevolent in intent.

The rapid pace of growth in the life sciences and its associated technologies—as described in Chapters 1 and 3—can lead to the unexpected emergence of new techniques and entirely new disciplines (e.g., RNA interference) in a very short period of time. Scientists working in the life sciences are best suited to recognize the dual-use implications of these newly emerging technologies and fields of knowledge, but they must develop a broadly distributed culture of awareness and responsibility if they are to recognize and shed light on potentially dangerous activities as they occur.

Because of the key features of this proposed "bottom-up" culture of awareness and responsibility—its globally distributed and decentralized adaptability—the committee likened it to the mammalian immune system, arguably the most spectacular example of a spatially distributed, decentralized, adaptive system. The hallmark of the mammalian immune system is its ability to respond to transgressions by microorganisms in ways that limit the growth of the transgressor and afford protection against its detrimental consequences. The responses of the immune system include both specific (adaptive immune system) and nonspecific (innate immune system) components. These are intricately linked but react in different ways to structures (antigens) that are foreign to the host. The innate immune system includes components that are present and pre-programmed for action even before an antigen challenge is encountered. The adaptive immune system, on the other hand, involves components that react to an antigen challenge with a high degree of specificity but only after some delay.[92] Through a complex network of local mechanisms involving both the innate and the adaptive immune systems, the essential global functions of immune surveillance, recognition, response, learning, and memory are constantly adapting to new microbial threats without central direction. Perhaps the global scientific community could fruitfully mimic this system.

The analogy between the global scientific community and the mammalian immune system is intended to be merely illustrative, not strict. The concept proceeds from two salient facts. First, as argued throughout the previous chapters of this report and under Recommendation 2, life science technologies with potential for dual use are developing and diversifying very rapidly. Any controlling mechanisms must therefore be dynamic and adaptive to the rapid pace of technological change. Second, as argued throughout Chapter 2 and above, the global decentralized nature of the problem demands that strategies for anticipating, identifying, and mitigating potential future threats must necessarily have global reach. Despite the existence of international conventions and related national legislation, no "top-down" solution presents itself at the moment with respect to the global regulation of dual-use agents and knowledge.

Given that unanticipated threats are virtually certain to emerge, decentralized and adaptive solutions, while potentially limited in effectiveness, are nonetheless of substantial interest. Their usefulness may be limited to their ability to engender public opprobrium, but active steps to promote the development of distributed, decentralized networks of scientists will at the least heighten awareness while potentially enhancing surveillance. These networks might be linked through a system analogous to the Program for Monitoring Emerging Diseases, which hosts the ProMED-mail Web site (see Box 4-3).[93] ProMED-mail was established in 1994 with the support of the Federation of American Scientists and SatelLife. Since October 1999, it has operated as an official program of the International Society for Infectious Diseases, a non-profit professional organization with 20,000 members worldwide. The ProMED-mail Web site has become an extremely useful locus for the posting of reports of infectious disease outbreaks by any concerned infectious disease specialist or expert, or lay person, including press reports, from around the globe. Such reports, while

**BOX 4-3
ProMED-mail**

ProMED-mail—the Program for Monitoring Emerging Diseases—a spinoff of the nonprofit Program for Monitoring Emerging Diseases, and now a project of the International Society for Infectious Diseases—is an Internet-based reporting system dedicated to rapid global dissemination of information on outbreaks of infectious diseases and acute exposures to toxins that affect human health, including those in animals and in plants grown for food or animal feed. Electronic communications enable ProMED-mail to provide up-to-date and reliable news about threats to human, animal, and food plant health around the world, seven days a week.

Among the outbreaks first reported on ProMED-mail were the early reports of SARS in both China and Toronto in 2003; Venezuelan equine encephalitis in Venezuela in 1995; H5N1 influenza in Indonesia November 2003; and the 2005 outbreak of human disease in China attributed to *Streptococcus suis*.

By providing early warning of outbreaks of emerging and reemerging diseases, public health precautions at all levels can be taken in a timely manner to prevent epidemic transmission and to save lives.

ProMED-mail is open to all sources and is free of political constraints. Sources of information include media reports, official reports, online summaries, local observers, and others. Reports are often contributed by ProMED-mail subscribers. A team of expert human, plant, and animal disease moderators screen, review, and investigate reports before posting to the network. Reports are distributed by e-mail to direct subscribers and posted immediately on the ProMED-mail Web site. ProMED-mail currently reaches over 30,000 subscribers in at least 150 countries.

A central purpose of ProMED-mail is to promote communication among the international infectious diseases community, including scientists, physicians, veterinarians, plant pathologists, epidemiologists, public health professionals, and others interested in infectious diseases on a global scale. ProMED-mail encourages subscribers to participate in discussions on infectious disease concerns, to respond to requests for information, and to collaborate in outbreak investigations and prevention efforts. ProMED-mail also welcomes the participation of interested persons outside the health and biomedical professions.

often uncertain in their accuracy or significance early on, prompt the attention of recognized infectious disease experts who moderate and help facilitate an international Web-based dialogue, including comments on what is and is not known about the suspect disease. Although supported by a specific organization operating a centralized Web site, the reporting system represented by ProMED-mail is essentially decentralized, distributed, and adaptive. It has no direct investigative or public health authority, but it serves as an early warning system, capable of earlier recognition of disease outbreaks than established institutional systems of public health surveillance and free of the potential political constraints on reporting infectious diseases that may be felt by national epidemiological and public health reporting systems.

A useful parallel to ProMED-mail would seem possible in the creation of a similarly distributed system for reporting potential inappropriate applications of emerging life sciences research and technologies. Candidate activities for reporting might include, for example, (1) experiments leading to the insertion of certain genes (e.g., interleukin-4) into known pathogens (e.g., orthopox viruses) *for no identifiable therapeutic or scientific reason;* (2) directed evolution (breeding) of novel pathogens *for no identifiable therapeutic or scientific reason*; or (3) the acquisition of supplies, equipment, or biological reagents by groups or individuals *in the absence of any identifiable appropriate scientific aim.*

Unanticipated results that generate a new and substantial dual-use threat need not be considered indicative of malevolent intent by the individuals involved in such a distributed reporting system. It is possible, indeed likely, that novel pathogens or other dual-use technologies of security concern will emerge through sheer serendipity in the course of legitimate research—that is, research undertaken and funded explicitly for identifiable therapeutic or bone fide scientific purposes; neither incompetence, idle curiosity, nor intentional malevolence need be involved. The research may actually be consistent with what had been initially proposed, peer reviewed, and funded by government or not-for-profit agencies. For example, the introduction of interleukin-4 into ectromelia virus, an experiment that was supported by the Australian government, aimed to improve vaccine responses but achieved quite different and unexpected results.[94] Nonetheless, a reporting system similar to ProMED-mail can call attention to the hazards of such experiments, and thereby sensitize the scientific community to their potential implications.

Unlike ProMED-mail, however, where the adversary is Mother Nature (often abetted by human activities impacting on the environment for infectious disease transmission), it is possible that the posting of certain information concerning dual-use applications of life sciences technology on a public Web site could have unintended negative consequences, perhaps informing those with purposeful malevolent intent. Such postings would thus need to be screened by a group of informed and concerned moderators, as they are today for ProMED-mail. However, the main intent of such a distributed reporting system would be to promote the free flow of information in real time, with the view, as expressed in Recommendation 1, that the open and free exchange of information may be one of the most important means of ensuring that risks are considered, appropriate countermeasures are developed, and possible consequences are mitigated in a timely fashion.

In the event that it is a colleague or superior who is engaged in the questionable activity, a scientist may need a way to report the suspect activity anonymously (so that fears of reprisal do not deter reporting). One possibility is that a Web site could be maintained by the science community (like Linux[95]). Even with the most secure technology, however, in any country where penalties may be grave, there will still be deterrence against reporting questionable activities if there is any perception that it might lead to the identification of the reporting person. Efforts would also need to be taken to ensure that inappropriate allegations are not made against scientists in situations where the reporter may be trying to right a perceived wrong or to "get even" with an individual with whom they have a personal or professional dispute. The committee acknowledges these issues yet believes that an open global forum of the type envisioned may be able to overcome some of these problems by shining the light of public attention on them and that such a forum will prove useful despite such obvious limitations.

The committee was under no illusions that interventions and responses by the global scientific community that do not involve responses by law enforcement agencies—for example, the threat of professional ostracism and/or academic sanction—would deter potential terrorists or determined state actors. Presumably, few terrorists worry about their stature in the scientific community or tenure at an institution. The distributed reporting and response network described above would be directed primarily at the *embedded community* of legitimate scientists, its aggregate aim being to stimulate creativity in anticipating activity that could be malicious and to stimulate vigilance in detecting and reporting such activity. The collective experience of the entire scientific community would be accumulated into one online memory, available to participants in the network.

The existence of such a network could profoundly alter the risk calculus for potential offenders. That is, they would know that the embedding community is alert to anomalous behavior; and, when appropriate, can alert enforcement agencies that are capable of formal investigation, at least in those countries that have enacted appropriate national legislation. Indeed, it is probable that security agencies in multiple countries would monitor this reporting network, for both good and possibly also inappropriate reasons. Again, the aim is to self-organize a body of norms and a climate of vigilance across the global community of legitimate scientists in order to change the risk calculus of potential offenders.

Admittedly, there is a thin line between vigilance and vigilantism. The former is the state of being watchful (i.e., without necessarily acting), whereas the latter refers to a reactive behavior. The presence of vigilantism could be as devastating as the absence of vigilance. Frivolous charges will need to be deterred and censured as surely as legitimate ones need to be followed up. In the social sciences, this involves the notion of a metanorm.[96] Metanorms already exist in the scientific community. For example, it is a central meta-norm of researchers to report falsification of data or abuse of human subjects, and it is seen as a violation of the meta-norm to *not* do so, given knowledge. Frivolous witch hunts and overreporting are considered violations as well. The search for a balanced strategy between under- and overreporting may take time and effort, but is probably worth the investment.

Other methodologies, in addition to these Internet-based approaches, may contribute to the development of a globally dispersed sense of awareness and responsibility on the part of legitimate scientists. Social norms, conventions, and institutions of many sorts emerge without central direction and are maintained by local conformity effects.[97] Educational efforts, scientific exchanges, international conventions, and codes of conduct and ethics—in effect, all of the measures suggested above—can contribute to the development of such norms for the global life sciences community over time. Once in effect, social scientists would view these norms as stable equilibria—social configurations from which no individual has any incentive to depart. Recognizing departures from the scientific norm will require subtle discrimination. Yet humans are capable of developing a very finely tuned sense of those behaviors that fall within a social norm and those that do not.

CONCLUSION 5: REGARDLESS OF THE STEPS TAKEN TO PREVENT SUCH EVENTS, THE COMMITTEE CONCLUDES THAT THERE IS A NEED TO RECOGNIZE THE VIRTUAL INEVITABILITY OF THE MALEVOLENT APPLICATION OF NEW TECHNOLOGIES AND AN OVERARCHING NEED FOR A RAPID, AND EFFECTIVE RESPONSE TO MITIGATE THE CONSEQUENCES SHOULD SUCH AN EVENT OCCUR.

Human history is replete with applications of technology for hostile purposes, and indeed the committee can think of no major category of technology that has not been used for such. The life sciences are no different, and it is only reasonable to expect that a technology derived from the life sciences will be used for malevolent purposes in the future. This likelihood dictates the need to be prepared for a rapid and effective response.

Recommendation 5

The committee notes with urgency the need to enhance public health infrastructure, achieve greater coordination among responsible federal agencies, and substantially strengthen existing response and recovery capabilities.

The committee recognizes that all of its recommended measures, taken together, provide no guarantee that continuing advances in the life sciences—and the new technologies they spawn—will not be used with the intent to cause harm. No simple or fully effective solutions exist where there is malevolent intent, even in cases where only minimal resources are available to individuals, groups, or states. Thus, the committee's recommendations recognize a critical need to strengthen the public health infrastructure and our existing response and recovery capabilities. In keeping with the focus of this report, the committee urges that the insights and potential benefits gained through advances in the life sciences and related technologies be fully utilized in the development of new public health defenses. It must be noted, however, that many of the concepts and suggestions embodied in these recomendations were articulated in the 2002 National Research Council Report, *Making the Nation Safer: The Role of Science and Technology in Countering Terrorism* (see 69-79) and remain as relevant and needed today as they were then.

An effective civil defense program will require a well-coordinated public health response, and this can only occur if there is strong integration of well-funded, well-staffed, and well-educated local, state, and federal public health authorities. Despite substantial efforts since September 11, 2001, few if any experts believe that the United States has achieved even a minimal level of success in accomplishing this goal, which is as important for responses to naturally emerging threats, such as pandemic influenza, as for the threat of a deliberate biological attack against one or more population centers. Current efforts to accomplish these aims have been woefully ineffective and have not provided the nation with the infrastructure it needs to deal rapidly, effectively, and in a clearly coordinated manner when faced with a catastrophic event such as an overwhelming tropical cyclone, a rapidly spreading pandemic, or a large-scale bioterror attack. These efforts need to be enhanced and expanded.

In making the recommendations that follow below, the committee recognizes that similar ones have been made in many different settings in response to the challenges of bioterrorism.[98] However, it is not dissuaded by the lack of novelty in such recommendations, given their overriding importance, and given the fact that, despite efforts to accomplish many of these goals, much remains to be done before the nation can be considered to be protected by the best possible public health infrastructure. In keeping with the focus of this report, the committee urges that the insights and potential benefits gained through advances in the life sciences and related technologies be fully utilized in the development of new public health defenses.

> Recommendation 5a. Strengthen response capabilities and achieve greater coordination of state, local, and federal public health agencies.

It remains unclear how the country's response to a future biological attack will be managed. The committee remains concerned about how the responses of many different federal departments (e.g., the Departments of Homeland Security, Health and Human Services, Justice, Defense, and the myriad agencies within them) will be effectively integrated and who will control operations and ensure that they are adequately interfaced with local and state governments and public health agencies. Although well beyond the scope of the committee's charge, the development of an effective means of integrating the responses by multiple governmental agencies would provide the nation with perhaps the most necessary of "tools" with which to meet any future challenge. Even current efforts to develop preventive measures are poorly coordinated and inappropriately placed into administrative "silos" with inadequate cross-fertilization and communication (e.g., environmental pathogen detection in Homeland Security versus disease diagnosis in Health and Human Services, or human infections in Health and Human Services and animal and zoonotic diseases in Agriculture and Homeland Security). Such an arrangement does not serve the nation's needs well.

With the profusion of federal public health, environmental, law enforcement, defense, and security agencies now engaged in various aspects of prevention, response, mitigation, and attribution in the event of a putative bioterrorist attack, the need for better integration and a clear command and control structure is critical. Rather than considering agents of biological origin as simply another form of weapons of mass destruction, such agents should be placed within the context of naturally emerging infectious diseases, and the public health measures needed to combat them.[99] "Defense" in the case of biological security means, above all, improvements in domestic and international disease surveillance and response.[100] Current efforts to accomplish these aims should be enhanced and expanded, and federal, state, and local governments (working with the best advice of the scientific community) should carry out a variety of communications activities, through both targeted and mass media efforts, to inform members of the public as to what they may expect during a biological event and what realistic and practical steps they could take to protect themselves.

Recommendation 5b. Strengthen efforts related to the early detection of biological agents in the environment and early population-based recognition of disease outbreaks, but deploy sensors and other technologies for environmental detection only when solid scientific evidence suggests they are effective.

Efforts are needed to improve the abilities of both the health care and public health communities to quickly detect disease outbreaks in human, plant, and animal populations caused by the intentional release of a biological agent. Ideally, surveillance systems should be sensitive enough to identify the emergence of an outbreak, categorize its nature, and identify those populations affected, so that an outbreak can be quickly and effectively contained. There are a variety of possible approaches today; some of them are based on the collection and analysis of population-based clinical, epidemiological, and even sociological data, including the number and nature of emergency room visits, types of prescriptions, calls to physicians, and so forth, and the careful application of public health informatics and computational modeling of epidemics. Other approaches in the future might be based on real-time monitoring of biological markers in individuals, on a massively parallel scale, including molecular markers of host response and profiles of indigenous microbial communities. It was beyond the scope of the committee's charge to develop specific recommendations concerning how current epidemic surveillance efforts could be enhanced, but the committee recognizes a clear need to accelerate current efforts to do so.

There is also a need to enhance present capabilities for detecting the presence of a biological agent in the environment, measuring its abundance, and determining the level of associated risk to the health of the potential target host (human, animal, plant, etc.). Given that the number of infectious agents may be exceptionally low following their dispersal, there is a need to develop and evaluate new technologies to improve currently available monitoring and detection systems, as well as a need to characterize a wide variety of environments over time, in the absence of a health threat (i.e., the "background"). The difficulties of accomplishing this should not be underestimated. However, advances in the life sciences and biotechnology will aid in this task—providing yet another reason to promote the general advance of research.

Efforts are now under way at present to develop a variety of new biosensors that can rapidly detect one or more potential bioweapon agents. Questions remain about how such sensors can best be deployed to provide maximum surveillance capability at a cost that will be affordable. Communications efforts also will be needed to explain to the public the merits—and limitations—of such detection systems and to prepare the public so that people can respond appropriately in the event that detection systems trigger an alert. Again, specific recommendations are beyond the scope of this committee's charge. However, while recognizing the importance and potential utility of such sensor systems, the committee cautions against implementing monitoring activities with such devices without compelling data to support their effectiveness.

> **Recommendation 5c. Improve capabilities for early detection of host exposure to biological agents of disease and early diagnosis of disease caused by them.**

Establishing a specific diagnosis is critical to implementing an appropriate public health response to a bioterrorism-related event, since the diagnosis will guide the use of specific therapies, immunizations, and other interventions. Efforts should thus be aimed at increasing the awareness of primary care and specialist clinicians to the potential for disease outbreaks initiated by the release of biological agents. As indicated in Recommendation 2, a broader perspective on the range of potential threats is essential, particularly in this age when relatively simple genetic engineering might easily change the pathogenicity of a relatively harmless microorganism. There is a strong need to improve the ability of clinicians to detect, report, and respond appropriately to patients who present with symptoms or signs consistent with a biological attack.

By making a specific diagnosis before the alarm has been raised, or by recognizing that a patient's clinical presentation lies outside the expected norm, a physician confronting an early case in an epidemic or a biological attack can make a uniquely important contribution to the timeliness of the public health response.[101] Better training should also be coupled with the availability of enhanced diagnostic tools that will provide physicians with a "real-time" bedside capacity to identify unusual infectious agents in patients with suspicious clinical signs and symptoms.

Early disease diagnosis, even prior to the onset of typical symptoms, should be the goal of research and development efforts. While it is reasonable to hope that improved diagnostic tests will be developed as a result of current federal biodefense research efforts, it is not clear that adequate attention, prioritization, or investment has been devoted to this important area, or that all of the potentially useful approaches (e.g., comprehensive monitoring of host-associated molecular biological markers) have been adequately explored. There is a similar need for early recognition and diagnosis of animal and plant diseases. As with Recommendation 5b, many of the concepts and suggestions mentioned above, were articulated in the 2002 National Research Council Report, *Making the Nation Safer: The Role of Science and Technology in Countering Terrorism* and remain as relevant and needed today as they were then.

Recommendation 5d. Provide suitable incentives for the development and production of novel classes of preventative and therapeutic agents with activity against a broad range of biological threats, as well as flexible, agile, and generic technology platforms for the rapid generation of vaccines and therapeutics against unanticipated threats.

No credible defensive effort can move forward without accelerating the rate at which vaccines and other preventatives and therapeutic agents are developed. Having effective vaccines available not only will help protect U.S. citizens and military personnel, but limiting the efficacy of biological weapons will reduce the attractiveness of such weapons and thereby offer some means of deterring their use. Continued research is needed to develop, or in some cases improve, vaccines against specific biological agents that are already of concern (e.g., anthrax, smallpox, influenza) as well as to develop the capacity to design and produce new vaccines rapidly in response to new threats, including threats that might emerge from advances in the life sciences. A particularly desirable goal would be to develop a single vaccine or biological response modifier capable of providing protection against a relatively large class of diseases. To date, well-established companies in the pharmaceutical and vaccine industries have had little financial incentive to develop new vaccines or therapeutics for biological threat agents for which the market is extremely uncertain and dependent ultimately on government procurement decisions. Therefore, the government's accomplishments in these areas have fallen far short of the goals regarding development of new vaccines and therapeutics.

The Bush administration's $5.6 billion BioShield initiative sought to solve this problem by placing large sums of money at the finish line, as it were, allowing purchase following the development of an effective countermeasure. However, there is no evidence to date that this has succeeded, due to a variety of concerns on the part of "big pharma," including the reliability of the government as a development partner, its previous threat to invoke eminent domain when concerned about the price of ciprofloxacin proposed by Bayer during the anthrax attacks of 2001, and (particularly in the case of vaccines) continuing worries about liability exposure.[102] For small biotech companies with limited venture capital funding, BioShield represents a potential windfall, but for the large publicly traded industry giants that are most capable of delivering these products, the opportunity costs are unacceptable related to the diversion of corporate research resources.[103]

Many pathogens are becoming resistant to today's antibiotics, and no new classes of drugs have been developed in recent years.[104] Already a problem for naturally occurring diseases, the dearth of new antibiotics may prove especially troublesome in the event of a biological attack with an engineered bacterial agent. In the initial phases of such an event, the level of antibiotic susceptibility will not be known, and more than one agent may be released simultaneously. Thus, new classes of broad-spectrum antibiotics are urgently needed, both for naturally acquired infections and to guard against the possibility of attacks with microbial agents resistant to current therapeutics. Advances in the fields of genomics, cell biology, structural biology, and combinatorial chemistry have resulted in the rapid development of some new antiviral agents, but no broad-spectrum antiviral agents are on the market, and there are no specific antivirals that are effective against the majority of the RNA and DNA viruses of concern. Expanded efforts are needed to develop new antiviral agents for specific diseases, but there is also a need to consider novel ways in which broad-spectrum antiviral agents could be developed. This might include the development of novel classes of immunomodulators for those agents for which there are no available therapeutics or vaccines.

Finally, in an age that bears witness to many newly emerging infections as well as the growing threat associated with the inadvertent or intentional creation of novel agents of biological origin, it is critical that the time to develop and license new therapeutics and vaccines be substantially shortened. The many years required for successful development and licensure of either drugs or vaccines is inconsistent with the flexible, agile responses required. The use of RNA silencing technologies offers promise for the rapid, sequence-specific development of therapeutic and possibly preventative antiviral compounds. Although many questions remain about the ultimate safety and efficacy of such approaches, the ability of this technology, as an example, to serve as a platform for rapid development of needed drugs makes it very attractive. Similarly, there is a need to develop vaccine platforms that are capable of being rapidly utilized to express novel immunogens and to elicit protective immunity against a newly appearing biological threat.

Again, it is not clear who might address these goals or how successful any such attempts will be. The committee believes that these are very important goals, however, and that their success will depend on the ability of science to continue to advance without being unduly fettered by overrestrictive laws and regulations, as well as novel approaches providing appropriate financial incentives to the industry entities most able to meet these challenges.

SUMMARY

Because it believes that continuing advances in the life sciences and related technologies are essential to countering the future threat of bioterrorism, the committee's recommendations affirm policies and practices that promote the free and open exchange of information in the life sciences. It also recognizes the need to adopt a broader perspective on the nature of the threat spectrum, and to strengthen the scientific and technical expertise available to the national security communities so that they are better equipped to anticipate and manage a diverse array of novel threats. Moreover, due to the global dispersion of life sciences knowledge and technological expertise, the committee recognizes the international dimensions of these issues, and makes recommendations that call for the global community of life scientists to adopt a common culture of awareness and a shared sense of responsibility and include specific actions that would promote such a culture.

No single recommendation by itself can provide a guarantee against the eventual successful use of the life sciences and related technologies for malevolent purposes. Rather, the actions and strategies that the committee recommends are intended to be complementary and synergistic. An effective system for managing the threats that face society will require a broad array of mutually reinforcing actions in a manner that successfully engages the variety of different communities that share stakes in the outcome. As in fire prevention, where the best protection against the occurrence of and damage from catastrophic fires comprises a multitude of interacting preventive and mitigating actions (e.g., fire codes, smoke detectors, sprinkler systems, fire trucks, fire hydrants, fire insurance), rather than any single "best" but impractical or improbable measure (e.g., stationing a fire truck on every block), the same is true here. The committee envisions a broad-based, intertwined network of steps—a *web of protection*—for reducing the likelihood that the technologies discussed in this report will be used successfully for malevolent purposes. The committee believes that the actions suggested in its recommendations, taken in aggregate, will likely decrease the risk of inappropriate application or unintended misuse of these increasingly widely available technologies.

Nonetheless, the committee recognizes that all of its recommended measures, taken together, cannot guarantee that continuing advances in the life sciences and the new technologies they spawn will not be used with the intent to cause harm. No fully effective solution exists where there is malevolent intent. The committee therefore reaffirms previous calls to strengthen the public health infrastructure and the nation's existing response and recovery capabilities, as it believes such steps will be essential for the early detection of malevolent applications and for mitigating the loss of life or other damage sustained by society in both the short and the long term should the worst-case scenario occur.

ANNEX 4-1

June 16, 2005

Secretary Carlos M. Gutierrez
Office of the Secretary
U.S. Department of Commerce
Room 5516
14th Street and Constitution Avenue, N.W.
Washington, DC 20230

Dear Secretary Gutierrez:

We appreciate this opportunity to provide comments on the advanced notice of proposed rule-making (ANPR) on "Revisions and Clarification of Deemed Export Related Regulatory Requirements. One of the key roles of the National Academies, consistent with our 1863 Congressional Charter, is to advise the nation on important issues involving science, engineering, and medicine such as this one. The members of our three honorary academies—the National Academy of Sciences, the National Academy of Engineering, and the Institute of Medicine—and the scientific experts who serve on the study committees of our operating arm, the National Research Council, are working at industrial, academic, and governmental institutions that are potentially affected by the proposed regulatory changes. We provide these comments in light of our background and experience with the U.S. scientific, engineering, and medical enterprise.

Our most important observation is the following: We believe <u>the rule changes that are being recommended by the Inspector General and the interpretation of existing regulations that are now being widely disseminated will serve to *weaken* both national security and the economic competitiveness of the United States.</u> The impact will likely be to dramatically hinder American scientific, engineering, and health care research and innovation, factors that have been so vital to our quality of life.

The clearest problem now is that universities and industry are unable to specify the expected impact of attempting to comply with these rules. We believe that the Department needs to address the following issues in the existing and proposed rules before we can provide you with a categorical response and before the Department determines which interpretations and rule changes to the Export Administration Regulations, if any, will make the nation safer.

First, the problems that these rule changes and new interpretations are attempting to address, as well as the costs and benefits of different regulatory approaches, need to be clarified. It is not simply that the affected communities will be more accepting of the need to tighten rules if they understand why (although that will help), but that complex problems require focused and tailored solutions. The measures being contemplated by the Department could be too broad, too narrow, or possibly irrelevant depending on whether one defines the challenge as primarily countering terrorist activities, political adversaries, or economic competitors.

Second, the new interpretations and proposed changes could eviscerate the Fundamental Research Exemption as enunciated in National Security Decision Directive (NSDD)-189 and reconfirmed by Secretary of State Rice and former Energy Secretary Spencer Abraham in November 2001 and May 2003, respectively. We favor a crisply defined regulatory "safe harbor" for fundamental research, so that universities can have confidence that activities within the "safe harbor" are in compliance, and so that the vital importance to national security of open fundamental research is reaffirmed as a matter of national policy. The new regulatory machinery could then be focused on university activities, if any, occurring OUTSIDE the "safe harbor." Such activities might be conducted in separate facilities, or even off campus. And if the regulatory "safe harbor" is properly defined and constructed, a number of universities might not even have any such activities.

Third, it is necessary to determine whether the perceived national security benefits are worth the cost that universities and industry will incur to implement these proposed changes. While the financial costs would be a burden, both sectors would find ways to manage them over time. Of much greater concern is that these measures will pose an irretrievable cost to our nation—especially our competitiveness and national security which have relied so heavily for the last sixty years on the fruits of technology derived from basic science, and bringing the "best and brightest" people from other countries to the U.S. Losing the "best and the brightest" foreign students and researchers to other countries because they feel unwelcome here will have very serious consequences for the future of America. Eleven of the last 45 winners of the Nobel Prize in science[105] from 1999-2004 were foreign-born Americans. In the same timeframe, fifteen of the last 51 recipients of the National Medal of Science, an annual award made by the U.S. President, were also immigrants to the United States.

Fourth, it is necessary to assess whether these particular measures will not *in fact* staunch the flow of scientific information to potential terrorists, adversaries, and/or competitors. In a world where access to information is increasingly global, those who intend to do harm to the United States will simply go elsewhere for the scientific or technological information they seek; the U.S. is far from the only advanced, research-capable country.

These four issues are manifestations of a single principle of U.S. policy concerning classified information: "Construct high fences around narrow areas." This refers to maintaining stringent security around sharply defined and narrowly circumscribed areas of *critical* importance in order to be able to maintain simultaneously the highest levels of national security and of scientific research. This principle was originally articulated in *A Review of the Department of Energy Classification: Policy and Practice* (1995)[106] and acknowledges that an attempt to protect *everything* in fact dilutes attention and protects *nothing*. It is our sense that the recommendations expressed by C.D. Mote, President of the University of Maryland, at the National Academies' May 6th workshop on the Department of Commerce Inspector General's Report on deemed export policy, could help to operationalize this principle in the area of deemed exports. We urge you to give them serious consideration as a first step:

1.   Greatly narrow the scope of controlled technologies requiring deemed export licenses and ensure the list remains narrow going forward.

2.   Delete all controlled technology from the list whose manuals are available in the public domain, in libraries, over the Internet, or from the manufacturers.

3.   Delete all equipment from the list that is available for purchase on the open market overseas from foreign or U.S. companies.

4.   Clear international students and postdoctoral fellows for access to controlled equipment when their visas are issued or shortly thereafter so that their admission to a university academic program is coupled with their access to use of export-controlled equipment.

5.   Do not change the current system of license requirements for use of export-controlled equipment in university basic research until the above four recommendations have been implemented.

To date, the Commerce Department has gained substantial goodwill within the science, engineering, and medical communities through its policy of openness in discussing and seeking comments on these rules. We give considerable credit to you and other responsible officials, such as Peter Lichtenbaum of the Bureau of Industry and Security, who have openly and willingly embarked on a dialogue that will ultimately make the research community more aware of how to secure our most advanced technologies from hostile entities. At the same time, we strongly recommend the Department embark on responses to the communities' concerns before implementing regulations that may chill ongoing research of critical importance to the future of the U.S.

Sincerely,

Bruce Alberts	Wm. A. Wulf	Harvey V. Fineberg
President	*President*	*President*
National Academy	National Academy	Institute
of Sciences	of Engineering	of Medicine

cc: Peter Lichtenbaum, Assistant Secretary of Commerce for Export Administration, Department of Commerce
Bureau of Industry and Security, Regulatory Policy Division, ATTN: RIN 0694-AD29

ANNEX 4-2

Recommendations for Enhancing the U.S. Visa System to Advance America's Scientific and Economic Competitiveness and National Security Interests[107]

May 18, 2005

Following the terrorist attacks of September 11, 2001, the U.S. government put in place new safeguards in the nation's visa system that made it extremely challenging for bona fide international students, scholars, scientists, and engineers to enter this country. While intended to correct weaknesses exposed by the attacks, the changes proved to be significant barriers for legitimate travelers and created a misperception that these visitors were no longer welcome here.

Other countries have used this opportunity to attract these individuals to their own educational, scientific, and technical institutions. In addition, key sending countries have enhanced their higher education systems in an effort to keep their best students at home.

Despite significant recent improvements to the U.S. visa system, considerable barriers remain that continue to fuel the misperception that our country does not welcome these international visitors, who contribute immensely to our nation's economy, national security, and higher education and scientific enterprises. These misperceptions must be dispelled soon, or we risk irreparable damage to our competitive advantage in attracting international students, scholars, scientists, and engineers, and ultimately to our nation's global leadership.

One year ago, most of the undersigned organizations of higher education, science, and engineering, in an effort to enhance national security and international exchange made a joint commitment to work with the federal government to make sensible changes to the visa system. We recommended several improvements, some of which have been adopted in the past year. Today we come together again to express gratitude and support for the changes that have been made, to continue to urge approval of those that have not, and to recommend additional improvements, so that America can continue to compete for and welcome the world's best minds and talents. We offer the following recommendations in the spirit of cooperation that has already resulted in improvements to the visa system:

- **Extend the validity of Visas Mantis security clearances for international scholars and scientists from the current two-year limit to the duration of their academic appointment.** While we appreciate that the limit has already been extended from one year to two years, this further extension would be comparable to that already provided for international students and would prevent redundant security checks that can waste resources and cause unnecessary delays and hardships.

- **Allow international students, scholars, scientists, and engineers to renew their visas in the United States.** Allowing individuals to complete, or at least initiate, the visa revalidation process before leaving the country to attend academic conferences or to visit family would reduce, and in many cases eliminate, visa delays, thus permitting them to continue their studies and research uninterrupted.

- **Renegotiate visa reciprocity agreements between the United States and key sending countries, such as China, to extend the duration of visas each country grants citizens of the other and to permit multiple entries on a single visa.** We applaud the State Department's initial efforts to achieve this and encourage continued efforts. Improved reciprocity would allow the federal government to focus its visa screening resources by reducing the number of visa renewals that must be processed.

- **Amend inflexible requirements that lead to frequent student visa denials.** The Immigration and Nationality Act of 1952 should place greater emphasis on student visa applicants' academic intent and financial means to complete a course of study in the United States, instead of their ability to demonstrate evidence of a residence and employment in their home country and their intent to return home. Up to 40 percent of student visa applicants from key sending countries are rejected because they are unable to demonstrate to the satisfaction of consular officials their intent and ability to return home after completing their studies. The United States is losing too many top students to this policy, and the act should be revised.

- **Develop a national strategy to promote academic and scientific exchange and to encourage international students, scholars, scientists, and engineers to pursue higher education and research opportunities in the United States.** In addition to visa reforms, this strategy should include a plan to counter prevailing negative perceptions of studying and conducting research in the United States and should promote study abroad by American students.

The following recommendation, while not related to visa issuance, addresses a potential barrier to international scientists and engineers seeking to study and conduct research in the United States.

• **The federal government should not require that export licenses be obtained for international scientists and engineers to use equipment required to conduct unclassified, fundamental research in the United States.** The Department of Commerce is considering expanding existing regulations to require that licenses be obtained before certain foreign nationals are permitted access to specialized scientific equipment required for unclassified, fundamental research. Requiring such licenses would further discourage top international scientists and engineers from making the United States their destination, prompting them to seek research opportunities overseas.

Lastly, it is essential that adequate resources continue to be provided by Congress and the Administration to administer an effective visa system and to implement the above recommendations.

We reiterate our commitment to work with the federal government to improve the visa system. That system should maintain our nation's security by preventing entry by those who pose a threat to the United States and encouraging the entry of the brightest and most qualified international students, scholars, scientists, and engineers to participate fully in the U.S. higher education and research enterprises. Such a system will foster American scientific and economic competitiveness. We commend the Administration for the improvements made to the visa system to date, and we look forward to continuing to work together for these further needed changes.

[signed]

Nils Hasselmo
President
Association of American
   Universities

Alan I. Leshner
President
American Association for the
   Advancement of Science

Bruce Alberts
President
National Academy of Sciences

David Ward
President
American Council on Education

C. Peter Magrath
President
National Association of State
   Universities and Land Grant Colleges

Wm. A. Wulf
President
National Academy of Engineering

Harvey V. Fineberg

Deborah L. Wince-Smith

President
Institute of Medicine

Marlene M. Johnson
Executive Director and CEO
NAFSA: Association of
  International Educators

Debra W. Stewart
President
Council of Graduate Schools
Education

Constantine W. Curris
President
American Association of State
  Colleges and Universities

Jerry P. Draayer
President and CEO
Southeastern Universities
  Research Association

Gerard A. Alphonse
2005 President
The Institute of Electrical and
  Electronics Engineers—USA

Eugene Arthurs
Executive Director
SPIE—The International Society
  for Optical Engineering

Rev. Charles L. Currie
President
Association of Jesuit Colleges
  and Universities

Judith Bond
President
American Society for Biochemistry
  and Molecular Biology

President
Council on Competitiveness

Marvin L. Cohen
President
American Physical Society

Allan E. Goodman
President and CEO
Institute of International

James M. Tiedje
President
American Society for
  Microbiology

Paul W. Kincade
President
Federation of American Societies
  for Experimental Biology

David L. Warren
President
National Association of
  Independent Colleges and
  Universities

Stephen Dunnett
President
Association of International
  Education Administrators

Sally T. Hillsman
Executive Officer
American Sociological Association

Katharina Phillips
President
Council on Governmental
  Relations

George R. Boggs
President and CEO
American Association of
  Community Colleges

Marc H. Brodsky
Executive Director and CEO
American Institute of Physics

Felice J. Levine
Executive Director
American Educational Research
  Association

James E. Morley, Jr.
President and CEO
National Association of College
  and University Business Officers

Roger Bowen
General Secretary
American Association of
  University Professors

Norman B. Anderson
Chief Executive Officer
American Psychological
  Association

Richard S. Dunn
Co-Executive Officer
American Philosophical Society

Mary Maples Dunn
Co-Executive Officer
American Philosophical Society

Richard L. Ferguson
CEO and Chairman of the Board
  ACT

John A. Orcutt
President
American Geophysical Union

Jerome H. Sullivan
Executive Director
American Association of Collegiate
  Registrars and Admissions Officers

Steven Block
President
The Biophysical Society

Elizabeth A. Rogan
CEO
Optical Society of America

Richard W. Peterson
President
American Association of Physics
  Teachers

Alyson Reed
Executive Director
National Postdoctoral Association

Robert P. Kirshner
President
American Astronomical Society

Stephen J. Otzenberger
Executive Director
College & University Professional
Organization for Human Resources

ENDNOTES

[1] It should be noted that the committee is in fundamental agreement with the findings and recommendations of the National Research Council's report *Biotechnology Research in an Age of Terrorism* (2004).

[2] As defined by the United Nations Commission on Human Security, human security means "to protect the vital core of all human lives in ways that enhance human freedoms and human fulfillment. Human security means protecting fundamental freedoms—freedoms that are the essence of life. It means protecting people from critical (severe) and pervasive (widespread) threats and situations. It means using processes that build on people's strengths and aspirations. It means creating political, social, environmental, economic, military and cultural systems that together give people the building blocks of survival, livelihood and dignity." UN Commission on Human Security. 2003. *Human Security-Now*. Available online at www.humansecurity-chs.org/finalreport/English/FinalReport.pdf [accessed January 5, 2006].

[3] As defined by the Human Development Reports of the United Nations Development Programme, "human development is a process of enlarging people's choices. Enlarging people's choices is achieved by expanding human capabilities and functionings. At all levels of development the three essential capabilities for human development are for people to lead long and healthy lives, to be knowledgeable and to have a decent standard of living. If these basic capabilities are not achieved, many choices are simply not available and many opportunities remain inaccessible. But the realm of human development goes further: essential areas of choice, highly valued by people, range from political, economic and social opportunities for being creative and productive to enjoying self-respect, empowerment and a sense of belonging to a community." Available online at hdr.undp.org/hd/glossary.cfm [accessed January 5, 2006].

[4] Order of December 10, 2001—Designation Under Executive Order 12958. *Federal Register* 66 (December 12):64345-64347.

[5] Order of September 26, 2002—Designation Under Executive Order 12958. *FederalRegister* 67 (September):61463-61465.

[6] Order of May 6, 2002—Designation Under Executive Order 12958. *Federal Register* 67(90):31109.

[7] National Research Council. 2004. *Biotechnology Research in an Age of Terrorism*. Washington, DC: The National Academies Press, see Chapter 3, reference 24, page 103.

[8] Department of Defense Security Directive 106. 2002. Mandatory Procedures for Research and Technology Protection Within the DOD (March). Available online at www.fas.org/sgp/news/2002/04/dod5200_39r_dr.html [accessed January 5, 2006].

[9] Knezo, G.J. 2003. 'Sensitive but Unclassified' and Other Federal Security Controls on Scientific and Technical Information: History and Current Controversy. Washington, DC: Congressional Research Service (April 2).

[10] The PATRIOT Act makes it illegal in the United States for anyone to possess any biological agent, including any genetically engineered organism created by

using recombinant DNA technology, of a type or in a quantity that, under the circumstances, is not reasonably justified by a prophylactic, protective, bona fide research, or other oeaceful purpose. The Act also prohibits the transfer or possession of a listed biological agent or toxin by a "restricted person."

[11] The Bioterrorism Response Act added new requirements for the secretaries of the Departments of Agriculture and Health and Human Services to consider in listing agents and in preventing unlawful access to agents during transfers; established new requirements for registration with the appropriate secretary concerning possession and use of select agents and toxins; and required the establishment of rules for appropriate physical security requirements for listed agents and for the Department of Justice—through the Federal Bureau of Investigation—to conduct background investigations on individuals who are permitted access to select agents or who work in a facility where select agents are stored.

[12] Center for Strategic and International Studies. 2005. Security Controls on Scientific Information and the Conduct of Scientific Research. Available online at thefdp.org/CSIS_0506_cscans.pdf [accessed January 5, 2006].

[13] No one knows precisely what nuclear secrecy cost the United States in monetary terms because the government has never tracked such costs. But Department of Energy officials routinely estimate that classified programs are 20 percent more expensive than unclassified ones. Using that rule of thumb, it is possible that up to $1 trillion of the $5.8 trillion in actual and anticipated nuclear weapons expenditures since 1940 was spent just on keeping things secret. Schwartz, S. 1998. *Atomic Audit: The Costs and Consequences of U.S. Nuclear Weapons Since 1940.* Washington, DC: Brookings Institution Press. Prepared for the 2004 Teaching Nonproliferation Summer Institute, University of North Carolina, Asheville, June 11-15, 2004. On an average annual basis, this would be equivalent to the total budget of the National Institutes of Health per year.

[14] Schwartz, S.I. 1995. Four trillion dollars and counting. *Bulletin of the Atomic Scientists* 51(6):32-52.

[15] National Academy of Sciences. 1982. *Scientific Communication and National Security.* Washington, DC: National Academy Press. Available at www.nap.edu/books/0309033322/html.

[16] Carlson, R. Briefing remarks before the Committee, February 2004.

[17] Wein, L.M. and Y. Liu. 2005. Analyzing a bioterror attack on the food supply: The case of botulinum toxin in milk. *Proceedings of the National Academy of Sciences* 102(28):9984-9989.

[18] Alberts, B. 2005. Modeling attacks on the food supply. *Proceedings of the National Academy of Sciences* 102(28):9737-9738.

[19] Wein, L.M. 2005. Got Toxic Milk? *New York Times* (May 30).

[20] USAMRIID states publicly that it does not conduct classified research.

[21] Gusterson, H. 1996. *Nuclear Rites: A Weapons Laboratory at the End of the Cold War.* Berkeley, CA: University of California Press, Chapter 4.

[22] A very thin line separates offense and defense bioweapons research. Also biodefense research can be problematic as in many cases defensive work generates an offensive capability. For more on this issue see Choffnes, E. 2002 Bioweapons: New Labs, More Terror? *Bulletin of the Atomic Scientists* 58(5):28-32.

[23] Ad Hoc Committee on Access to and Disclosure of Scientific Information. *In the Public Interest*. Massachusetts Institute of Technology, June 12, 2002.

[24] See NIAID Strategic Plan for Biodefense Research. February 2002. Available online at www3.niaid.nih.gov/biodefense/research/strategic.pdf [accessed January 5, 2006]. See also NIAID Biodefense Research Agenda for CDC Category A Agents. February 2002; and NIAID Biodefense Research Agenda for CDC Category A Agents, Progress Report, August 2003. Available online at www2.niaid.nih.gov/Biodefense/Research/strat_plan.htm [accessed January 5, 2006].

[25] National Research Council. 2004. *Biotechnology Research in an Age of Terrorism*. Washington, DC: The National Academies Press, see Chapter 4.

[26] National Research Council. 2004. *Biotechnology Research in an Age of Terrorism*. Washington, DC: The National Academies Press.

[27] Alberts, B. 2005. Modeling attacks on the food supply. *Proceedings of the National Academy of Sciences* 102(28):9737-9738. Center for Strategic and International Studies. 2005. Security Controls on Scientific Information and the Conduct of Scientific Research. Available online at www.csis.org/hs/0506_cscans.pdf. American Civil Liberties Union. 2005. *Science Under Seige: The Bush Administration's Assault on Academic Freedom and Scientific Inquiry*. Available online at www.aclu. org/Privacy/Privacy.cfm?ID=18534&c=39. Donohue, L.K. 2005. Censoring science won't make us any safer. *Washington Post* (June 26).

[28] Malakoff, D. and K. Drennan. 2004. Butler gets 2 years for mishandling plague samples. *Science* 303(5665)(19):1743-1745. Available online at www.sciencemag.org/cgi/reprint/303/5665/1743a.pdf [accessed January 5, 2006].

[29] The PATRIOT Act makes it illegal in the United States for anyone to possess any biological agent, including any genetically engineered organism created by using recombinant DNA technology, of a type or in a quantity that, under the circumstances, is not reasonably justified by a prophylactic, protective, bona fide research, or other peaceful purpose. The Act also prohibits the transfer or possession of a listed biological agent or toxin by a "restricted person."

[30] The Bioterrorism Response Act added new requirements for secretaries of the Departments of Agriculture and Health and Human Services to consider in listing agents and in preventing unlawful access to agents during transfers; established new requirements for registration with the appropriate secretary concerning possession and use of select agents and toxins; and required the establishment of rules for appropriate physical security requirements for listed agents and for the Department of Justice—through the Federal Bureau of Investigation—to conduct background investigations on individuals who are permitted access to select agents or who work in a facility where select agents are stored.

[31] See olpa.od.nih.gov/legislation/107/pendinglegislation/6bioterroism.asp for a list of all 17 bills introduced [accessed May 25, 2005].

[32] Indeed, such scientific exchanges and collaborations also increases the awareness within the United Sates of the extent and nature of technological capabilities of scientists from other countries.

[33] See www.genome.gov/11006939 for a list of the 20 sequencing centers [accessed May 25, 2005].

[34] Institute of Medicine. 2004. *Learning from SARS: Preparing for the Next Disease Outbreak*. Washington, DC: The National Academies Press.

[35] National Research Council. 2004. *Seeking Security: Pathogens, Open Access, and Genome Databases.* Washington, DC: The National Academies Press.

[36] General Accounting Office. 2004. Border Security: Improvements Needed to Reduce Time Taken to Adjudicate Visas for Science Students and Scholars. GAO-04-371 (February). As referenced in Brown, H.A. and P.D. Syverson. 2004. Findings from U.S. Graduate Schools on International Graduate Student Admission Trends. Available online at www.cgsnet.org/pdf/Sept04FinalIntlAdmissions SurveyReport.pdf [accessed January 4, 2006].

[37] General Accounting Office. 2005. Border Security: Streamlined Visas Mantis Program has Lowered Burden on Foreign Science Students and Scholars, but Further Refinements Needed. GAO-05-198 (February):7. Available online at www. gao.gov/new.items/d05198.pdf [accessed January 5, 2006].

[38] The question of shifting from country of citizenship to country of birth has been raised in the context of deemed exports. The Inspector General of the Department of Commerce has recommended consideration be given to changing to country of birth. At the meeting on this topic at NAS (2005) it seemed apparent that the Department of Commerce was listening but not pressing for this change. The responses by NAS and various academic organizations all speak to limiting changes and reducing the impact of deemed export regulations.

[39] See www7.nationalacademies.org/rscans/IG_Workshop_Transcripts.pdf [accessed January 5, 2006].

[40] For example, programs organized within the Department of Homeland Security's Science and Technology Directorate.

[41] The terms "bona fide" and "legitimate" are not defined in the statute.

[42] A "restricted person" is defined as "anyone who: is under indictment for or has been convicted in any court of a crime punishable by imprisonment for a term exceeding one year; is a fugitive from justice; is an unlawful user of any controlled substance; is an alien illegally or unlawfully in the United States; has been adjudicated as a mental defective or has been committed to any mental institution; is an alien who is a national of a country which is currently designated by the Secretary of State as a supporter of terrorism; or has been dishonorably discharged from U.S. armed forces." Currently there are seven countries on the State Department's List of State Sponsors of Terrorism: Cuba, Libya, Iran, Iraq, North Korea, Sudan, and Syria.

[43] There are four basic criteria used to evaluate whether an agent or toxin should be listed: 1) The effect on human health of exposure to the agent or toxin (or, in the case of the USDA list of plant and animal agents and toxins, the effect of an agent or toxin on animal or plant health or products). 2) The degree of contagiousness and the methods by which transfer of the agent or toxin to humans can occur (or, in plants and animals, the virulence of an agent or degree of toxicity of the toxin and the methods by which the agents or toxins are transferred to animals or plants). 3) The availability and effectiveness of pharmacotherapies and immunizations to treat and prevent any illness resulting from exposure (or, in the case of USDA agents, the availability and effectiveness of medicines and vaccines to treat and prevent any illness caused by an agent or toxin) 4) Any other criteria, including the needs of children and other vulnerable populations that the Secretary considers appropriate (or, in the case of USDA agents, other criteria that the

Secretary considers appropriate to protect animal or plant health, or animal or plant products).

[44] U.S. Congress. Antiterrorism and Effective Death Penalty Act of 1996, P.L. 104-132 (April 24), sec. 511.

[45] Malakoff, D. 2003. Security Rules Leave Labs Wanting More Guidance. *Science* 299(5610):1175.

[46] See www.asm.org/Policy/index.asp?bid=8648 [accessed May 25, 2005].

[47] See www.wmd.gov/report/.

[48] "A senior National Security Council official is said to have praised Bio-Chem 2020 but was quick to note that it is a 'cottage program,' not part of a broader Intelligence Community endeavor." From Commission on the Intelligence Capabilities of the United States Regarding Weapons of Mass Destruction. Report to the President of the United States, March 31, 2005. Chapter 13 "The changing proliferation threat and the intelligence response." See www.globalsecurity .org/intell/library/reports/2005/wmd_report_25mar2005_chap13.htm [accessed January 6, 2006]. The Committee believes that there may be current on-going discussions among officials in the national security and intelligence communities about the need for such an advisory group.

[49] Based on a personal conversation with Dr. David Sencer, former director of the CDC, the ACIP was informally created in 1964 by combining several small ad hoc committees into one, unified committee to advise the CDC on topics related to vaccines (personal communication with Dr. David Sencer, January 9, 2006). The ACIP was formally created in 1993 under the statutory authority of 42 U.S.C. 217a, Section 222 of the Public Health Service Act, as amended. The committee is governed by the provisions of Public Law 92-463, the Federal Advisory Committees Act of 1972, as amended (5 U.S.C. A 2), which sets forth standards for the formation and use of advisory committees. In addition, the ACIP was given a statutory role, and budget, under Section 13631 of the Omnibus Budget Reconciliation Act of 1993, Public Law 103-66 (42 U.S.C. 1396s(c)(2)(B)(i) and (e), subsections 1928(c)(2)(B)(i) and 1928(e) of the Social Security Act) to provide advice and guidance to the Secretary; the Assistant Secretary for Health, HHS; and the Director, CDC, regarding the most appropriate application of antigens and related agents for effective communicable disease control in the civilian population. The committee develops written recommendations for routine administration of vaccines to the pediatric and adult populations, along with schedules regarding the appropriate periodicity, dosage, and contraindications applicable to the vaccines. ACIP is the only entity in the federal government which makes such recommendations. For more information regarding the structure and functions of the ACIP please see www.cdc.gov/nip/acip/charter.htm.

[50] Federal Advisory Committee Act. - 483 -. Federal Advisory Committee Act. 5 USC a, as amended. See www.usdoj.gov/04foia/facastat.pdf.

[51] Under the Federal Advisory Committee Act, a federal advisory committee shall terminate in two years after it is established, unless *a statute authorizing the committee specifically provides for a different duration or the agency head renews its charter.* In addition, the President or agency head can terminate the advisory committee earlier if he or she determines that the committee has fulfilled its purpose, it is no longer carrying out its purpose, or the cost of operation is too much relative to

the benefits of the committee. See www.redlodgeclearinghouse.org/legislation/faca3.html [accessed January 6, 2006].

[52] The NSABB is specifically charged with guiding the development of a system of institutional and federal research review that allows for fulfillment of important research objectives while addressing national security concerns; guidelines for the identification and conduct of research that may require special attention and security surveillance; professional codes of conduct for scientists and laboratory workers that can be adopted by professional organizations and institutions engaged in life science research; and, materials and resources to educate the research community about effective biosecurity. The NSABB Charter was signed March 4, 2004. See www.biosecurityboard.gov/SIGNED% 20NSABB%20 Charter.pdf, emphasis added [accessed January 6, 2006].

[53] Steinbrook, R. 2005. Biomedical Research and Biosecurity. *New England Journal of Medicine* 353(21):2212-2214.

[54] Report of the Commission on the Intelligence Capabilities of the United States Regarding Weapons of Mass Destruction, March 31, 2005. Available online at www.wmd.gov/report/wmd_report.pdf [accessed January 6, 2006]; National Commission on Terrorist Attacks Upon the United States. 2004. The 9/11 Commission Report. Available online at www.9-11commission.gov/report/index.htm [accessed January 6, 2006].

[55] For more on this issue see The National Academies. 2005. *Rising Above the Gathering Storm.* Washington, DC: The National Academies Press.

[56] A regime comprises the multitude of cooperative and coercive measures—including international agreements, multilateral organizations, national laws, regulations, and policies—intended to prevent the spread of dangerous weapons and technologies.

[57] The Biological and Toxin Weapons Convention can be viewed online at www.opbw.org/convention/documents/btwctext.pdf. The fact that its prohibitions do not expressly extend to 'research' is sometimes cited as a loophole or as an obstacle to raising awareness of the BWC within the research community. It should be recalled, however, that the original negotiators of the BWC differentiated 'pure' and 'applied' research. Although the precise meaning of 'development' is unclear, at least some of the negotiators understood it to subsume end-item and process research; others, however, took a contrary view. This is a matter that states partiers could perhaps clarify at a future BWC review conference.

[58] BWC/MSP/2004/INF.2, 3 December 2004 See www.opbw.org.

[59] For a detailed discussion on the challenges that the BWC currently faces see National Research Council. 2005. *An International Perspective on Advancing Technologies and Strategies for Managing Dual-Use Risks.* Washington, DC: The National Academies Press. Available online at www.nap.edu/catalog/11301.html.

[60] The Chemical Weapons Convention can be viewed online at www.opcw.org/html/db/cwc/eng/cwc_frameset.html.

[61] General Accounting Office. 2004. Nonproliferation: Delays in Implementing the Chemical Weapons Convention Raise Concerns about Proliferation. Report to the Chairman, Committee on Armed Services, House of Representatives. GAO-04-361. Available online at www.gao.gov/new.items/d04361.pdf [accessed January 6, 2006].

[62] The Australia Group (AG) is an informal consultative group of nations (38 countries plus the European Commission) which meet annually with the objective "to ensure, through licensing measures on the export of certain chemicals, biological agents, and dual-use chemical and biological manufacturing facilities and equipment, that exports of these items from their countries do not contribute to the spread of CBW." See www.australiagroup.net. The group formed in 1985, in response to evidence that Iraq had used chemical weapons in the Iran-Iraq war and that Iraq had obtained many of the materials for their chemical weapons program from the international chemical industry. In 1990, the AG group expanded its efforts to address the increasing spread of bioweapons materials and technology.

[63] UN Security Council Resolution 1540 (2004) obliges all UN member states to have national laws in place to prohibit the proliferation of terrorism with biological materials. In effect, the resolution obliges all UN member states, not just BWC States Parties, to comply with Articles III and IV of the BWC, both of which apply to non-state actors. Article III creates a very clear obligation not to transfer to any recipient whatsoever any sort of material, equipment, or know-how for making biological weapons. Article IV obliges all States Parties to take national measures to fully implement these obligations and responsibilities, which means that all States Parties must enact legislation containing the prohibitions of the BWC and penalties for noncompliance.

[64] SCR 1540 was a resolution passed by the United Nations Security Council requiring all U.N. member states to establish effective domestic controls to prevent non-state actors from acquiring nuclear, chemical, or biological weapons, their means of delivery, and related materials.

[65] For more information, see Rappert, B. 2005. Towards a Life Science Code: Possibilities and Pitfalls in Countering the Threats from Bioweapons. Available online at www.ex.ac.uk/codesofconduct/ Publications/ Bradford %2012.7.4.doc [accessed January 6, 2006].

[66] See www.asm.org/ASM/files/CCLIBRARYFILES/FILENAME/0000000656/ Council%20approved%20Code%20of%20Ethics2.pdf [accessed January 6, 2006].

[67] See www.ex.ac.uk/codesofconduct/Chronology/index.htm [accessed January 6, 2006].

[68] See www.theasm.com.au/ [accessed January 6, 2006].

[69] Somerville, M.A. and R.M. Atlas. 2005. Ethics: A weapon to counter bioterrorism. *Science* 307(5717):1881-1882.

[70] See www.biotech.ca/EN/ethics.html [accessed January 6, 2006].

[71] www.europabio.org/documents/corevalues.pdf.

[72] Institute of Medicine/National Research Council. 2005. *An International Perspective on Advancing Technologies and Strategies for Managing Dual-Use Risks*. Washington, DC: The National Academies Press. Available online at www.nap.edu/catalog/11301.html.

[73] For a discussion of what happened at this experts group meeting—MX 2005—see *The CBW Conventions Bulletin*, No. 68, June 2005 online at www.sussex.ac.uk/Units/spru/hsp/CBWCB68.pdf [accessed January 6, 2006].

[74] The recommendation reads as follows: Relevant United Nations offices should be tasked with producing proposals to reinforce ethical norms, and the

creation of codes of conduct for scientists, through international and national scientific societies and institutions that teach sciences or engineering skills related to weapons technologies, should be encouraged. Such codes of conduct would aim to prevent the involvement of defence scientists or technical experts in terrorist activities and restrict public access to knowledge and expertise on the development, production, stockpiling and use of weapons of mass destruction or related technologies. Available online at www.un.org/terrorism/a57273.htm [accessed January 6, 2006].

[75] Institute of Medicine/National Research Council. 2005. *An International Perspective on Advancing Technologies and Strategies for Managing Dual-Use Risks.* Washington, DC: The National Academies Press. Available online at www.nap.edu/catalog/11301.html.

[76] Ibid.

[77] Iverson M., M. Frankel, and S. Siage. 2003. Scientific societies and research integrity: what are they doing and how well are they doing it? *Science and Engineering Ethics* 9(2):141-158.

[78] Doig, A. and J. Wilson. 1998. The effectiveness of codes of conduct. *Journal of Business Ethics* 7(3):140-149.

[79] Higgs-Kleyn, N. and D. Kapelianis. 1999. The role of professional codes in regulating ethical conduct. *Journal of Business Ethics* 19:363-374.

[80] Luegenbiehl, C. 1991. Codes of ethics and the moral education of engineers. In: Johnson, D. 1991. *Ethical Issues in Engineering.* Upper Saddle River, NJ: Prentice Hall:136-154.

[81] Davis, M. 1998. *Thinking Like an Engineer.* Oxford: Oxford University Press.

[82] Meselson M. 2000. Averting the exploitation of biotechnology. *FAS Public Interest Report* 53:5.

[83] Unger, S. 1991. Code of engineering ethics. In: Johnson, D. 1991. *Ethical Issues in Engineering.* Upper Saddle River, NJ: Prentice Hall:105-130.

[84] Reiser, S. and Bulger, R. 1997. The social responsibilities of biological scientists. *Science and Engineering Ethics* 3(2):137-143.

[85] The NSABB is charged specifically with guiding the development of: A system of institutional and federal research review that allows for fulfillment of important research objectives while addressing national security concerns; Guidelines for the identification and conduct of research that may require special attention and security surveillance; Professional codes of conduct for scientists and laboratory workers that can be adopted by professional organizations and institutions engaged in life science research; and Materials and resources to educate the research community about effective biosecurity. For more information on the NSABB, see www.biosecurityboard.gov/.

[86] As of this writing, Duke University, MIT, Princeton, and the University of California, San Diego have education modules; The University of California, Berkeley and SUNY Stonybrook are in the process of developing education modules. The Arms Control Association has also developed an education module on the history of biological weapons, arms control treaties and the "dual use" dilemma. The Federation of American Scientists is also developing an interactive teaching module to promote awareness of biosecurity issues among bioscience researchers." See www.fas.org/main/content.jsp?formAction=297&contentId=146.

[87] For a discussion of the laws and regulations in the United States governing the handling and control of biological materials and the rules governing who may or may not work with these materials, please see National Research Council, 2004, *Biotechnology Research in an Age of Terrorism*. Washington DC: The National Academies Press; Chapter 2.

[88] Daar, A.S. and P.A. Singer. 2005. Biotechnology and Human Security. In: *Helsinki Process Papers on Human Security*. Foreign Ministry's Publications: Helsinki:120-162. Available online at www.utoronto.ca/jcb/home/documents/Biotech_human_security.pdf [accessed January 6, 2006].

[89] UN Millennium Project Task Force on Science, Technology and Innovation. 2005. *Innovation: Applying knowledge to development*. London: Earthscan.

[90] Kellman, B. The global bargain for biosecurity. Unpublished manuscript distributed to committee (June 2004).

[91] Ibid.

[92] This description of the mammalian innate and adaptive immune systems is adapted from Kathryn Nixdorff, briefing to the Committee at the Committee's International Workshop. Institute of Medicine/National Research Council. 2005. *An International Perspective on Advancing Technologies and Strategies for Managing Dual-Use Risks*. Washington, DC: The National Academies Press; 44-49. Available online at www.nap.edu/catalog/11301.html.

[93] See www.promedmail.org.

[94] Jackson, R.J. et al. 2001. Expression of mouse interleukin-4 by a recombinant ectromelia virus suppresses cytolytic lymphocyte responses and overcomes genetic resistance to mousepox. *Journal of Virology* 75(3):1205-1210.

[95] Linux is a free Unix-type operating system. See www.linux.org.

[96]Young, H.P. 1998. *Individual Strategy and Social Structure: An Evolutionary Theory of Institutions*. Princeton: Princeton University Press; Axelrod, R. 1984. *The Evolution of Cooperation*. New York: Basic Books; Axelrod. R. 1986. An Evolutionary Approach To Norms. *American Political Science Review* 80(4):1095-1111; Epstein, J.M. 2001. Learning to be Thoughtless: Social Norms and Individual Competition. *Computational Economics* 18:9-24.

[97] Fox, J.A. and A.R. Piquero. 2003. Deadly demographics: Population characteristics and forecasting homicide trends. *Crime & Delinquency* 49(3):339-359.

[98] National Research Council. 2004. *Biotechnology Research in an Age of Terrorism*. Washington, DC: The National Academies Press. Institute of Medicine. 2002. *Biological Threats and Terrorism: Assessing the Science and Response Capabilities*. Washington, DC: The National Academies Press; Institute of Medicine. 2003. *Microbial Threats to Health: Emergence, Detection, and Response*. Washington, DC: The National Academies Press. NIAID Blue Ribbon Panel on Bioterrorism and Its Implications for Biomedical Research, February 2002. More information on the NIAID Blue Ribbon Panel is available online at www.niaid.nih.gov/publications/btbluribbon.htm [accessed January 6, 2006].

[99] Chyba, C.F. 2001. Biological Terrorism and Public Health. *Survival* 43(Spring):93-106.

[100] Prevention is a cornerstone of public health. Just as mosquito netting can be used to prevent the spread of malaria, the built environment can be used to minimize risks of exposure to airborne biological agents. See www.cdc.gov/niosh/

bldvent/2002-139.html#foreward; www.cdc.gov/niosh/docs/2003-136/2003-136.html; and www.ashrae.org/content/ASHRAE/ASHRAE/Article AltFormat/20053810917_347.pdf.

[101] October 2, 2001: infectious-disease specialist Dr. Larry Bush found a high white blood cell count and rod-shaped bacilli in Robert Stevens, 63, photo editor at the supermarket tabloid *The Sun*. He soon was convinced Stevens had contracted anthrax. He then notified the Palm Beach County Health Department. See en.wikipedia.org/wiki/Timeline_of_the_2001_anthrax_attacks_in_Florida [accessed January 6, 2006]. Feb 28, 2003: World Health Organization officer Carlo Urbani, MD, examines an American businessman with an unknown form of pneumonia in a French hospital in Hanoi, Vietnam. March 10, 2003: Urbani reports an unusual outbreak of the illness, which he calls severe acute respiratory syndrome or SARS, to the main office of the WHO. He notes that the disease has infected an usually high number of healthcare workers (22) at the hospital. March 29, 2003: Carlo Urbani, who identified the first cases of SARS, dies as a result of the disease. Researchers later suggest naming the agent that causes the disease after the infectious disease expert. See my.webmd.com/content/article/63/72068.htm [accessed January 6, 2006].

[102] Wysocki, B. 2005. U.S. Struggles for Drugs to Counter Biological Threats. *Wall Street Journal* (July 11).

[103] The failure of the government to get the countermeasures it needs to protect its citizens is a major problem. BioShield gives HHS more flexibility to purchase countermeasures but there is a critical piece missing—funding of initial product development, the so-called "Valley of Death" for new drugs. BioShield does not provide sufficient financial incentives for pharmaceutical companies to invest years of research into a product. For a detailed description of what BioShield does and does not do, as well as the difficulties in getting countermeasures for biodefense, see Borio, L.L. and Gronvall, G.K. 2005. Anthrax countermeasures: current status and future needs. *Biosecurity and Bioterrorism: Biodefense Strategy, Practice, and Science* 3(2):102-112.

[104] Institute of Medicine. 2003. *The Resistance Phenomenon in Microbes and Infectious Disease Vectors*. Washington, DC: The National Academies Press.

[105] The areas of science reflected in the Nobel Prize include chemistry, medicine and physiology, and physics. Areas of science for which the National Medal of Scinece is awarded include biology, chemistry, engineering, and math and physics.

[106] This reference can be found online at www.nap.edu/books/0309053382/html/89.html.

[107] See *www4.nationalacademies.org/news.nsf/isbn/s05182005?OpenDocument*

Acronyms and Abbreviations

ACIP	Advisory Committee on Immunization Practices
ADMET	Absorption, Distribution, Metabolism, Excretion, Toxicity
AIDS	Acquired Immunodeficiency Syndrome
ART	Antiretroviral Therapy
ASM	American Society of Microbiology
BERD	Business Enterprise Research and Development
BIA	BioIndustry Association
BIO	Biotechnology Industry Organization
BIS	Bureau of Industry and Security
BWC	Biological and Toxin Weapons Convention
CDC	Centers for Disease Control and Prevention
cDNA	Complementary Deoxyribonucleic Acid
CFC	Chlorofluorocarbon
CoV	Coronavirus
CWC	Chemical Weapons Convention
DHS	Department of Homeland Security
DIA	Defense Intelligence Agency
DNA	Deoxyribonucleic Acid
DOD	Department of Defense
DPI	Dry Powder Inhaler
dsRNA	Double-stranded Ribonucleic Acid

EGFR	Epidermal Growth Factor Receptor
EPO	European Patent Office
FDA	Food and Drug Administration
GDP	Gross Domestic Product
GFP	Green Fluorescent Protein
GM	Genetically Modified
GMO	Genetically Modified Organism
HA	Haemagglutinin
HHS	Department of Health and Human Services
HIV	Human Immunodeficiency Virus
HPLC	High-performance Liquid Chromatography
HTS	High-throughput Screening
HUGO	Human Genome Organization
IAEA	International Atomic Energy Agency
ICGEB	International Center for Genetic Engineering and Biotechnology
ICT	Information and Computer Technology
INPI	National Patent Office of Brazil
ISAAA	International Service for the Acquisition of Agri-Biotech Applications
MAMP	Microbe-associated Molecular Pattern
MDI	Metered-dose Inhaler
MEMS	Microelectromechanical Systems
miRNA	Micro-ribonucleic Acid
mRNA	Messenger Ribonucleic Acid
NA	Neuraminidase
NIH	National Institutes of Health
NASSCOM	National Association of Software and Services Companies
NMR	Nuclear Magnetic Resonance
NPT	Nuclear Nonproliferation Treaty
NRC	National Research Council
NSABB	National Science Advisory Board for Biosecurity
NSF	National Science Foundation
NSDD	National Security Decision Directive
NSG	Nuclear Suppliers Group

OECD	Organisation for Economic Co-operation and Development
OPCW	Organization for the Prohibition of Chemical Weapons
PCR	Polymerase Chain Reaction
PCT	Patent Cooperation Treaty
PDB	Protein Data Bank
pMDI	Propellant Metered-dose Inhaler
qPCR	Quantitative Polymerase Chain Reaction
REDI	Regional Emerging Diseases Intervention Center, Singapore
RISCS	RNA-induced Silencing Complexes
RNA	Ribonucleic Acid
RNAi	Ribonucleic Acid Interference
SAR	Structure-Activity Relationships
SARS	Severe Acute Respiratory Syndrome
SARS-CoV	SARS Coronavirus
SCF	Supercritical Fluid
SCNT	Somatic Cell Nuclear Transfer
SIPI	State Intellectual Property Office of the People's Republic of China
siRNA	Small Interfering Ribonucleic Acid
SNP	Single Nucleotide Polymorphisms
TLR	Toll-like Receptor
UNICEF	United Nations Children's Fund
USDA	U.S. Department of Agriculture
USPTO	U.S. Patent and Trademark Office
VEGF	Vascular Endothelial Growth Factor
WMD	Weapons of Mass Destruction
WTO	World Trade Organization

APPENDIX B

Committee Meetings

February 23-24, 2004
Washington, DC

Guest Speakers
Dr. Robert Carlson, University of Washington
Dr. James B. Petro, Defense Intelligence Agency

April 27-28, 2004
Washington, DC

Guest Speakers
Dr. Pim Stemmer, Avidia
Dr. Charlie Rice, Rockefeller University
Dr. Drew Endy, Massachusetts Institute of Technology
Dr. Herb Lin, The National Academies
Sonia Miller, SE Miller Law Firm

June 23-24, 2004
Washington, DC

Guest Speakers
Dr. John Steinbruner, University of Maryland
Barry Kellman, DePaul University
Michael Moodie, Chemical and Biological Arms Control Institute
Terence Taylor, International Institute for Strategic Studies
Dr. David Lipman, National Center for Biotechnology Information/
    National Library of Medicine

Dr. Charles Jennings, *Nature*
Dr. Phillip Campbell, *Nature*
Dr. Jonathan Tucker, Center for Nonproliferation Studies/Monterey Institute of International Studies
Dr. Gerald Epstein, Center for Strategic and International Studies
Dr. Jerrold Post, George Washington University

September 22-23, 2004
Cuernavaca, Mexico

Guest Speakers
Terence Taylor, International Institute for Strategic Studies
Dr. David Banta, consultant
Decio Ripandelli, International Centre for Genetic Engineering and Biotechnology
Dr. Charles Arntzen, Arizona State University
Miguel Gomez Lim, CINVESTAV
Luis Herrera-Estrella, National Polytechnic Institute
Dr. Rosiceli Barreto Gonçalves Baetas, Biomanguinhos
Dr. Jacques Ravel, The Institute for Genomic Research
Dr. Patrick Tan Boon Ooi, Genome Institute of Singapore
Dr. Abdallah Daar, University of Toronto
Gerardo Jimenez-Sanchez, National Institute of Genomic Medicine
Ambassador Tibor Tòth, Hungarian Embassy, Geneva
Dr. Amy Sands, Monterey Institute of International Studies
Robert Mathews, Australian Department of Defence
Jerome Amir Singh, Centre for the AIDS Programme of Research in South Africa
Peter Herby, International Committee of the Red Cross
Dr. Nadrian Seeman, New York University
Michael Morgan, The Wellcome Trust
Dr. Kathryn Nixdorff, University of Darmstadt
Elliott Kagan, Department of Defense

January 25-26, 2005
Washington, DC

Discussion of draft report. No guest speakers.

March 8-9, 2005
Washington, DC

Discussion of draft report. No guest speakers.

APPENDIX C

Biographical Sketches of Committee Members

Dr. Stanley M. Lemon, M.D. *co-chair,* is the John Sealy Distinguished University Chair and Director of the Institute for Human Infections and Immunity at the University of Texas Medical Branch (UTMB) at Galveston. He received his undergraduate A.B. degree in biochemical sciences from Princeton University *summa cum laude,* and his M.D. with honor from the University of Rochester. He completed postgraduate training in internal medicine and infectious diseases at the University of North Carolina at Chapel Hill, and is board certified in both. From 1977 to 1983, he served with the U.S. Army Medical Research and Development Command, followed by a 14 year period on the faculty of the University of North Carolina School of Medicine. He moved to UTMB In 1997, serving first as chair of the Department of Microbiology & Immunology, then as dean of the School of Medicine from 1999 to 2004. Dr. Lemon's research interests relate to the molecular virology and pathogenesis of positive-strand RNA viruses responsible for hepatitis. He has had a longstanding interest in antiviral and vaccine development, and has served previously as chair of the Anti-Infective Drugs Advisory Committee, and the Vaccines and Related Biologics Advisory Committee, of the U.S. Food and Drug Administration. He is past chair of the Steering Committee on Hepatitis and Poliomyelitis of the World Health Organization Programme on Vaccine Development. He presently serves as a member of the U.S. Delegation of the U.S.-Japan Cooperative Medical Sciences Program, and chairs the Board of Scientific Councilors of the National Center for Infectious Diseases of the Centers for Disease Control and Prevention. He is chair of the Forum on Microbial Threats of the Institute of Medicine, and recently chaired an Institute of Medicine study committee related to vaccines for the protection of the military against naturally occurring infectious disease threats.

David A. Relman, M.D. *co-chair,* is an associate professor of medicine (infectious diseases and geographic medicine) and of microbiology and immunology at Stanford University School of Medicine, Stanford, California, and chief of the Infectious Diseases Section at the Veterans Affairs Palo Alto Health Care System, Palo Alto, California. Dr. Relman received his B.S. degree in biology from Massachusetts Institute of Technology, Cambridge, Massachusetts, and his medical degree from Harvard Medical School. He completed his residency in internal medicine and a clinical fellowship in infectious diseases at Massachusetts General Hospital, Boston, after which he moved to Stanford as a research fellow and postdoctoral scholar. He joined the Stanford faculty in 1994. His major focus is laboratory research directed toward characterizing the human endogenous microbial flora, host-microbe interactions, and identifying previously-unrecognized microbial pathogens, using molecular and genomic approaches. He has described a number of new human microbial pathogens. Dr. Relman's lab (relman.stanford.edu) is currently exploring human oral and intestinal microbial ecology, sources of variation in host genome-wide expression responses to infection and during states of health, and how *Bordetella* species (including the agent of whooping cough) cause disease. He has published over 150 peer-reviewed articles, reviews, editorials and book chapters on pathogen discovery and bacterial pathogenesis. Dr. Relman has served on scientific program committees for the American Society for Microbiology (ASM) and the Infectious Diseases Society of America (IDSA), and advisory panels for NIH, CDC, the Departments of Energy and Defense, and NASA. He is a member of the Board of Directors of the IDSA, the Board of Scientific Counselors at NIDCR/NIH, and the Forum on Microbial Threats at the Institute of Medicine. He received the Squibb Award from IDSA in 2001, the Senior Scholar Award in Global Infectious Diseases from the Ellison Medical Foundation in 2002, and is a fellow of the American Academy of Microbiology.

Roy Anderson, Ph.D., FRS, is professor of Infectious Disease Epidemiology and Head of the Department of Infectious Disease Epidemiology at Imperial College Faculty of Medicine, University of London. Roy Anderson is a fellow of the Royal Society and a Foreign Member of the Institute of Medicine at the US National Academy of Sciences. He has published over 400 scientific papers on the epidemiology, population biology, evolution and control of a wide range of infectious disease agents, including HIV, BSE, vCJD, parasitic helminths and protozoa, and respiratory tract viral and bacterial infections. His principal research interests are epidemiology, biomathematics, demography, parasitology, immunology, and health economics. He also has a keen interest in science policy and the public understanding of science. He has held a wide variety of advisory and consultancy posts with government departments, pharmaceutical companies and international aid agencies. Professor Anderson has been a member of SEAC since January 1998.

Steven M. Block, Ph.D., is a biophysicist at Stanford University, where he holds a joint appointment as a professor in the Departments of Biological Sciences and Applied Physics. He is also a Senior Fellow of the Stanford Institute for International Studies, and a member of the JASONs, a group of academicians who consult for the U.S. government and its agencies on technical matters related to national security. Prior to joining the Stanford faculty in 1999, Professor Block held positions at Princeton University (1994-1999), Harvard University (1987-1994), and the Rowland Institute for Science in Cambridge, MA (1987-1994). He received his undergraduate training in both physics and biology at Oxford University, earned his doctorate from the California Institute of Technology (1983), and conducted postdoctoral research at Stanford. Professor Block's technical interests are in interdisciplinary science, particularly the biophysics of motor proteins. His laboratory pioneered the use of laser-based optical traps ("optical tweezers") to study the nanoscale motions of these mechanoenzymes at the level of single molecules, and his group was the first to develop instrumentation able to resolve the individual steps taken by single kinesin motors moving along microtubules. Other biological systems currently under study in his laboratory include RNA polymerase, exonuclease, and helicase, enzymes that move processively along DNA. Professor Block is a strong proponent of nanoscience, but he is also an outspoken critic of the "futurist" element of the nanotechnology movement.

Christopher Chyba, Ph.D., is professor of astrophysical sciences and international affairs at Princeton University. Until July 2005, he was associate professor in the Department of Geological and Environmental Sciences at Stanford University, and co-director of the Center for International Security and Cooperation, Stanford Institute for International Studies. He holds the Carl Sagan Chair for the Study of Life in the Universe at the SETI Institute. His security-related research focuses on nuclear proliferation and biological terrorism. His planetary science and astrobiology research focuses on the search for life elsewhere in the solar system. A graduate of Swarthmore College, Chyba studied as a Marshall Scholar at the University of Cambridge and received his Ph.D. in planetary science from Cornell University in 1991. He served on the White House staff from 1993 to 1995, entering as a White House Fellow on the National Security Council staff and then serving in the National Security Division of the Office of Science and Technology Policy (OSTP). After leaving the White House, he drafted the President's decision directive on responding to emerging infectious diseases, and authored a report for OSTP in 1998 on preparing for biological terrorism. He received the Presidential Early Career Award, "for demonstrating exceptional potential for leadership at the frontiers of science and technology during the 21st century." He chaired the Science Definition Team for NASA's Europa Orbiter mission and served on the executive committee of NASA's Space Science Advisory Committee, for which he chaired the Solar System Exploration Subcommittee. Dr. Chyba currently serves on the National Academy of Sciences' Committee for International Security and Arms Control, on the Monterey Nonproliferation Strategy Group, and chairs the National Research Council's Committee on Preventing the Forward Contamination of Mars. In October 2001, he was named a MacArthur Fellow for his work in astrobiology and international security.

Nancy Connell, Ph.D., is vice chair for research, department of medicine, and professor of microbiology and molecular genetics, and has been appointed director of the NJMS-Center for Biodefense. She is an NIH-funded basic scientist, a permanent member of the NIH Study Section on Bacteriology and Mycology-1, and serves as director of the Biosafety Level Three Facility of the NJMS-Center for Emerging and Re-emerging Pathogens. She is a graduate of Harvard Medical School and has been a faculty member at NJMS since 1992.

Freeman Dyson is now retired, having been for most of his life a professor of physics at the Institute for Advanced Study in Princeton. He was born in England and worked as a civilian scientist for the Royal Air Force in World War II. He graduated from Cambridge University in 1945 with a BA degree in mathematics. He went on to Cornell University as a graduate student in 1947 and worked with Hans Bethe and Richard Feynman. His most useful contribution to science was the unification of the three versions of quantum electrodynamics invented by Feynman, Schwinger and Tomonaga. Cornell University made him a professor without bothering about his lack of Ph.D. He subsequently worked on nuclear reactors, solidstate physics, ferromagnetism, astrophysics and biology, looking for problems where elegant mathematics could be usefully applied. He has written a number of books about science for the general public. *Disturbing* *the Universe* (1974) is a portrait-gallery of people he has known during his career as a scientist. *Weapons of Hope* (1984) is a study of ethical problems of war and peace. *Infinite in All Directions* (1988) is a philosophical meditation based on Dyson's Gifford Lectures on Natural Theology given at the University of Aberdeen in Scotland. *Origins of Life* (1986, second edition 1999) is a study of one of the major unsolved problems of science. *The Sun, the Genome and the Internet* (1999) discusses the question of whether modern technology could be used to narrow the gap between rich and poor rather than widen it. Dyson is a fellow of the American Physical Society, a member of the U.S. National Academy of Sciences, and a fellow of the Royal Society of London. In 2000, he was awarded the Templeton Prize for progress in Religion.

Joshua M. Epstein, Ph.D., is a Senior Fellow in Economic Studies at the Brookings Institution, a member of the Brookings-Johns Hopkins Joint Center on Social and Economic Dynamics, and a member of the External Faculty of the Santa Fe Institute. He holds a Ph.D. in Political Science from MIT and is a member of the New York Academy of Sciences. He is also a member of the editorial boards of the journal *Complexity*, and of the Princeton University Press Studies in Complexity book series. His primary research interest is in the modeling of complex social, economic, and biological systems using agent-based computational models and nonlinear dynamical systems. He has taught computational and mathematical modeling at Princeton and the Santa Fe Institute Summer School. He has published widely in the modeling area, including recent articles on the dynamics of civil violence, the demography of the Anasazi (both in the *Proceedings of the National Academy of Sciences*) and the epidemiology of smallpox (in the *American Journal of Epidemiology*). His two most recent books are *Growing Artificial Societies: Social Science from the Bottom Up*, with co-author Robert Axtell, (MIT Press, 1996); and *Nonlinear Dynamics, Mathematical Biology, and Social Science* (Addison-Wesley/Santa Fe Institute, 1997). His book, *Generative Social Science: Studies in Agent-Based Computational Modeling*, is forthcoming from Princeton University Press.

Stanley Falkow, Ph.D., (NAS, IOM) is professor of microbiology and immunology and professor of medicine at Stanford University. Dr. Falkow is recognized internationally for his research related to the molecular mechanisms of bacterial pathogenesis. Dr. Falkow is the former president of the American Society for Microbiology and has been elected to the American Academy of Arts and Sciences, the National Academy of Sciences, and the Institute of Medicine. He has received the Squibb Award from the Infectious Diseases Society of America (1978), the Paul Erhlich Award from Germany (1980), the Brisol-Myers-Squibb Award for Infectious Diseases Research (1997), and the Robert Koch Prize from Germany (2000). Dr. Falkow holds a B.S. in Bacteriology from the University of Maine, an M.S. in Biology from Brown University, and a Ph.D. in Biology from Brown University.

Stephen S. Morse, Ph.D., is Founding Director of the Center for Public Health Preparedness at the Mailman School of Public Health of Columbia University, and Associate Professor in the Epidemiology Department. Dr. Morse recently returned to Columbia from 4 years in government service as Program Manager at the Defense Advanced Research Projects Agency (DARPA), where he co-directed the Pathogen Countermeasures program and subsequently directed the Advanced Diagnostics program. Before coming to Columbia, he was Assistant Professor (Virology) at The Rockefeller University in New York, where he remains an adjunct faculty member. Dr. Morse is the editor of two books, *Emerging Viruses* (Oxford University Press, 1993; paperback, 1996) (selected by *American Scientist* for its list of "100 Top Science Books of the 20th Century"), and *The Evolutionary Biology of Viruses* (Raven Press, 1994). He currently serves as an editor of the CDC journal *Emerging Infectious Diseases* and was formerly an editor-in-chief of the Pasteur Institute's journal *Research in Virology*. Dr. Morse was chair and principal organizer of the 1989 NIAID/NIH Conference on Emerging Viruses (for which he originated the term and concept of emerging viruses/infections); served as a member of the Institute of Medicine's Committee on Emerging Microbial Threats to Health (and chaired its Task Force on Viruses), and was a contributor to its report, Emerging Infections (1992); was a member of the IOM's Committee on Xenograft Transplantation; currently serves on the steering committee of the Institute of Medicine's Forum on Emerging Infections, and has served as an adviser to WHO (World Health Organization), PAHO (Pan American Health Organization), FDA, the Defense Threat Reduction Agency (DTRA), and other agencies. He is a fellow of the New York Academy of Sciences and a past chair of its Microbiology Section. He was the founding chair of ProMED (the nonprofit international Program to Monitor Emerging Diseases) and was one of the originators of ProMED-mail, an international network inaugurated by ProMED in 1994 for outbreak reporting and disease monitoring using the Internet. Dr. Morse received his Ph.D. from the University of Wisconsin-Madison.

Randall S. (Randy) Murch, Ph.D., received a B.S. degree in Biology from the University of Puget Sound, Tacoma, Washington in 1974, an M.S. degree in Botanical Sciences from the University of Hawaii in 1976, and a Ph.D. degree in Plant Pathology from the University of Illinois in 1979. After 23 years of service as a special agent, he retired from the FBI in

November 2002. During his FBI career, he was assigned to the Indianapolis, Los Angeles and New York field divisions, and to the national security, (forensic) laboratory and investigative technology (engineering) divisions at FBI Headquarters and Quantico, Virginia. He served as a department head and deputy division head in the FBI Laboratory, as well as a deputy division head of the FBI's electronic surveillance division (investigative technology). He has extensive experience in counterintelligence, counterterrorism, forensic science, electronic surveillance, WMD threat reduction, and outreach to those communities. He created the FBI's WMD forensic investigation/S&T response program in 1996, and served as the FBI's science advisor to the 1996 Olympics. From December 1999 to June 2001, he was detailed to the Defense Threat Reduction Agency as the director of DTRA's advanced systems and concepts office. He has participated in National Academy of Sciences/National Research Council, Defense Science Board and DTRA Threat Reduction Advisory Committee studies and panels and other senior review panels. He joined the Institute for Defense Analyses in December 2002, and now works to deliver creative solutions for difficult national security problems across a range of operational, science and engineering disciplines.

Paula Olsiewski, Ph.D., is leading the Alfred P. Sloan Foundation's program to reduce the threat of bioterrorism. Since joining the Foundation in 2000, she has created a collaborative network from the public, private and government sectors that has become critical to the nation's civilian biodefense movement. Among the many projects Dr. Olsiewski has facilitated is the Department of Homeland Security's READY campaign, a public education effort that empowers Americans to prepare for potential terrorist attacks. Another important grant to the Center for Law and the Public's Health at Georgetown and Johns Hopkins Universities produced model legislation for dealing with bioterrorism and catastrophic infectious diseases. Thirty-three states and the District of Columbia have enacted legislation based on the Model State Emergency Health Powers Act. A grant to the National Academies resulted in the Fall 2003 NRC Report *Biotechnology Research in an Age of Terrorism* and led to the establishment of the National Science Advisory Board for Biosecurity by the US Department of Health and Human Services in March 2004. During the 1990s, Dr. Olsiewski founded and directed a consulting practice, Neo/Tech Corp., providing expertise in structuring, implementing, and directing technology development programs. Before that, she was vice president of commercial development at Enzo Biotech, Inc. where she was responsible for overall management of product development, technology licensing and transfer programs. Dr. Olsiewski serves on numerous advisory committees and boards. She is a member of the MIT Corporation and was the president of the MIT Alumni/ae Association 2003-2004. She is chairman of the Board of Trustees of Asphalt Green, Inc., a not-for-profit organization dedicated to assisting individuals of all ages and backgrounds achieve health through a lifetime of sports and fitness. Dr. Olsiewski received a B.S. in Chemistry from Yale College, and a Ph.D. in Biological Chemistry from MIT.

Chandra Kumar N. Patel, Ph.D., a member of the National Academy of Engineering and National Academy of Sciences, is chief executive officer and chairman of the board of Pranalytica, Inc. and professor of physics and former vice chancellor of research at the University of California at Los Angeles. Until 1993, Dr. Patel served as executive director of the Research, Materials Science, Engineering and Academic Affairs Division at AT&T Bell Laboratories. Dr. Patel has an extensive background in several fields, to include materials, lasers, and electro-optical devices. During his career at AT&T, which began in 1961, he made numerous seminal contributions in several fields, including gas lasers, nonlinear optics, molecular spectroscopy, pollution detection and laser surgery. Dr. Patel has served on numerous government and scientific advisory boards and he is past president of Sigma Xi and the American Physical Society. In addition, Dr. Patel has received numerous honors, including the National Medal of Science, for his invention of the carbon dioxide laser.

Clarence J. (CJ) Peters, M.D., is the John Sealy Distinguished University Chair in Tropical and Emerging Virology at the University of Texas Medical Branch in Galveston and is Director for Biodefense in the Center for Biodefense and Emerging Infectious Diseases at that institution. Before moving to Galveston in 2001, he worked in the field of infectious diseases for three decades with NIH, CDC, and the U.S. Army. He has been Chief of Special Pathogens Branch at the Centers for Disease Control and Prevention in Atlanta, Georgia and previous to that, Chief of the Disease Assessment Division and Deputy Commander at USAMRIID. He was the head of the group that contained the outbreak of Ebola at Reston, Virginia and led the scientists who identified hantavirus pulmonary syndrome in the southwestern U.S. in 1993. He has worked on global epidemics of emerging zoonotic virus diseases including Bolivian hemorrhagic fever, Rift Valley fever, and Nipah virus. He received his M.D. from Johns Hopkins University and has more than 275 publications in the area of virology and viral immunology. Dr. Peters is currently also a member of the National Academy of Sciences Committee on Research Standards and Practices to Prevent the Destructive Application of Biotechnology.

George Poste, D.V.M., Ph.D., is chief executive of Health Technology Networks, a consulting group based in Scottsdale, Arizona, and suburban Philadelphia specializing in the application of genetics and computing in healthcare and bioterrorism defense. From 1992 to 1999 he was chief science and technology officer and president, Research and Development of SmithKline Beecham (SB). During his tenure at SB he was associated with the successful registration of 29 drug, vaccine and diagnostic products. He is chairman of diaDexus and Structural GenomiX in California and Orchid Biosciences in Princeton. He serves on the Board of Directors of AdvancePCS and Monsanto. He is an advisor on biotechnology to several venture capital funds and investment banks. In May 2003, he was appointed as Director of the Arizona Biodesign Institute at Arizona State University. This is a major new initiative combining research groups in biotechnology, nanotechnology, materials science, advanced computing and neuromorphic engineering. He is a fellow of Pembroke College Cambridge and distinguished fellow at the Hoover Institution and Stanford University. He is a member of the Defense Science Board of the U.S. Department of Defense and in this capacity he chairs the Task Force on Bioterrorism. He is also a member of the National Academy of Sciences Working Group on Defense Against Bioweapons. Dr. Poste is a Board Certified Pathologist, a fellow of the Royal Society and a fellow of the Academy of Medical Sciences. He was awarded the rank of Commander of the British Empire by Queen Elizabeth II in 1999 for services to medicine and for the advancement of biotechnology. He has published over 350 scientific papers, co-edited 15 books on cancer, biotechnology and infectious diseases and serves on the editorial board of multiple technical journals. He is invited routinely to be the keynote speaker at a wide variety of academic, corporate, investment, and government meetings to discuss the impact of biotechnology and genetics on healthcare and the challenges posed by bioterrorism.

C. Kameswara Rao, Ph.D., initially taught at the Department of Botany, Andhra University, Waltair, and served the Bangalore University from 1967 to 1998. He received the B.Sc. (Hons.), M.Sc., and Ph.D. degrees from the Andhra University, and a D.Sc., (honoris causa) from the Medicina Alternativa Institute, Open International University for Complementary Medicines, Colombo. He was a professor of botany and the chairman of the department of botany, and the department of sericulture at the Bangalore University. Currently, he is executive secretary for the Foundation for Biotechnology Awareness and Education. On a Commonwealth Academic Staff Fellowship and a Royal Society and Nuffield Foundation Bursary, Professor Kameswara Rao worked on the computer applications in plant systematics, at the Natural History Museum, London, and the Royal Botanic Gardens, Kew, in the UK, besides some other institutions. Professor Kameswara Rao was the president of the Indian Association for Angiosperm Taxonomy for 1999. He is a member of the Indian Subcontinent Plant Specialist Group of the Species Survival Commission, IUCN. He is a member of the Programme Advisory Committee of the Botanical Survey of India and the Zoological Survey of India, Ministry of Forests and Environment, Government of India. He is the executive secretary of the Foundation for Biotechnology Awareness and Education. Professor Rao's research interests are, applications of computers and phytochemistry in plant systematics, and databases of medicinal plants. Recently, he was awarded a Certificate of Merit by the World Peace Foundation, Beijing, an affiliate of the UN, for his research work on Indian medicinal plants.

Julian Robinson, Ph.D., a chemist and patent lawyer by training had, previously held research appointments at the Stockholm International Peace Research Institute (SIPRI), the Free University of Berlin, and the Harvard University Center for International Affairs. He has been active in the Pugwash Conferences on Science and World Affairs since 1968. He has served as an advisor or consultant to a variety of national and international organizations, governmental, and nongovernmental, including the World Health Organization, other parts of the United Nations system, the International Committee of the Red Cross, and the UK National Authority for the Chemical Weapons Convention. In association with the Belfer Center for Science and International Affairs of the Kennedy School of Government at Harvard, he directs the UK end of the Harvard Sussex Program (HSP), which is a collaborative research, teaching, and publication activity focused on chemical/biological-warfare armament and arms limitation. This is a subject on which he has published some 400 papers and monographs since 1967, including much of the six volume SIPRI study The Problem of Chemical and Biological Warfare (1971-76), Effects of Weapons on Ecosystems (1979), Chemical Warfare Arms Control (1984), NATO Chemical Weapons Policy and Posture (1986), and The Problem of Chemical-Weapon Proliferation in the 1990s (1991). Since 1988, he has been editing, with Matthew Meselson of Harvard University, one of the few journals in the field, The *CBW Conventions Bulletin*, now published quarterly from the Sussex end of HSP.

Peter A. Singer, M.D., MPH, FRCPC, is Sun Life Financial Chair in Bioethics and Director of the University of Toronto Joint Centre for Bioethics and Professor of Medicine at the University of Toronto and University Health Network. He also directs the World Health Organization Collaborating Centre for Bioethics and the Canadian Program on Genomics and

Global Health at the University of Toronto. He studied internal medicine at the University of Toronto, medical ethics at the University of Chicago, and clinical epidemiology at Yale University. Singer is the recipient of awards that include the Nellie Westerman Prize in Ethics of the American Federation for Clinical Research, Young Educator Award of the Association of Canadian Medical Colleges, American College of Physicians George Morris Piersol Teaching and Research Scholar, Canadian Life and Health Insurance Association Medical Scholarship, NHRDP National Health Research Scholar, CIHR Investigator, and CIHR Distinguished Investigator, Senior Fellow at Massey College, and the Award for Excellence from Yale University School of Public Health. He has published over 200 articles, held over $20 million in research grants, and trained over 50 graduate students and fellows. He is a member of the Scientific Advisory Board of the Bill & Melinda Gates Foundation Grand Challenges for Global Health Initiative, a Director of BIOTECanada, and board chair of Branksome Hall School for Girls. His contributions have included improvements in quality end of life care, fair priority setting in healthcare organizations, and teaching bioethics. His current research focus is global health, in particular harnessing genomics and nanotechnology to improve health in developing countries.

Christopher L. Waller, Ph.D., received his Ph.D. in Medicinal Chemistry and Natural Products from the University of North Carolina in Chapel Hill in 1992. His graduate research efforts were directed at the design, synthesis, and biological evaluation of antiedema agents. Following graduation, Dr. Waller accepted a postdoctoral fellowship under the direction of Dr. Garland Marshall at Washington University in St. Louis where he focused his efforts on the design HIV protease inhibitors. In 1993, Dr. Waller accepted a position with the U.S. EPA in which he was responsible for the development of structure-activity relationship and pharmacokinetic models as a research chemist and leader of a team of analytical, computational, and synthetic organic chemists, toxicologists, and biomedical engineers. From 1996-1999, Dr. Waller served as a Research Manager at OSI Pharmaceuticals. In this role, he managed a group of computational chemists, scientific application developers, and robotics engineers. In early 1999, Dr. Waller joined Eli Lilly- Sphinx Laboratories as a computational chemist and Head of Cheminformatics in the Discovery Chemistry group. Since 2001, Dr. Waller has been Associate Director of Research Informatics for Pfizer Global Research And Development, Ann Arbor Laboratories. Dr. Waller has published over 25 peer-reviewed articles and has received numerous honors and awards including The Board of Publications Award for the Best Paper in Toxicology and Pharmacology in 1996.